Towards Net-Zero Carbon Initiatives

Carbon Initiatives

A Life Cycle Assessment Perspective

Towards Net-Zero Carbon Initiatives

Carbon Initiatives

A Life Cycle Assessment Perspective

Editors

Hsien Hui Khoo
A*STAR, Singapore

Reginald B H Tan
A*STAR, Singapore
National University of Singapore, Singapore

World Scientific

W JERSEY · LONDON · SINGAPORE · BEIJING · SHANGHAI · HONG KONG · TAIPEI · CHENNAI · TOKYO

Published by

World Scientific Publishing Co. Pte. Ltd.

5 Toh Tuck Link, Singapore 596224

USA office: 27 Warren Street, Suite 401-402, Hackensack, NJ 07601

UK office: 57 Shelton Street, Covent Garden, London WC2H 9HE

Library of Congress Control Number: 2023047031

British Library Cataloguing-in-Publication Data
A catalogue record for this book is available from the British Library.

TOWARDS NET-ZERO CARBON INITIATIVES
A Life Cycle Assessment Perspective

ISBN 978-981-12-7620-0 (hardcover)
ISBN 978-981-12-7566-1 (ebook for institutions)
ISBN 978-981-12-7621-7 (ebook for individuals)

For any available supplementary material, please visit
https://www.worldscientific.com/worldscibooks/10.1142/13393#t=suppl

Desk Editors: Gregory Lee/Amanda Yun

Typeset by Stallion Press
Email: enquiries@stallionpress.com

Preface

In the context of global warming, "low carbon economy" or "low carbon technology" has become the interest in the worldwide research arena. Industrialization has brought great emissions of carbon dioxide, which needs to be reduced. With the development of industrialized nations, various types of energy demands, along with carbon dioxide emissions, are expected to increase over time. Low carbon technologies play a vital role in generating less carbon dioxide than traditional processes, and will play a vital role in the transition to a low carbon economy.

This book explores various initiatives and potential methods to achieve Net-Zero Carbon targets, with a broad system-based scope encompassing energy and resource use. As industrialized nations look into emerging new technologies focusing on renewable or efficient energy use, along with the move towards Sustainable Development Goals, issues and challenges related to achieving low-carbon economy projects have gained much global attention.

In this area, Life Cycle Assessment (LCA) plays a critical role as an effective and comprehensive method to analyze potential carbon footprint, along with other unintended environmental impacts of any low-carbon technological process. LCA is applied as a holistic and system-wide scientific method that can be used to quantify impact metrics chosen to evaluate any emerging Net-Zero Carbon technologies of interest, and it will reveal environmental trade-offs or further research opportunities that are required for balancing carbon dioxide emissions. The LCA's perspectives of Net-Zero Carbon technologies can also be used to outline decision-making strategies for a nation's shift towards low-carbon economic development. Comprehensive guidelines of an LCA for low-carbon technologies are also presented.

Hsien Hui, Dr. KHOO
ISCEE, A*STAR

About the Editors

Hsien Hui KHOO is a Research Scientist at the Institute of Sustainability for Chemicals, Energy and Environment (ISCEE), Agency for Science, Technology and Research (A*STAR), Singapore. She graduated with an M.Eng degree (Master of Engineering) from Nanyang Technological University (NTU) in 1998, and completed a research internship for environmental management and supply chain studies at CQUniversity, Australia. In 2007, she obtained her Ph.D. from the National University of Singapore (NUS), specializing in Life Cycle Assessment (LCA) and sustainability management. With a passion for research, she has over 17 years of research experience in the areas of green chemistry and environmental studies. Her research topics cover carbon capture and utilization, carbon dioxide (CO_2)-based products, bio-derived products, and national plastic recycling methods. She obtained the certificate of Business Sustainability Management from the Cambridge Institute of Sustainability Leadership (CISL), UK, in 2017. Dr. Khoo was also awarded the Singapore Commonwealth Fellowship in Innovation in 2017 by The Royal Commonwealth Society and the Enterprise Merit Award in 2018 by Singapore's Quality and Standards Board. In 2023, Dr. Khoo was awarded an appreciation certificate for her work in the LCA of CO_2-based fuel done for Aerospace Technology, Singapore.

Reginald B.H. TAN is Executive Director at the Science and Engineering Research Council at A*STAR, and concurrently Professor at the Department of Chemical and Biomolecular Engineering at NUS. Professor Tan has been a keen educator and advocate of science-based and evidence-based solutions to environmental issues and is recognized as a highly respected LCA mentor. He has applied the technique of environmental LCA to case studies in mineral extraction, energy options, carbon capture and sequestration, material selection, and waste management, resulting in several significant publications in the field. With his expertise in environmental sustainability, Prof. Tan has been an active standards volunteer over the past 25 years. He has served in many leadership roles both internationally (ISO TC207 SC5 on LCA and ISO Coordinating Committee on Climate Change) and nationally (Chair of Environmental and Resources Standards Committee). He was awarded the Singapore Standards Council Distinguished Partner Award in 2019.

Contents

Chapter 1

Trends of Emerging Zero-Carbon Technologies: The Role of the Life Cycle Assessment for Evaluating Carbon Dioxide Reduction Targets

Hsien Hui KHOO[a]* and Reginald B.H. TAN[b]

[a]*Institute of Sustainability for Chemicals,
Energy and Environment (ISCE²),
Agency for Science, Technology and Research (A*STAR),
Singapore 627833*
[b]*Department of Chemical and Biomolecular Engineering,
National University of Singapore,
Singapore 117585*

*khoo_hsien_hui@isce2.a-star.edu.sg

This chapter presents an introduction of emerging zero-carbon technologies and processes envisaged to reduce global warming impacts. Worldwide, a growing field of research sciences and technological developments has advanced towards the goal of decarbonization in the areas of energy efficiency, Carbon Capture and Utilization (CCU) and other associated areas. As aims to reduce climate change impacts become a rising global concern, it has also been well recognized that the Life Cycle Assessment (LCA) perspective adds significant scientific value by highlighting environmental trade-offs and/or further research opportunities. Overall, the LCA plays a crucial role in performing the overall evaluation of the actual environmental performance of any processes employed to achieve zero carbon emissions. Along with new avenues of science and research geared towards low-carbon hydrogen (H_2) methods, process recycling involving circularity, and chemical transformation methods (carbon dioxide (CO_2) to chemicals/fuels), it was suggested that scientific approaches in LCA data selection and compilation be evaluated to avoid uncertainties in results.

1. Introduction

The growing levels of atmospheric greenhouse gas (GHG) emissions have spurred worldwide concerns of impending global climate change impacts among international organizations, governments and environmental scientists. The Paris Agreement with the UN Framework Convention on Climate Change is an international agreement that aims to tackle climate change and maintain global average temperatures well below 2°C above pre-industrial levels and ideally below 1.5°C [1]. To achieve the goals of the Paris Agreement, significant reductions in carbon dioxide (CO_2) emissions are required. Targets set to achieve net-zero CO_2 emissions will require a fundamental change in the way energy and resources are being used, as demonstrated in the United Nations Sustainable Development Goals (SDGs). The success of a variety of fuels and technologies to reach net-zero CO_2 emissions depends on process-specific circumstances.

The chapters compiled in this book present various emerging technologies or processes that are ongoing, with the aim to achieve GHG reduction targets or reach zero carbon emission strategies. The list of scientific topics included are: Carbon Capture and Utilization (CCU), low-carbon hydrogen (H_2) energy systems, waste management or recycling coupled with the circular economy, and CO_2 accounting methodologies. Aligned with the need to evaluate net CO_2 and other potential environmental impacts, the important role of the Life Cycle Assessment (LCA) will also be introduced to identify areas for improvements in research developments and used to support informed decision-making [2, 3].

2. Carbon Capture and Utilization: Brief Overview

Carbon Capture and Utilization (CCU) offers the possibility of carbon recycling, where CO_2 is converted into fuel or chemicals that can be fed

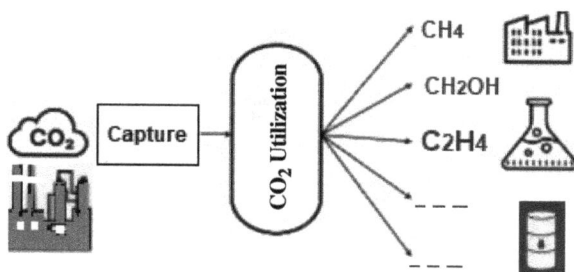

Fig. 1.   From CO_2 to chemicals and fuels. (Adapted from Ref. [6].)

back into the production cycle [4, 5]. CCU is regarded by both science and industry as a promising approach to circumvent resource depletion concerns for the chemical industry. Besides reducing CO_2 concentrations in the atmosphere, the use of CO_2 offers opportunities to manufacture a wide range of chemicals. This scheme is illustrated in Fig. 1 [6].

Recent research advancements led to the production of a number of new CO_2-based products. As a notable example, a factory known as Carbon Recycling International (CRI) in Iceland makes use of CO_2 sourced from a geothermal power plant to produce methanol on a large scale [7]. In another case, Cuéllar-Franca and Azapagic [8] stated in their work that the utilization of CO_2 for producing dimethyl carbonate (DMC) could lower global warming impacts by 4.3 times, as compared to conventional production.

In the academic field, an evaluation of CCU concepts covers a wide range of sustainability and environmental subjects. Chauvy and colleagues [9] applied a multi-criteria assessment to classify viable, emerging CO_2-based products according to technical, economic, energetic, environmental, and market considerations. Apart from the economical value versus costs consideration, the industrial concerns of utilizing CO_2 as raw material for large-scale production are Technology Readiness Level (TRL) ranks. TRLs are scales ranging from 1 (observation of basic principles) to 9 (demonstration in operational environment).

Among the trends found in CCU developments, dimethyl ether (DME) is a promising clean fuel that can substitute diesel and be a valuable intermediate in the petrochemical industry. Manufacturing it from CO_2, although expensive due to high electricity costs, has been proven as technically feasible, which is promising due to the projected increase of DME's market value and demand [10]. Patricio and colleagues [11]

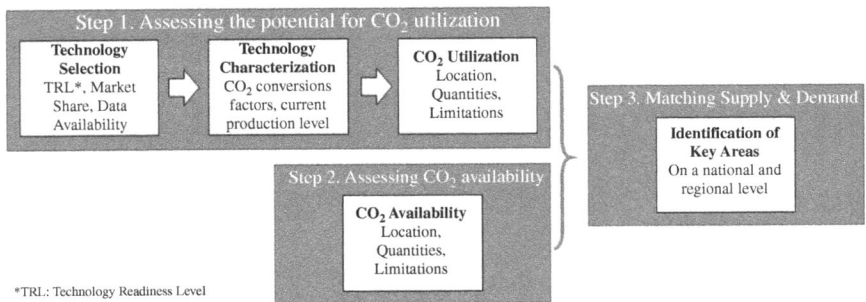

Fig. 2. A top-down methodological approach for Carbon Capture and Utilization. (Reproduced from Ref. [11] with permission from Elsevier.)

recommended a methodological approach for CCU research arenas based on a top-down approach (Fig. 2), where CO_2-based value chains can be developed for different types of carbon-derived products. Various CO_2 capture and sequestration technologies will be presented in Chap. 15.

According to Müller and colleagues [2], carbon footprints of different CO_2 sources differ due to their energy demands of CO_2 capture methods. The carbon footprint of CO_2 feedstock varies from -0.99 to -0.98 kg CO_2-eq. in a low carbon economy. This implies that for an overall sustainability assessment of CCU options, CO_2 sources have to be taken into account; it is therefore necessary to take a step back in the carbon production chain and consider the wide variety of capture technologies.

2.1. *Sources of CO_2 — Capture technologies*

Sources of CO_2 can span a wide range of power plants, refineries and industries. There is a vast array of technologies that exist to capture CO_2 from power plants. Four different types of methods are post-combustion, pre-combustion, oxy-combustion, and chemical looping combustion. Each CO_2 capture technology has its own advantages and disadvantages and is at different stages of development [12].

One of the mature methods for the post-combustion capture of CO_2 is the use of chemical absorption with amine solvents [13]. Post-combustion capture with solvents has already been commercialized

Table 1.   Summary chemical absorption with monoethanolamine.

Capture Efficiency (%)	Energy Penalty (MJ/kg CO_2)	Brief Description	Ref.
91.7	3.93	Optimization of CO_2 capture case study; process model simulated with UniSim.	[16]
~95	4.3	Simulation results of CO_2 capture using ASPEN Plus.	[17]
95	3.8–4.2	Energy analysis carried out using GateCycle software (for NGCC) + ChemCAD software.	[18]
99	2.3–2.8	Development and testing of new solvent. (*CANSOLV DC-201*) at NSE-Kimitsu pilot site, Tokyo, Japan.	[19]
~80	3.9–4.2	Based on various literature and simulation studies.	[20]

and is considered a robust technology fit for large-scale CO_2 capture from point sources such as power plants and industrial flue gases [14]. Among all the other methods, CO_2 capture with monoethanolamine (MEA) is also rated as an industrial benchmark due to its relatively lower cost and high CO_2 capture rate. Various investigations have already been carried out to identify key process variables and design parameters pertaining to MEA used for CO_2 capture [15]. A summary of energy used for CO_2 captured from power plants is given in Table 1.

Sources of CO_2 can also be recovered from large-scale industrial processes. As an example, De Lena and colleagues [21] presented the integration of the calcium looping (CaL) process into a cement plant for post-combustion CO_2 capture. This process is based on the reversible reaction between calcium oxide (CaO) and CO_2 at high temperatures and is considered one of the most promising CO_2 capture technologies in cement plants. Petroleum refineries are also classified as being part of the CO_2-intensive industries. A fluid catalytic cracking unit (FCCU), which is an important refinery process, generates a large amount of CO_2 emissions. Wei and colleagues [22] introduced the capture of CO_2 from the refinery FCCU flue gas with a solvent-based carbon-capture technology. Simulation studies via Aspen HYSYS/

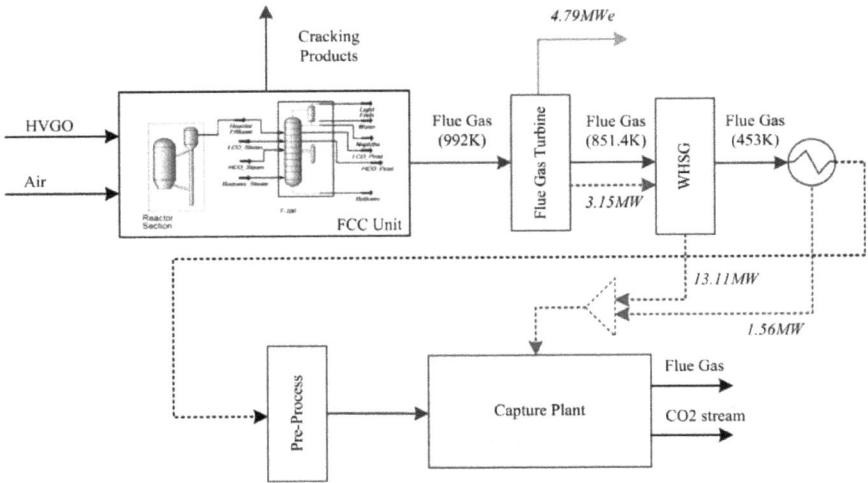

Fig. 3. CO_2 capture from a refinery fluid catalytic cracking unit. (Reproduced from Ref. [22] with permission from Elsevier.)

Petroleum Refining submodels were carried out to project the operating parameters and associated energy input for the carbon capture plant with MEA solvent from an actual industrial-scale refinery (illustrated in Fig. 3).

While CCU aims to improve economic benefits for industrial applications and lower global warming potential or fossil resource depletion, the overall reduction of environmental impacts cannot be taken for granted. The LCA plays an important role in this area to perform the overall evaluation of the actual environmental performance of any processes employed to achieve zero carbon emissions.

3. Life Cycle Assessment

An LCA will play an important role as an effective and comprehensive method to analyze potential GHG emissions and other environmental impacts of a technology or system. As aims to reduce climate change impacts become a rising global concern, it has also been well recognized that the LCA perspective adds significant scientific value by highlighting environmental trade-offs and/or further research opportunities.

3.1. *Life Cycle Assessment methodology*

In order to achieve the aims of reducing CO_2 emissions — or overall environmental sustainability goals — of the evolving processes targeted to reduce GHGs, an environmental tool is needed for evaluating these technologies. The LCA is commonly described as a compilation and evaluation of the inputs, outputs and potential environmental impacts of a product system. It is a valuable tool in various fields, e.g., product or process design, decision-making in industry, policy as well as economics [23], and can also be conducted to meet the objectives of decision-makers considering climate-related sustainability roadmap strategies.

The LCA methodology was standardized by the International Standardization Organization (ISO) [24–25]. Updates and revisions of ISO 14040 and 14044 LCA standards have been made to reaffirm the validity of the technical contents of LCA procedures and to serve as core reference documents for the users and practitioners of the LCA [26]. As

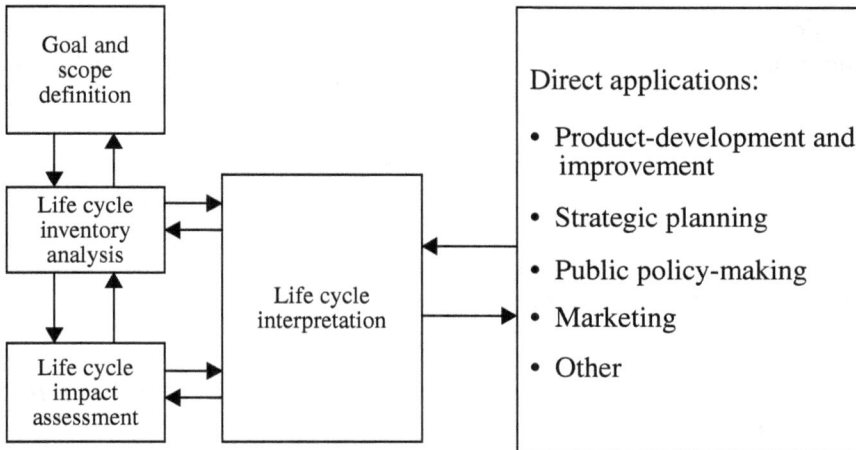

Fig. 4.   Phases of a Life Cycle Assessment. (Reproduced from the International Organization for Standardization.)

displayed in Fig. 4, the ISO 14040 standards define the technical as well as the organizational aspects of LCA phases. The description of each phase of the LCA is as follows:

I. **Goal definition and scope** — the product, process or service is described in terms of quantity and function/quality. The scope or perimeter of the LCA system boundary is next decided.

II. **Inventory analysis** — input-output material flows and resource use (energy, chemicals, water, etc.) for the product/process defined in the first phase are recorded and compiled. This phase is also known as the Life Cycle Inventory (LCI). It is also the most time-consuming aspect of the LCA.

III. **Impact assessment** — the compiled LCI is converted into a range of environmental impacts for analyzing the process/product of interest.

IV. **Interpretation** — results are interpreted, and any unexpected "environmental hotspots" of the life cycle stages involved within the defined LCA system boundary/scope are identified.

Further descriptions of the LCA methodology will also be described in Chap. 15.

For an overall sustainability assessment of CCU options, the production chain from the CO_2 source to products, including capture technologies, energy and resource requirements, and their associated emissions, have to be accounted for [8, 27]. In a case study involving CO_2-based ethylene production via the electrochemical process, Khoo and colleagues [28] elaborated how the LCA offers a more holistic approach than a single-stage environmental assessment. The system boundary took into account small and large-scale additional emissions that arose from the energy use and indirect resources required.

3.2. *Life Cycle Assessment Applications in Carbon Capture and Utilization*

Various other cases of LCA-based investigations of CCU have been reported. Thonemann and Pizzol [29] introduced a generalized LCA scheme for the investigation of CCU systems (Fig. 5). Two separate studies were done comparing different CCU technologies for the production of

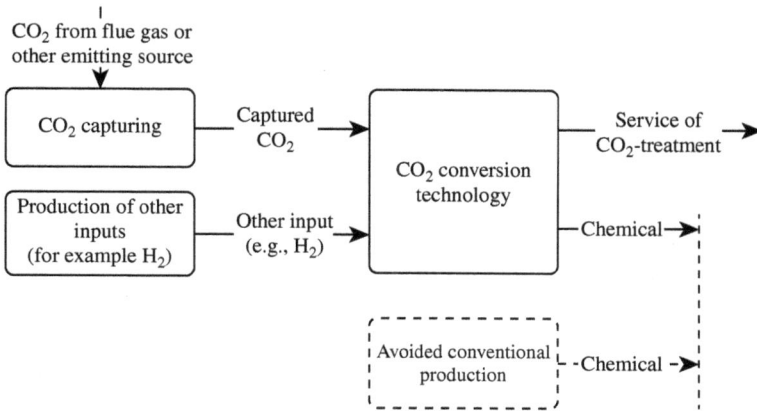

Fig. 5. Generalized Life Cycle Assessment scheme of the investigated product system. (Reproduced from Ref. [29] with permission from the Royal Society of Chemistry.)

CO_2-based formic acid [30] and methane [31]. Both studies concluded that significant environmental impact reductions can be achieved, compared to their conventional production pathways. In other CCU examples, the results of the LCA is highly dependent on the selections of functional units (FUs) (i.e., the relative basis for which environmental impacts are assessed) or system boundaries and can result in different outcomes. For example, separate LCA studies in the literature report cradle-to-gate carbon footprints between −1.7 and +9.7 kg of CO_2-eq. per kg of CO_2-based methanol [2]. For the purpose of enhancing justification effects of comparability among studies, a decision tree to define a suitable FU was derived for each class of CCU technologies (Fig. 6).

Process parameters in LCA model variations can alter the inventory values used in the LCA (e.g., energy source, resource use, product yield, wastes, etc.), as well as its associated emissions. Khoo and colleagues [28] indicated how different scales of similar CCU concepts may result in dissimilar net CO_2 outcomes. Their work displayed that the expected CO_2 reduction, or environmental benefits, from small-scale experimental designs (FU = 1 gram) cannot be directly transferred — or linearly projected — to large-scale models (FU = ton output). This is mainly due to different sources of CO_2 feedstock. The potential net 0.98–3.7 g CO_2-eq. is achievable in the small-scale setup, while 0.65–3.0 t CO_2-eq. is achievable in the large-scale model (Fig. 7(a) and 7(b)).

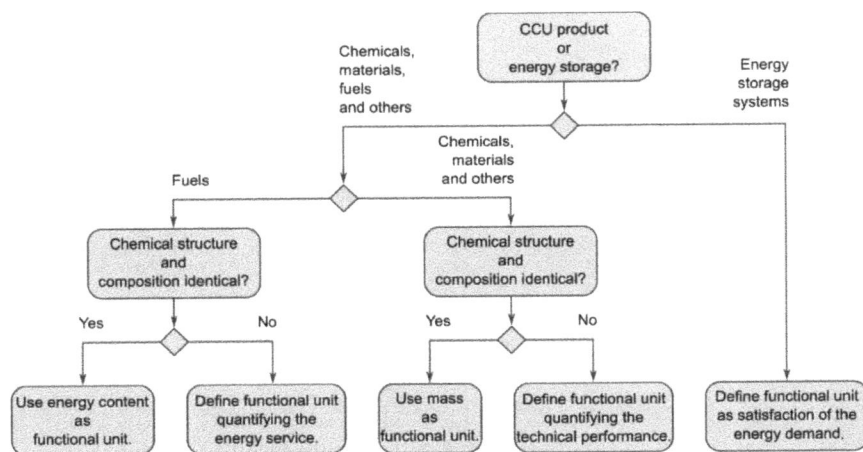

Fig. 6.   Decision tree for suitable functional units in Life Cycle Assessment studies. (Reproduced from Ref. [2] with permission from the Royal Society of Chemistry.)

In another LCA investigation, the production of CO_2-based polyether carbonate polyols in a real industrial pilot plant was carried out by von der Assen and Bardow [32]. A cradle-to-gate system of polyol production and all upstream processes, such as the provision of energy and feedstocks, were considered. They considered a lignite power plant equipped with a pilot plant for CO_2 capture. It was concluded that the production of polyols with 20 wt% CO_2 in the polymer chains causes GHG emissions of 2.65–2.86 kg CO_2-eq. kg^{-1}. The results are displayed in Fig. 8.

Apart from the focus on global warming potential reduction, other environmental consequences should also be carefully considered (e.g., Refs. [8, 28, 33]). In order to provide a comprehensive evaluation of any overall benefits and drawbacks of CCU developments, the LCA can be applied to generate other impact category results such as Acidification, Eutrophication, Freshwater Ecotoxicity, Human Toxicity, etc. As a classical example, the environmental impacts of various post-conversion CO_2 capture methods [34] from fossil-based power plants were reviewed and compared by Cuéllar-Franca and Azapagic [8]; the results are illustrated in Fig. 9.

GWP results per 1 g ethylene

(a) GWP results for small-scale (1 g)

GWP results per 1 t ethylene

(b) GWP results for large-scale (1 ton)

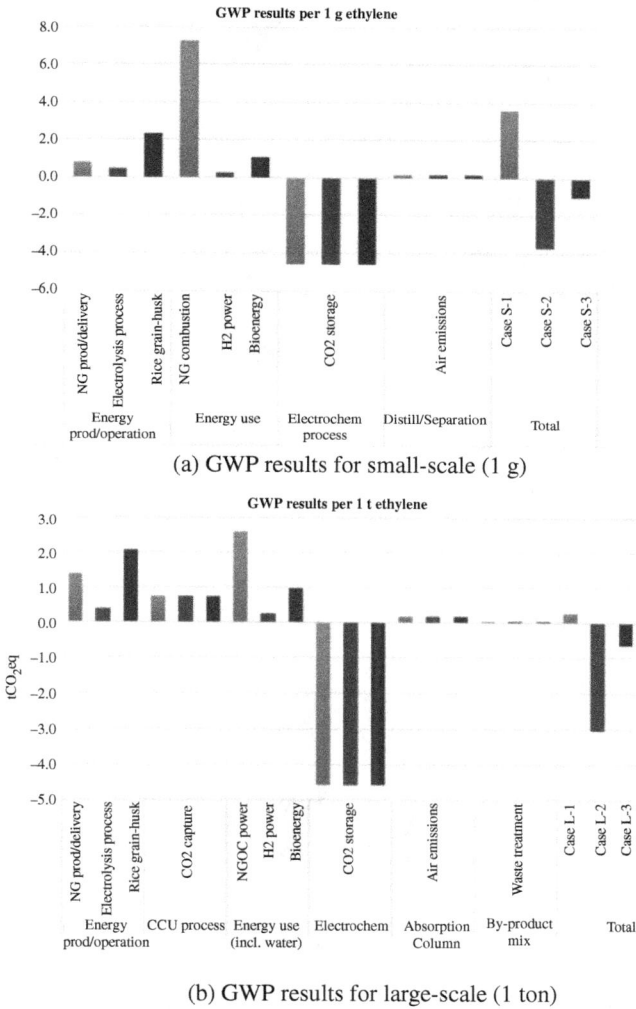

Fig. 7. Global warming potential (GWP) results for two different scales. (Reproduced from Ref. [28] with permission from Elsevier.)

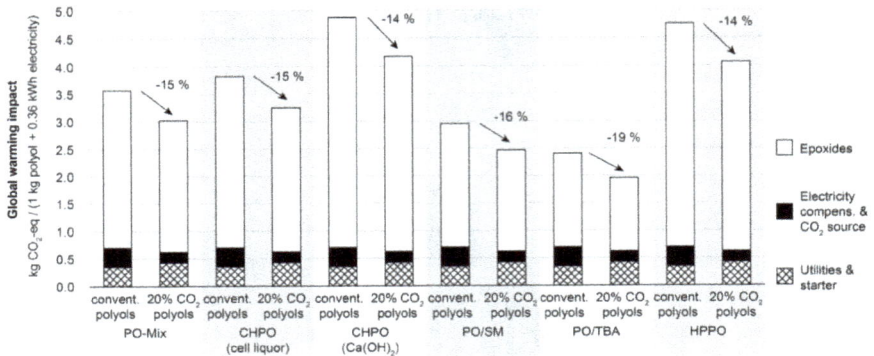

Fig. 8.   Global warming impact considering alternative propylene oxide (PO) production technologies for a product system of conventional polyether polyols (left bar for each PO technology) and CO_2-based polyethercarbonate polyols with 20 wt% CO_2 (right bar for each PO technology). (Reproduced from Ref. [32] with permission from the Royal Society of Chemistry.)

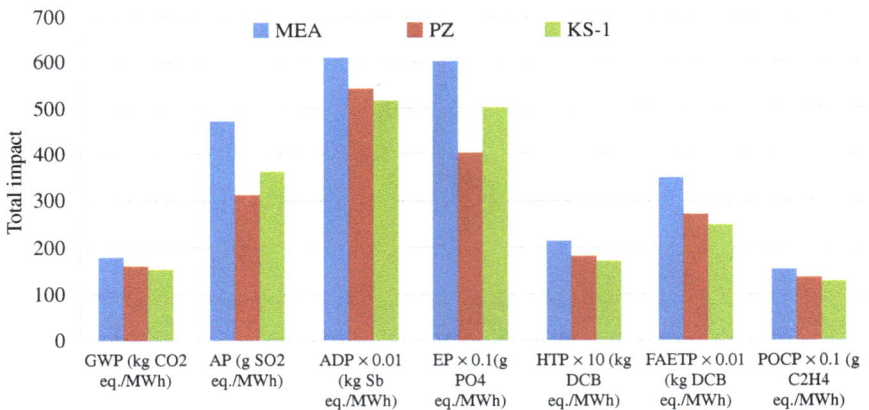

Fig. 9.   Environmental impacts of a pulverized-coal power plant with post-conversion capture via monoethanolamine (MEA), piperazine with potassium carbonate (PZ) and hindered amine (KS-1). (Reproduced from Ref. [8] under the terms of the Creative Commons CC BY license.)

4. Alternative Energy Resources: Trends Towards Low-Carbon Energy

With the aim of achieving long-term decarbonization of power generation systems and, at the same time, increasing energy security, highly developed nations worldwide are looking into alternative energy resources to replace fossil fuels. A wide range of renewable energy technologies are on the rise. In this chapter, the two selected renewable energy resources presented are: bio-energy and H_2.

4.1. *Bio-energy*

The utilization of diverse forms of bioenergy is expected to expand in various economic sectors [35]. As an example, Muhammad and colleagues [36] analyzed the potential of utilizing the abundant supply of rice husk as an alternative energy provider. Scientific work concerning the sustainability of bioenergy use in terms of feedstock production, planned targets and policies, and socio-economics has been done [37]. Although numerous biomass feedstocks for bioenergy have been explored, challenges and other types of environmental issues (e.g., land use, water consumption, biodiversity loss) still exist.

Various LCA studies have already been applied to aid in the identification of bioenergy systems to reach low carbon targets adopted by national policy organizers and other decision-makers [38, 39]. However, different aspects used in LCA models (e.g., system boundary limitations, co-product allocation, land use) will generate dissimilar results for the same product [40]. The US Environmental Protection Agency (EPA) [41] has stated that environmental performance evaluation of feedstock production used in biofuels should include domestic and international agricultural/forestry sector-wide impacts (Fig. 10).

The sum of the following three lifecycle emissions for renewable fuel pathways are considered for comparison to the direct emissions from the baseline petroleum fuel it displaces [41]:

- Agricultural sector impacts such as increases or decreases in feedstock and livestock production,
- Impacts from using biofuel co-products in the agricultural sector, such as the use of distillers grains as livestock feed, and

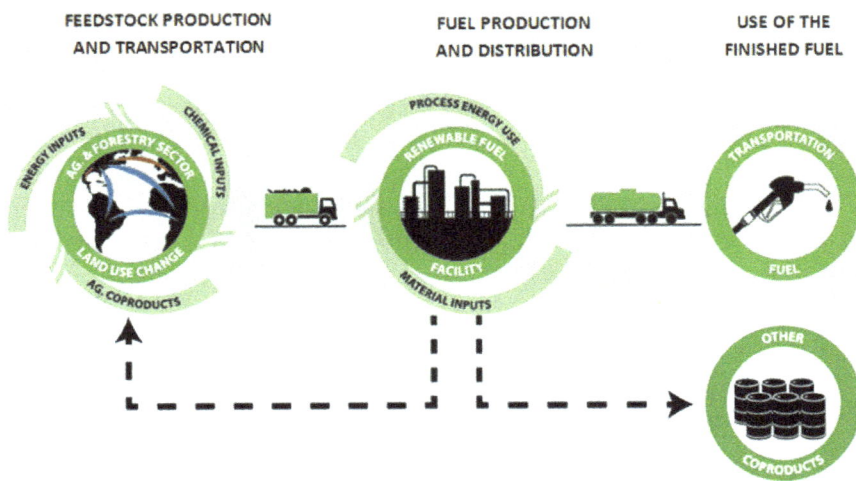

Fig. 10.   The US Environmental Protection Agency's Life Cycle Assessment for the Renewable Fuel Standard [41].

- Significant emissions from land use changes, such as cultivating new land for feedstocks.

Additionally, Brandão and colleagues [38] recommend that in LCA studies concerning bioenergy, emissions from land use changes are a major determining parameter that should not be omitted. Land use changes may lead to altered soil organic carbon (SOC) and changes in a host of ecosystem services. Aligned with a similar notion for LUC, Koellner and colleagues [42] presented guidelines from UNEP-SETAC on environmental indicators for global land use to be considered in LCA system boundaries (Fig. 11).

O'Keeffe and colleagues [43], on the other hand, introduced "RELCA" as a regional life cycle inventory method applied for assessing the environmental burdens of producing a bioenergy product. RELCA can be applied to explore the regional variability of direct regional environmental burdens as well as the indirect non-regional burdens associated with the production of regional bioenergy products (biodiesel, bioethanol, and biogas). Figure 12 demonstrates the results of the RELCA simulations. The results indicate that the overall regional average of 1 MJ of biodiesel is *ca.* 35.4 g CO_2-eq. MJ^{-1}.

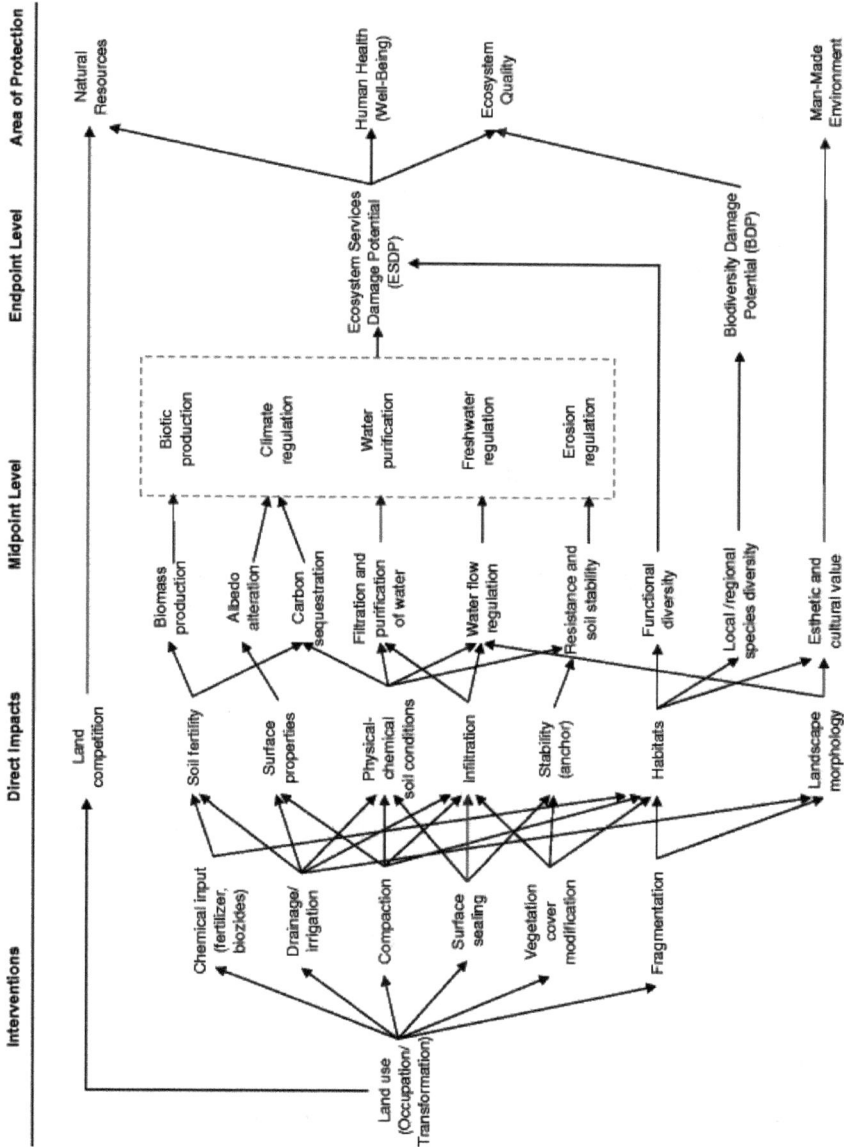

Fig. 11. The UNEP-SETAC guideline on global land use impact assessment on biodiversity and ecosystem services in the Life Cycle Assessment. (Reproduced from Ref. [42] under the terms of the Creative Commons CC BY license.)

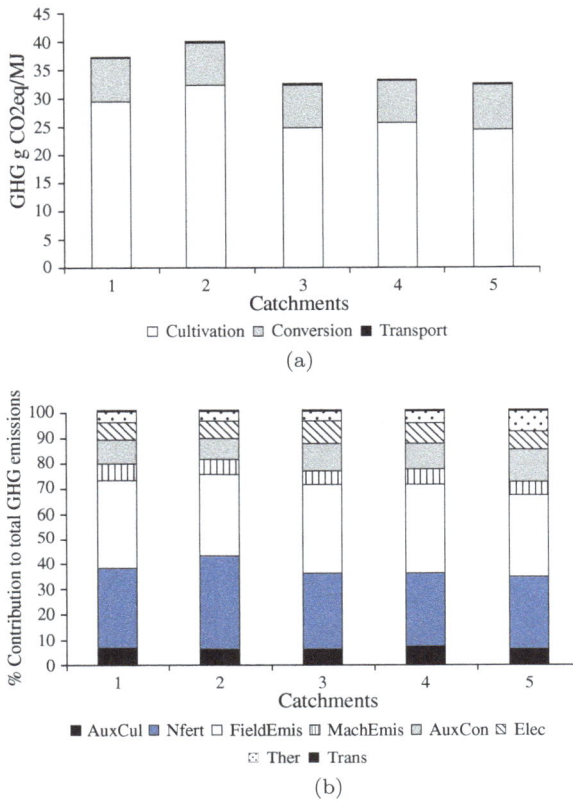

Fig. 12.   (a) g CO_2-eq./MJ results of process steps (cultivation, conversion and transport), and (b) Contribution of different parameters to the total greenhouse gas emissions. (Reproduced from Ref. [43] under the terms of the Creative Commons CC BY license.)

A case study of biodiesel production considering the techno-economic assessment (TEA) and LCA will be presented in Chap. 11.

4.2. *Hydrogen as a source of clean energy*

Among alternative fuels, H_2 is viewed as an important energy carrier of the future. The long-term goal for the "hydrogen economy" is generally the production of H_2 from sustainable renewable energy sources. H_2 has the highest energy content per unit weight of any known fuel

(122 MJ/kg H_2), and it is the only fuel that is not chemically bound to carbon. Despite such initiatives, the actual environmental performance of any H_2 energy system depends on how it is being produced. Conventional production processes of H_2 utilize steam methane reforming and coal gasification — both these technologies result in the emission of a large amount of GHGs.

With the increased interest in H_2 as a potential low-carbon fuel, the environmental performance of its production via the LCA has gained importance. In order to contribute to emission reduction targets, H_2 production has to be powered by renewable electricity sources such as PV systems [44]. Al-Breiki and Bicer [45] applied the LCA to compare the GHG emissions of LNG, ammonia, methanol, DME, and H_2 production and use. The LCA system is illustrated in Fig. 13. The results show that when natural gas is used as a feedstock, the total GHG emissions (from production, ocean transportation, and utilization/combustion) resulted in: 73.96 g CO_2-eq. MJ^{-1} for liquified natural gas, 95.73 g CO_2-eq. MJ^{-1} for methanol, 93.76 g CO_2-eq. MJ^{-1} for DME, 50.83 g CO_2-eq. MJ^{-1} for liquid H_2, and 100.54 g CO_2 eq. MJ^{-1} for liquid ammonia. The cleanest energy carrier was liquid

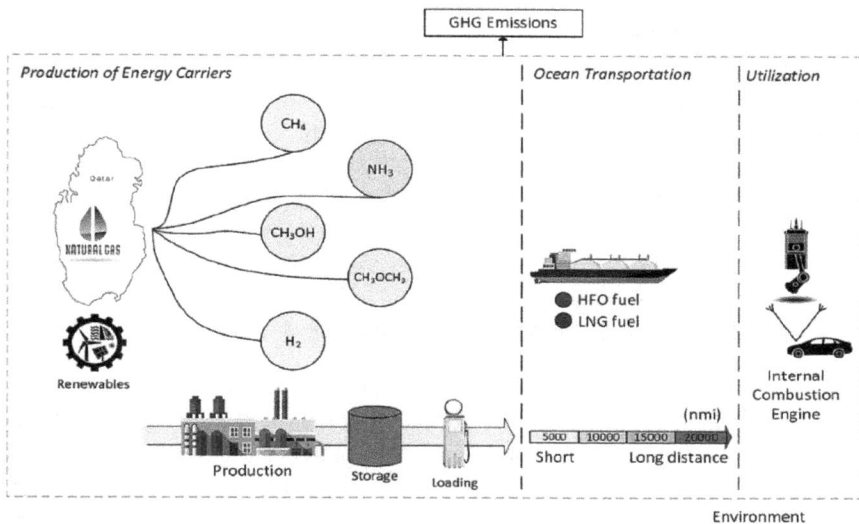

Fig. 13.   Life Cycle Assessment comparison of energy carriers. (Reproduced from Ref. [45] under the terms of the Creative Commons CC BY license.)

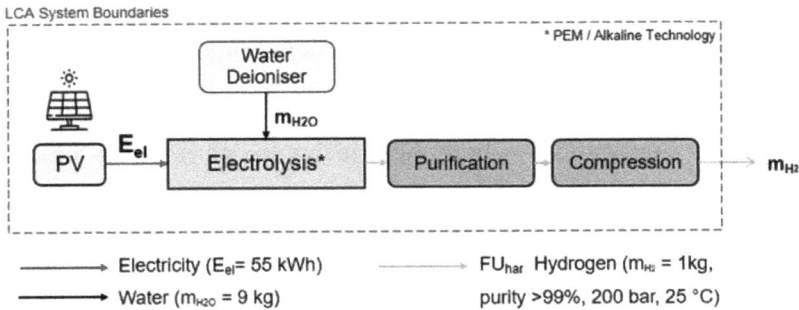

Fig. 14.   The system boundaries of reviewed LCA studies divided into electrolysis, inclusive of purification and compression subsystems. (Reproduced from Ref. [46] under the terms of the Creative Commons CC BY license.)

H_2 produced from solar-powered electrolysis, which resulted in 42.50 g CO_2 eq. MJ^{-1} fuel.

In a different viewpoint of investigation, Kanz and colleagues [46] review the LCA of studies of up to 13 sets of H_2 electrolysis systems using power from photovoltaic (PV) systems (Fig. 14). Within the LCA scope, 55–57.5 kWh of electricity and 9–10 L of water are required to produce 1 kg H_2. Their work discusses the assumptions, strengths and weaknesses of the LCA studies and identifies the causes of the varying environmental impacts, which range from 0.7 to 6.6 kg CO_2-eq/kg H_2. Overall, the main reasons were due to different efficiencies and thus the electricity consumption of the electrolyzer, PV technology and production, location of use, and operation time.

More research developments of clean H_2 will be demonstrated in Chaps. 5 to 7.

5. Further Developments: Circular Economy

The Paris Agreement calls for bold actions needed to mitigate climate change [1]. In order to achieve this, a major shift in the way products (resources or energy, chemicals, other materials) are being produced, used and recovered or recycled is needed. Circular Economy (CE) is a concept where wastes are recycled into resources through a series of processes or technological feedback mechanisms so that the stock of resources is constant or increasing over time. Various industries are beginning to recognize the benefits of CE. With the aim of optimizing how economies deliver societal demands, CE can play a vital role in realizing the transformational change needed to meet low-carbon economy targets. The benefits of CE could also reduce pressures on the environment or ecosystem as well as improve the efficiency and security of resources and raw materials. In this manner, the life cycle of products is extended.

According to Lee and colleagues [47], carbon-intensive industries such as the chemical and waste management sectors that are dependent on carbon feedstock for their production/operation are also increasingly under pressure to reduce their CO_2 emissions. The authors presented a concept to support the transformation from a linear *"one-way cradle to grave manufacturing model"* towards a circular carbon economy (Fig. 15). In their work, technological innovations and developments that are necessary to support a successful sector coupling of the energy, chemical and waste management sectors were identified. Technological recommendations were also highlighted as necessary steps towards the transformation from a linear to circular carbon economy.

Pires da Mata Costa and colleagues [48] illustrated how implementing CE can also be an important strategy to reduce the carbon footprint of the plastic chain, since circular strategies can keep the carbon circulating in

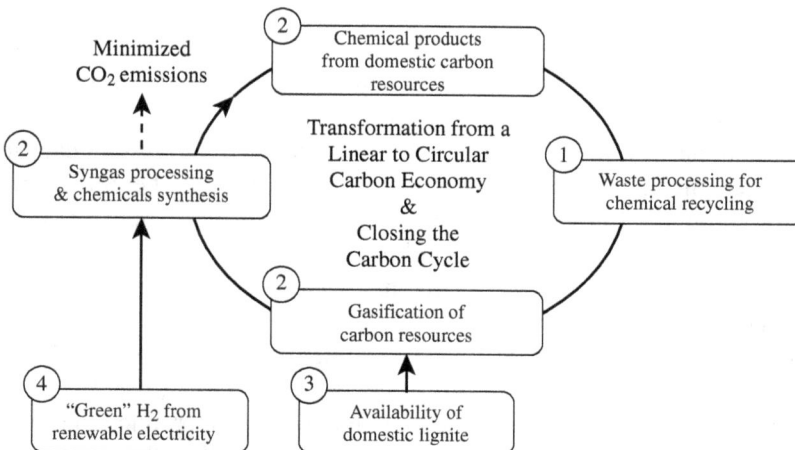

Fig. 15. Transformation from a linear to circular carbon economy and closing the carbon cycle. (Reproduced from Ref. [47] under the terms of the Creative Commons CC BY license.)

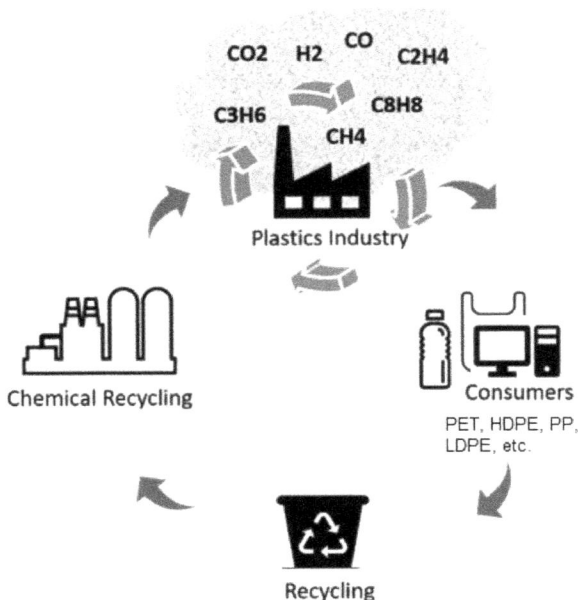

Fig. 16.   Circular Economy applied in plastic production and recycling. (Reproduced from Ref. [48] under the terms of the Creative Commons CC BY license.)

the production chain for long periods, extracting the maximum value of the raw materials and recovering and regenerating products at the end of their service life (illustrated in Fig. 16).

Additionally, a possible strategy to fully include the plastic chain into the CE scheme involves the sequestration and reutilization of emitted carbon dioxide through CCU techniques. In yet another example, Somoza-Tornos and colleagues [49] explored the benefits of the CE approach in the chemical industry by quantifying the economic and environmental benefits of ethylene recovery from polyethylene plastics using process modeling coupled with the LCA. The recovery of ethylene resulted in net emissions of −0.56 kg CO_2-eq./kg of waste PE. The results presented suggest that waste PE pyrolysis is an appealing route to close the loop in the ethylene production process, thereby promoting the adoption of CE principles within the plastics and chemicals sector. Jiang and colleagues [50] highlighted recent achievements of chemical recycling technologies with a focus on turning "waste to wealth". A CE approach was employed (Fig. 17). In their review, the authors

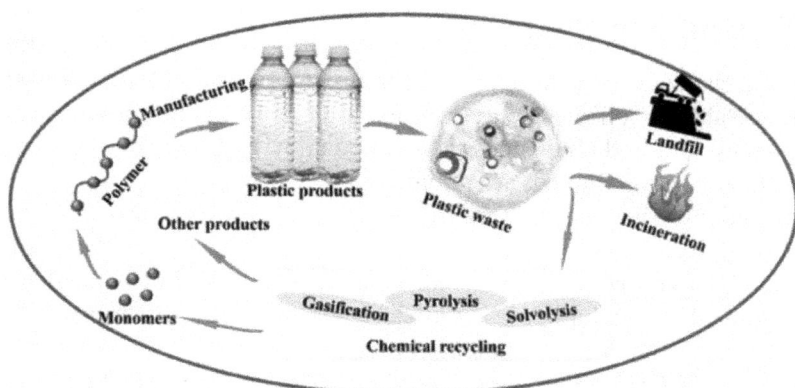

Fig. 17. Concept of the Circular Economy: "waste to wealth". (Reproduced from Ref. [50] with permission from Elsevier.)

evaluated the effects of reaction temperature, catalyst use, and reactor on recovery efficiencies. Related environmental performance during the treatment procedure was also critically discussed. The outcome of their work offers guidance for the development of industrial-relevant catalysts and recycling systems.

The CE for methanol will be demonstrated in Chap. 3. Details relating to sets of proper requirements of LCA models for chemical recycling strategies involving waste conversion to useful resources/products will be elaborated upon in Chap. 7. Carbon footprint reduction potentials for various polymer-recycling methods will be presented in Chap. 8.

Overall, LCA methodologies have evolved to serve different levels of climate sustainability objectives. Along with new avenues of science and research geared towards low carbon H_2 methods, process recycling involving circularity, and chemical transformation methods (CO_2 to chemicals/fuels), it would be worthy to exercise scientific approaches in LCA data selection and compilation. The flow of input-output data needed for LCA models becomes a challenge, especially in new experimental design setups involving complex molecular transformation. Khoo and colleagues [51] highlighted how LCA result uncertainties due to (ambiguous) data sources can be avoided, and recommended several steps involved to ensure correct information flows are given for LCA models to serve the needs of decision-makers, considering climate-related sustainability strategies.

6. Concluding Remarks

As industrialized nations look into emerging new technologies focusing on renewable or efficient energy use — along with the move towards SDGs — issues and challenges related to achieving low-carbon economy projects have gained much attention.

In order to ensure aims of reducing CO_2 emissions — or overall environmental sustainability goals — of the evolving processes targeted to reduce GHG can be well achieved, an environmental tool is needed for evaluating these technologies. As CCU attracts much attention in both industry and academia, the potential benefits — or any unintended drawbacks — of CCU applications should be analyzed in a holistic method. The LCA plays a crucial role to perform the overall evaluation of the actual environmental performance of any processes employed to achieve zero carbon emissions. In literature, various LCA studies of CCU have already been reported [29–32].

Besides the plethora of scientific fields involved in CCU, the decarbonization of energy systems is also seen as an opportunity to address the environmental impacts of conventional fossil-based power generation technologies. Various LCA studies have been carried out to explore the possibility of bioenergy systems to reach low carbon targets [39]. In this area, the need to look into land use change in the system boundary of LCA models has been highlighted [38, 42]. With increased research and developments in alternative energy resources, the environmental performance of H_2 production has gained much attention [40–43]. Explorations of other types of processes or recycling/recovery technologies, coupled with the LCA, are also ongoing. As a critical role in the path towards decarbonization, CE models involve designing products, processes and supply chains that can contribute to meeting low GHG emission targets [47–49]. As LCA approaches evolve to serve different levels of climate-related sustainability objectives, it would be worthy to exercise the scientific steps needed in LCA data needed to ensure the authenticity of LCA models [51].

References

1. Y. Gao, X. Gao and X. Zhang, The 2°C global temperature target and the evolution of the long-term goal of addressing climate change — from the United Nations Framework Convention on Climate Change to the Paris Agreement, *Engr.* **3**, 272–278 (2017).

2. L.J. Müller, A. Kätelhön, M. Bachmann, A. Zimmermann, A. Sternberg, *et al.* A guideline for Life Cycle Assessment of carbon capture and utilization, *Front. Energy Res.* **8**, 1–20 (2020).
3. F. Cherubini and G. Jungmeier, LCA of a biorefinery concept producing bioethanol, bioenergy, and chemicals from switchgrass, *Int. J. Life Cycle Assess.* **15**, 53–66 (2010).
4. S.P. Philbin, Critical analysis and evaluation of the technology pathways for carbon capture and utilization, *Clean Technol.* **2**, 492–512 (2020).
5. P. Markewitz, W. Kuckshinrichs, W. Leitner, J. Linssen, P. Zapp, *et al.*, Worldwide innovations in the development of carbon capture technologies and the utilization of CO_2, *Energy Environ. Sci.* **5**, 7281–7305 (2012).
6. G. Garcia-Garcia, M.C. Fernandez, K. Armstrong, S. Woolass and P. Styring, Analytical review of life-cycle environmental impacts of carbon capture and utilization technologies, *ChemSusChem* **14**, 995–1015 (2021).
7. Carbon Recycling International (CRI) (2016). http://www.carbonrecycling.is
8. R M. Cuéllar-Franca and A. Azapagic, Carbon capture, storage and utilisation technologies: A critical analysis and comparison of their life cycle environmental impacts, *J. CO_2 Util.* **9**, 82–102 (2015).
9. R. Chauvy, N. Meunier, D. Thomas and G. De Weireld, Selecting emerging CO_2 utilization products for short- to mid-term deployment, *Appl. Energy* **236**, 662–680 (2019).
10. S. Michailos, S. McCord, V. Sick, G. Stokes and P. Styring, Dimethyl ether synthesis via captured CO_2 hydrogenation within the power to liquids concept: A techno-economic assessment, *Energy Conv. Manag.* **184**, 695–703 (2019).
11. J. Patricio, A. Angelis-Dimakis, A. Castillo-Castillo, Y. Kalmykova and L. Rosado, Region prioritization for the development of carbon capture and utilization technologies, *J. CO_2 Util.* **17**, 50–59 (2017).
12. J.C. Abanades, B. Arias, A. Lyngfelt, T. Mattisson, D.E. Wiley, *et al.*, Emerging CO_2 capture systems, *Int. J. Greenh. Gas Control* **40**, 126–166 (2015).
13. H. Kim and K.S. Lee, Energy analysis of an absorption-based CO_2 capture process, *Int. J. Greenh. Gas Control* **56**, 250–260 (2017).
14. S.Y. Oh, M. Binns, H. Cho and J.K. Kim, Energy minimization of MEA-based CO_2 capture process, *Appl. Energy* **169**, 353–362 (2016).
15. B. Dutcher, M. Fan and A.G. Russell, Amine-based CO_2 capture technology development from the beginning of 2013 — review, *ACS Appl. Mater. Interfaces* **7**, 2137–2148 (2015).
16. S.Y. Oh, M. Binns, H. Cho and J.K. Kim, Energy minimization of MEA-based CO_2 capture process, *Appl. Energy* **169**, 353–362 (2016).
17. C. Han, K. Graves, J. Neathery and K. Liu, Simulation of the energy consumption of CO_2 capture by aqueous monoethanolamine in pilot plant, *Energ. Environ. Res.* **1**, 67–80 (2011).

18. M. Vaccarelli, R. Carapellucci and L. Giordano, Energy and economic analysis of the CO_2 capture from flue gas of combined cycle power plants, *Energy Procedia* **45**, 1165–1174 (2014).

19. P.E. Just, Advances in the development of CO_2 capture solvents, *Energy Procedia* **37**, 314–324 (2013).

20. B. Xue, Y. Yu, J. Chen, X. Luo and M. Wang, A comparative study of MEA and DEA for post-combustion CO_2 capture with different process configurations, *Int. J. Coal Sci. Technol.* **4**, 15–24 (2017).

21. E. De Lena, M. Spinelli, I. Martínez, M. Gatti, R. Scaccabarozzi, *et al.*, CO_2 capture in cement plants by "tail-end" calcium looping process, *Energy Procedia* **148**, 186–193 (2018).

22. M. Wei, F. Qian, W. Du, J. Hu, M. Wang, *et al.*, Study on the integration of fluid catalytic cracking unit in refinery with solvent-based carbon capture through process simulation, *Fuel* **219**, 364–374 (2018).

23. H.H. Khoo, R.M. Eufrasio-Espinosa, L.S.C. Koh, P.N Sharratt and V. Isoni, Sustainability assessment of biorefinery production chains: A combined LCA-supply chain approach, *J. Clean. Prod.* **235**, 1116–1137 (2019).

24. ISO 14040, Environmental management: Life cycle assessment — Principles and framework, International Organization for Standardization, Geneva (2006).

25. ISO 14044, Environmental management: Life cycle assessment — Requirements and guidelines, International Organization for Standardization, Geneva (2006).

26. M. Finkbeiner, A. Inaba, R.B.H. Tan, K. Christiansen and H.-J. Klüppel, The new international standards for life cycle assessment: ISO 14040 and ISO 14044, *Int. J. Life Cycle Assess.* **11**, 80–85 (2006).

27. A. Ghannadzadeh and A. Meymivand, Environmental sustainability assessment of an ethylene oxide production process through cumulative exergy demand and ReCiPe, *Clean Technol. Environ. Policy* **21**, 1765–1777 (2019).

28. H.H. Khoo, I. Halim and A.D. Handoko, LCA of electrochemical reduction of CO_2 to ethylene, *J. CO_2 Util.* **41**, 101229 (2020).

29. N. Thonemann and M. Pizzol, Consequential life cycle assessment of carbon capture and utilization technologies within the chemical industry, *Energy Environ. Sci.* **7** (2019).

30. A. Sternberg, C.M. Jens and A. Bardow, Life cycle assessment of CO_2-based C1-chemicals, *Green Chem.* **19**, 2244–2259 (2017).

31. W. Hoppe, N. Thonemann and S. Bringezu, Life Cycle Assessment of carbon dioxide-based production of methane and methanol and derived polymers, *J. Ind. Ecol.* **12**, 327–380 (2018).

32. N. von der Assen and A. Bardow, Life cycle assessment of polyols for polyurethane production using CO_2 as feedstock: Insights from an industrial case study, *Green Chem.* **16**, 3272–3280 (2014).

33. M. Rosental, T. Fröhlich and A. Liebich, Life Cycle Assessment of carbon capture and utilization for the production of large volume organic chemicals, *Front. Clim.* **2**, 586199 (2020).
34. A. Korre, Z. Nie and S. Durucan, Life cycle modelling of fossil fuel power generation with post-combustion CO_2 capture, *Int. J. Greenh. Gas Control* **4**, 289–300 (2010).
35. S. Sohni, N.A. Nik Norulaini, R. Hashim, *et al.*, Physicochemical characterization of Malaysian crop and agro-industrial biomass residues as renewable energy resources, *Ind. Crops Prod.* **111**, 642–650 (2018).
36. A. Muhammad, K. Ab Saman and N.A. Farid, The utilization potential of rice husk as an alternative energy source for power plants in Indonesia, *Adv. Mat. Res.* **845**, 494–498 (2014).
37. S. Kumar, P. Abdul Salam, P. Shrestha and E.K. Ackom, An assessment of Thailand's biofuel development, *Sustainability* **5**, 1577–1597 (2013).
38. M. Brandão, E. Azzi, R.M.L. Novaes and A. Cowie, The modelling approach determines the carbon footprint of biofuels: The role of LCA in informing decision makers in government and industry, *Clean. Environ. Sys.* **2**, 100027 (2021).
39. H. Rocha, R.S. Capaz, E.E.S. Lora, L.A.H. Nogueira, M.M.V. Leme, *et al.*, Life cycle assessment (LCA) for biofuels in Brazilian conditions: A meta-analysis, *Renew. Sust. Energ. Rev.* **37**, 435–459 (2014).
40. T. Prapaspongsa and S.H. Gheewala, Consequential and attributional environmental assessment of biofuels: Implications of modelling choices on climate change mitigation strategies, *Int. J. Life Cycle Assess.* **22**, 1644–1657 (2017).
41. US EPA, *Lifecycle Analysis of Greenhouse Gas Emissions under the Renewable Fuel Standard*, Renewable Fuel Standard Program (2022). https://www.epa.gov/renewable-fuel-standard-program/lifecycle-analysis-greenhouse-gas-emissions-under-renewable-fuel
42. T. Koellner, L. de Baan, T. Beck, M. Brandão, B. Civit, *et al.*, UNEP-SETAC guideline on global land use impact assessment on biodiversity and ecosystem services in LCA, *Int. J. Life Cycle Assess.* **18**, 1188–1202 (2013).
43. S. O'Keeffe, S. Wochele-Marx and D. Thrän, D. RELCA: A REgional Life Cycle inventory for Assessing bioenergy systems within a region, *Energy Sustain. Soc.* **6**, 12 (2016).
44. I. Dincer and C. Acar, Review and evaluation of hydrogen production methods for better sustainability, *Int. J. Hydrog, Energy* **40**, 11094–11111 (2015).
45. M. Al-Breiki and Y. Bicer, Comparative life cycle assessment of sustainable energy carriers including production, storage, overseas transport and utilization, *J. Clean. Prod.* **279**, 123481 (2021).
46. O. Kanz, K. Bittkau, K. Ding, U. Rau and A. Reinders, Review and harmonization of the life-cycle global warming impact of PV-powered hydrogen production by electrolysis, *Front. Electron.* **2**, 711103 (2021).

47. R.P. Lee, F. Keller and B. Meyer, A concept to support the transformation from a linear to circular carbon economy: Net zero emissions, resource efficiency and conservation through a coupling of the energy, chemical and waste management sectors, *Clean Energy* **1**, 102–113 (2017).
48. L. Pires da Mata Costa, D. M. Vaz de Miranda, A.C. Couto de Oliveira, L. Falcon, M.S.S. Pimenta, *et al.*, Capture and reuse of carbon dioxide (CO_2) for a plastics circular economy: A review, *Process.* **9**, 759 (2021).
49. A. Somoza-Tornos, A. Gonzalez-Garay, C. Pozo, M. Graells, A. Espuña, *et al.*, Realizing the potential high benefits of circular economy in the chemical industry: Ethylene monomer recovery via polyethylene pyrolysis, *ACS Sustain. Chem. Eng.* **8**, 3561–3572 (2020).
50. J. Jiang, K. Shi, X. Zhang, K. Yu, H. Zhang, *et al.*, From plastic waste to wealth using chemical recycling: A review, *J. Environ. Chem. Eng.* **10**, 106867 (2022).
51. H.H. Khoo, V. Isoni and P.N. Sharratt, LCI data selection criteria for a multidisciplinary research team: LCA applied to solvents and chemicals, *Sustain. Prod. Consum.* **16**, 68–87 (2018).

Chapter 2

From Carbon Dioxide to Renewable Fuels and Chemicals: The Important Role of Catalysis

Pin LIM[a,b]*, CHANG Jie[a], CHEN Luwei[a],
POH Chee Kok[a], LIM San Hua[a] and Hsien Hui KHOO[a]*****

[a]*Institute of Sustainability for Chemicals,
Energy and Environment (ISCE²),
Agency for Science, Technology and Research (A*STAR),
Singapore 627833*
[b]*Department of Chemical Engineering,
Imperial College London,
South Kensington Campus,
London SW7 2AZ, United Kingdom*

**p.lim21@imperial.ac.uk
**chen_luwei@isce2.a-star.edu.sg
***khoo_hsien_hui@isce2.a-star.edu.sg*

Recently, Carbon Capture and Utilization (CCU) has emerged as a key tool for the efforts to reduce net carbon emissions and combat global warming. CCU provides an advantage over traditional Carbon Capture and Storage (CCS), allowing for the valorization or upgrading of undesired carbon dioxide (CO_2) emissions to valuable fuels and chemicals. Advancements and developments of CCU methods can contribute considerably to supplying low carbon fuels and chemicals to various industries, along with reducing greenhouse gas emissions. To this end, catalysis research has become highly important in devising and developing alternative pathways for the production of renewable and sustainable fuels from CO_2. The main focus of this chapter is to review and discuss an overview of catalyst use in the viable pathways for CO_2 valorization, with some selected examples. A case study involving the heterogeneous catalysis of CO_2 to jet kerosene via a tandem catalytic modified Fischer–Tropsch process will be presented. Life Cycle Assessment (LCA) was applied to compare CO_2-based and conventional jet fuel.

1. Introduction: Catalyst Use for Carbon Capture and Utilization

To reduce the reliance on fossil fuels, alternative routes for producing low carbon liquid fuels and valuable chemicals from captured carbon dioxide (CO_2) have been a topic of great interest [1, 2]. Carbon Capture and Utilization (CCU) technology is a key tool in carbon mitigation and decarbonization strategies. CO_2 valorization strategies are highly valued, as converting undesired CO_2 emissions to fuels and chemicals will not only significantly contribute to greenhouse gas (GHG) reduction but also provide an avenue for recycling CO_2 through the production of CO_2-based fuels and chemical feedstocks for other industrial applications [3]. In the broad, emerging scientific arena of CCU, catalysis has taken up an important role in expanding the knowledge and possibilities in CO_2 conversion science and technology [4], encompassing a wide range of possible application areas, ranging from the replacement of conventional fuels with CO_2-based bulk and commodity chemicals [2].

However, many CO_2 valorizations pathways, especially for higher hydrocarbons products, are still in the stage of development. To date, most reported studies of CO_2 hydrogenation have focused on short-chained products such as CO, HCOOH, CH_3OH, CH_4, and C_2-C_4 olefins, while studies on efficient production of liquid (C_5+) hydrocarbons are rarely reported. This is mainly due to several inherent constraints that are present simply due to the intrinsic nature of the task at hand [3, 4]. Firstly, the CO_2 molecule is highly stable and unreactive, with the required deoxygenate molecule cleavage for activation hampered by the high bond strength (532 kJ/mol) [5].

In CCU research science, one of the main challenges faced lies in the efficient activation and cleavage of the small and relatively unreactive CO_2 molecule (O = C = O). This area of science and research has stimulated various research arenas covering chemistry, emerging new technologies, as well as process economics involved in product selectivity and potential yields with the use of available catalysts [3–6]. This is well-reflected in industry; only four major industrial processes presently utilize CO_2 as a raw material — urea synthesis, salicylic and para-hydroxy benzoic acid, polycarbonate synthesis, and methanol production [3]. Although the direct conversion of CO_2 to liquid fuels has received widespread attention and concerted efforts for the research and development of such catalysts, there is still significant potential for further improvement.

The emerging areas encompassing CCU, in retrospect, abate carbon emissions [7]. Homogeneous catalysis approaches have been applied to use CO_2, in particular for the synthesis of fine chemicals [8, 9]. The study of heterogeneous catalysis in CO_2 upgrading has also been keenly investigated in recent years — some important pathways are the synthesis of methanol, reverse water–gas shift, higher alcohols and hydrocarbons [5, 8–10].

A key pathway for the future is the modified Fischer–Tropsch (FT) synthesis route, which involves the direct conversion of CO_2 to C_{2+} hydrocarbons, removing the need for syngas feedstock. This will be discussed in detail in later sections as well as in the provided case study. A summary of the major reactions and pathways based on CO or CO_2 hydrogenation relevant to the energy and chemicals industries is provided in Fig. 1.

CO$_2$ based chemistries		CO based chemistries
$CH_4 + 2H_2O \rightarrow 4H_2 + CO_2$	Steam reforming	
$CO_2 + 3H_2 \rightarrow CH_3OH + H_2O$	Methanol synthesis	$CO + 2H_2 \rightarrow CH_3OH$
$CO_2 + 4H_2 \rightarrow CH_4 + 2H_2O$	Methanation	$CO + 3H_2 \rightarrow CH_4 + H_2O$
$CO_2 + H_2 \rightarrow CO + H_2O$	Reverse water gas shift	
$nCO_2 + (3n+1)H_2 \rightarrow$ $C_nH_{(2n+2)} + 2nH_2O$	Fischer Tropsch	$nCO + (2n+1)H_2 \rightarrow$ $C_nH_{(2n+2)} + nH_2O$

Fig. 1. Chemical pathways based on CO or CO_2 hydrogenation. (Reproduced from Ref. [8]; open access — Creative Commons CC BY license.)

2. Important CCU Pathways

2.1. *CO_2 to methane*

Methane (CH_4) is an important chemical feedstock and industrial fuel. The key challenge of a CO_2-to-methane (Sabatier) process is that the reduction of CO_2 to methane is the largest change of carbon's oxidation state, requiring an energy-intensive transformation [11]. The methanation process itself is also highly exothermic, requiring efficient energy removal to prevent thermodynamic bottlenecks during reactor operation. However, the advantage of methane is that it is a highly energetic fuel, combusting to release approximately 891 kJ mol^{-1}.

Heterogeneous catalysis of CO_2 methanation has been well-studied, with methanation processes typically employing nickel catalysts, which have good methane selectivity and high conversion efficiency [11–13]. In addition, Ru-based catalysts have also received significant attention recently [11]. Biological methanation is also an option, being less susceptible to the presence of impurities and disruptive components such as sulfur [13, 14]. Different reactor concepts can also aid process design while fixed-bed reactors are typically used for catalytic methanation; fluidized-bed reactors and novel concepts such as biological alternatives and three-phase methanation are also being studied [13]. Presently, several catalyst development efforts have progressed beyond the laboratory bench-testing stage, such as the Ni-based CO_2 methanation catalyst jointly developed by the Institute of Sustainability for Chemicals, Energy and Environment (ISCE^2, Singapore) and IHI Corporation (Japan) [15], which has undergone testing at a capacity up to of 500 kg CO_2/day. Another example of a catalyst that has reached the pilot plant stage is the proprietary CO_2 methanation catalyst being tested by the Institute for Chemical Processing of Coal (IChPW) at the Łaziska Power Plant (Poland) in a catalytic two-stage reactor system [16].

Other variants of methane production systems involving biomass-derived CO_2 as a feedstock have also been investigated [17]. Further catalytic improvements are required for efficient commercialization, especially with regard to lowering the required reaction temperature and reaction heat recovery [13].

2.2. *CO$_2$ to methanol*

Methanol is one of the top five commodity chemicals that can be directly employed as a fuel resource. Fossil-based methanol, referred to as MeOH or CH$_3$OH, is typically produced by the FT process, where pressurized synthesis gas (or syngas, a mixture of H$_2$ and CO) reacts in the presence of a catalyst [18]. In order to reduce the use of fossil fuels, as well as mitigate carbon emissions, there is a need to find alternative routes to synthesize MeOH. Ongoing research efforts focusing on the emerging technological conversion of methane to MeOH have been widely reported, including a biological process involving methanotrophic bacterial fermentation and plasma-aided synthesis processes [14]. Direct conversion of CO$_2$ to MeOH is another area of great interest. For example, Chou and Lobo [19] reported the investigation of supported indium oxide catalysts for CO$_2$ hydrogenation to methanol (illustrated in Fig. 2), significantly achieving near-perfect (100%) methanol selectivity at the low temperature range (~528 K) and a reaction pressure of 40 bar.

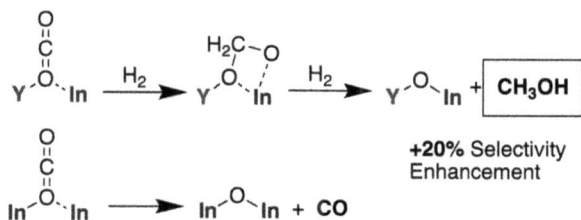

Fig. 2. Utilization pathway of indium oxide catalysts for CO$_2$ hydrogenation to methanol. (Reproduced from Ref. [19] with permission from Elsevier.)

Several chemical producers have already reported progress on their technology for CO$_2$ conversion to methanol; these catalysts are usually proprietary and undisclosed — however, modified variants of the conventional CuO/ZnO/Al$_2$O$_3$ type catalysts are known [20]. As mentioned in Chapter 1, one such example is Carbon Recycling International (CRI, Iceland), which has successfully made use of geothermal-sourced CO$_2$ feedstock to produce methanol at a commercial scale (3,000 t/y of methanol) [21].

2.3. *CO_2 to higher alcohols*

Direct conversion of CO_2 to higher alcohols (HA) is more challenging to achieve, mainly due to the increased reaction complexity and difficulty in controlling C-C coupling (Fig. 3(a)) [22]. The most versatile and useful HAs are within the range C_2–C_5, as these have the appropriate properties for blending into the gasoline pool, hence possessing great potential for the transportation industry. One method of ethanol production is via the catalytic conversion of syngas (a mixture of primarily CO and H_2), derived from biomass, coal or natural gas [8]. Butanol is also of industrial interest, as it has sufficient energy density to make up a significant portion of the gasoline blend pool [23]. Currently, these alcohols are primarily produced from biomass through microbial fermentation, which has inherent scale-up limitations [23, 24].

Direct syngas to alcohol production routes have been studied and reported for many decades. In 1985, Mo-KCl/SiO_2 catalyst was the first catalyst found to display reasonable activity for direct CO_2 to HA transformation [25]. More recently, CuFe catalysts developed by researchers at the Chinese Academy of Sciences have shown promising activity [26]. Additionally, modified Rh-based catalysts promoted by Li, Fe and V and supported on silica have also been found to be highly efficient [22]. Fe is an important promoter for this process, as FeO_x is an important RWGS facilitator [27], generating CO for the hydrogenation to HAs. Silica was found to be the most suitable support for Rh catalysts, but other reducible

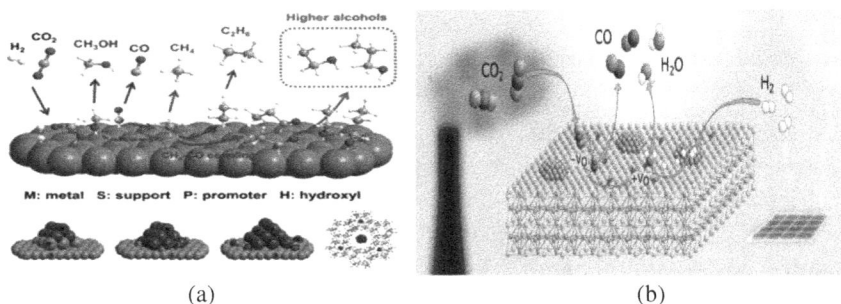

(a)   (b)

Fig. 3.   (a) Catalytic conversion of CO_2 to possible products, including higher alcohols. (b) Surface adsorption mechanism of the reverse water–gas shift pathway over novel perovskite catalysts. (Reproduced from Ref. [29] with a Creative Commons permission.) (Reproduced from Ref. [22] with permission from Elsevier.)

metal-oxide supports such as TiO_2 and of course Fe_3O_4 have also been found to be effective [22].

2.4. *CO_2 to higher hydrocarbons*

Key transformations of CO_2 into relevant higher hydrocarbon molecules have also been well studied. Specifically, the production of long-chained (C_{5+}) hydrocarbon molecules from CO_2 takes place primarily via the aforementioned FT process, which has syngas (CO) as a reactant [28]. Thus, this necessitates an additional step, to convert CO_2 into CO via the reverse water–gas shift pathway [27, 29] (see Fig. 3(b)). Alternatively, CO_2 can first be converted to methanol, then to hydrocarbons through the methanol-to-gasoline (MTG) process [30].

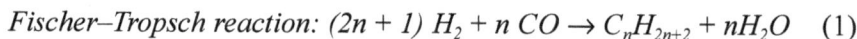

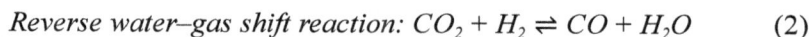

$$\text{\textit{Fischer–Tropsch reaction: }} (2n+1)\,H_2 + n\,CO \rightarrow C_nH_{2n+2} + nH_2O \quad (1)$$

$$\text{\textit{Reverse water–gas shift reaction: }} CO_2 + H_2 \rightleftharpoons CO + H_2O \quad (2)$$

FT synthesis is one of the most widely utilized processes for gas-to-liquid (GTL) conversion and is one of the primary production pathways for CO_2 to long-chained hydrocarbons [28, 31]. However, due to the lack of confinement effects on FT metal surfaces, C-C coupling is difficult to control. This leads to chain growth in a similar fashion to that of polymerization reactions, resulting in a hydrocarbon distribution of what is well-established as the Anderson–Schulz–Flory (ASF) distribution [7], which drastically limits the yield of higher carbon products.

As mentioned, in a traditional two-step process, conversion to CO is required before FT upgrading to hydrocarbon fuels can occur. This additional step, which can be achieved through the reverse water–gas shift (RWGS) reaction or thermochemical splitting using solar power [32], results in a two-step (cascade) process that inherently necessitates additional capital and utility requirements. Considering direct CO_2 conversion leads us to appraise the possibility of an alternative FT process that converts CO_2 to hydrocarbons directly. Known as the modified CO_2 FT process, this involves directly converting CO_2 to long-chained hydrocarbons by performing the tandem catalytic coupling of RWGS and FT with a bifunctional catalyst.

Direct CO_2 utilization via FT, while holding great potential, is more challenging due to its chemically inert nature of CO_2 [3, 4] and the selectivity challenges that arise with CO [33]. While the CO_2 to hydrocarbons pathway could be effectuated in a cascade system, there are several clear benefits to utilizing a tandem catalytic single-reactor system. A smaller footprint and greater energy efficiency can be achieved via direct CO_2 conversion combining the endothermic (RWGS) and exothermic (FT) reactions in one reactor. This also simplifies the process, as the knockout sections for H_2O removal, heat exchangers and compressor between two reactors can be omitted. Thus, the development of an effective catalyst and a suitably optimized process for a modified (two-step) CO_2 FT is likely to effectively reduce production costs, drastically improving the economic and environmental viability of the process [34].

Due to the potential benefits of such a tandem process, this particular strategy has also been investigated by researchers in recent years [28]. For instance, Wei and colleagues developed a high-performing Na-Fe_3O_4/ HZSM-5 catalyst for this modified FT process [35]. Gao and colleagues reported the development of a bifunctional In_2O_3/HZSM-5 catalyst, which produces hydrocarbons via the MTG route [36]. An organic combustion method was utilized by Yao and colleagues to synthesize a Fe-Mn-K catalyst, which achieved good hydrocarbon yields at relatively lower temperatures [34]. However, no commercial catalysts are currently available for the direct conversion of CO_2 to liquid fuels via this modified FT synthesis. The high stability of CO_2 and complex reaction pathways result in an inherently low CO_2 conversion and desired product selectivity, limiting potential applications at industry scale [28].

2.5. *CO_2 to light olefins*

Light olefins (LO) (C_{2-4}) are also highly valuable as they are important building block chemicals for industrial products such as synthetic rubbers and plastics [37, 38]. In fact, ethylene and propylene command major demand in the petrochemical industry, being produced in the highest capacities (200 Mt/a) [39]. Similar to the production of higher hydrocarbons, olefin production can also take place through the modified CO_2 FT route, as discussed earlier, and the methanol-to-olefins (MTO) route. As mentioned, the wide product selectivity and the ASF distribution limitations are barriers to olefin yield via the modified CO_2 FT [40]. The MTO route involves the utilization of a bifunctional catalyst for the hydrogenation of CO_2 to methanol and subsequent conversion to olefins [41].

High-performing catalysts for these production routes generally utilize zeolite supports or enclosures [42, 43]. Core-shelled or hollow-spherical catalysts have also been found to be beneficial for selectivity tuning to improve LO yield [44, 45]. Notable examples of commercially operational MTO processes are MTO (UOP/Norsk Hydro) and MTP (Lurgi), both of which are in use in China [46].

3. Applications of LCA Evaluations for CO_2 Valorization Processes

For CCU applications, the Life Cycle Assessment (LCA) can be used as a scientific tool to appraise the objective of global warming mitigation. From a systems approach perspective, apart from CO_2 utilization potential, the LCA accounts for additional (unexpected) emissions that may arise from intensive energy utilization and other fossil-based resources required. In the pursuit of ensuring the benefit of CCU, the application of LCA ensures decarbonization targets are met [47].

The scientific arena of CO_2 conversion covers a wide range of emerging application areas, spanning liquid fuels to bulk and commodity chemicals [39, 48]. In order to justify and ascertain the benefits of CO_2 utilization, a vigorous LCA evaluation is required to confirm CO_2 mitigation compared to the conventional (often, fossil) route, taking into account the whole value chain from the origin of the utilized CO_2 to the final usage of the product [8]. This will provide a comprehensive and accurate outlook on the viability of the proposed pathway.

As an example, Artz and colleagues reviewed selected state-of-the-art synthetic methodologies and processes for the potential to reduce the LCA environmental footprint of dimethyl carbonate (DMC) production [4]. As seen from Fig. 4, all reported global warming impacts for CO_2-based production via transesterification to DMC are lower.

Apart from DMC, the authors [4] also carried out a study of the use of CO_2 for making formic acid, a commodity chemical with a reported production rate of 720,000 t/a in 2013 [49]. Different routes (two different process concepts involving direct hydrogenation [50, 51]) for CO_2 to formic acid processing were studied, and the global warming impact of the CO_2-based processes were benchmarked against the commercial fossil-based pathway for formic acid production (Fig. 5). Zang and colleagues, on the other hand, conducted an LCA and technoeconomic analysis

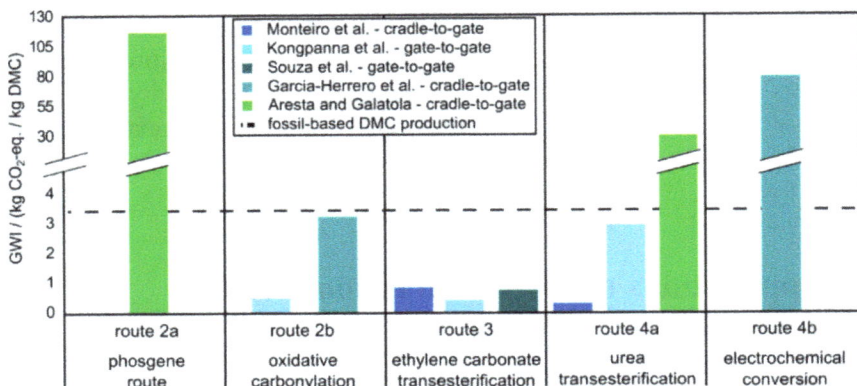

Fig. 4.   Reported global warming impacts for dimethyl carbonate (DMC) production. The global warming impacts reported by the five LCA studies vary largely due to diverging LCA methodologies, dissimilar background LCA data, and differing underlying process designs. (Reproduced from Ref. [4] with permission from the American Chemical Society.)

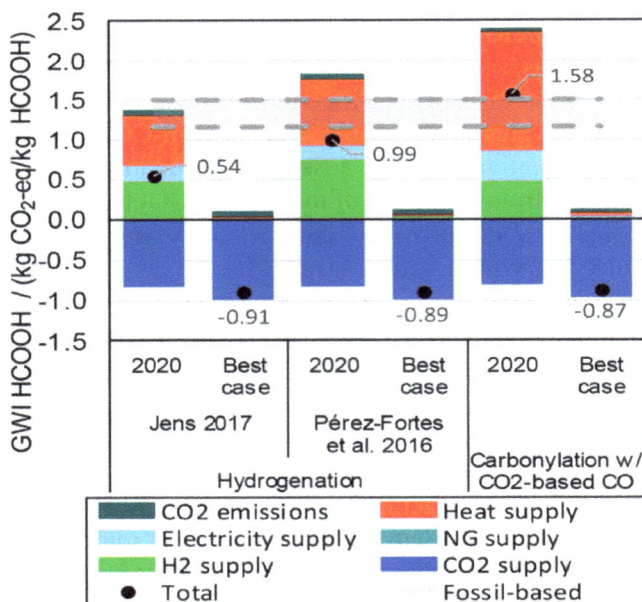

Fig. 5.   Global warming impact (GWI) for CO_2-based formic acid production as compared to the current industrial benchmark (cradle-to-gate). (Obtained from Ref. [4], with permission from the American Chemical Society.)

involving the following process systems [52]: integrated methanol–ethanol coproduction, integrated methanol–ammonia coproduction, and standalone methanol production. CO_2 feedstock was supplied from ethanol plants, ammonia plants, and the general CO_2 supply market. The cradle-to-gate GHG emissions of three considered methanol production systems were estimated with various H_2 production and electricity generation technologies and coproduct allocation methods, as shown in Fig. 6.

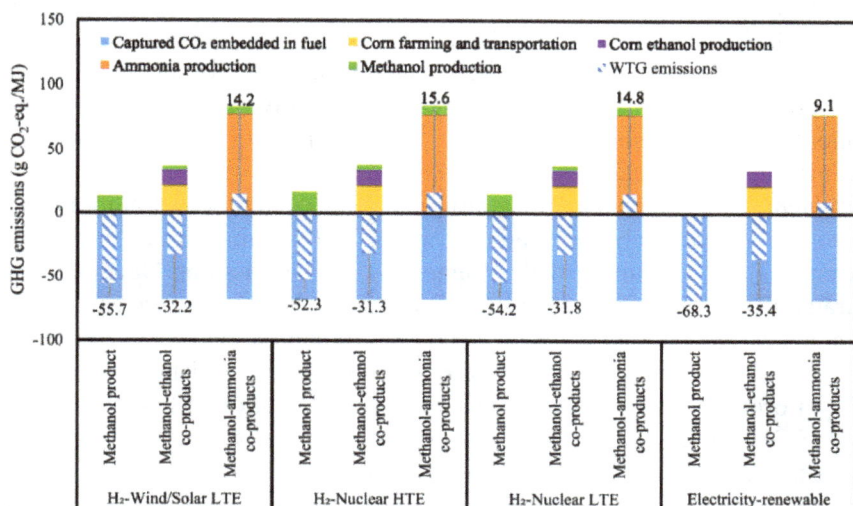

Fig. 6. Methanol cradle-to-gate GHG emissions with various H_2 sources, electricity types, and allocation methods. (Reproduced from Ref. [52] with permission from the American Chemical Society.)

Additionally, Rosental and colleagues performed a cradle-to-gate basis LCA for the production of large-volume organic chemicals such as methanol, ethylene, propylene and BTX [53]. Their investigation found that assuming the usage of electricity powered by renewable wind sources, producing these essential commodity chemicals via CCU production pathways will result in a reduction of GHG by up to 97%, along with other environmental benefits. Finally, two previous studies have evaluated the environmental impacts of CO_2-based higher hydrocarbon production [54, 55], something that is directly relevant to our provided case study. However, both were based on the conventional (cascade) FT process, where CO_2 was first converted into CO in a reactor situated upstream.

4. LCA Case Study: CO_2 to Kerosene via a Modified FT Process

4.1. *Background and significance*

As a global logistics hub, Singapore's aviation and maritime sectors are naturally major fossil fuel consumers and, thus, major emission contributors. Due to the inherent weight and space constraints, decarbonizing these sectors by electrification is likely to face considerable challenges [56, 57]. Specifically, carrying heavy batteries would simply be counterproductive, increasing fuel requirements to sustain the journey, and is not space-efficient. Similarly, hydrogen carriers simply do not have the sufficient energy density required for aviation journeys and would also require significant infrastructure and engine modifications.

Henceforth, finding a direct, drop-in replacement fuel for traditional jet kerosene appears to be the preferred way of solving this particular conundrum [57]. In alignment with Singapore's Low-Emissions Development Strategy, the objective of this project was to develop an efficient catalyst to produce sustainable jet kerosene from CO_2 through a viable CCU production pathway to reduce GHG emissions and fossil-fuel reliance.

In order to achieve the goal of sustainable aviation, a Fe-based catalyst (FCZA) was developed for the production of jet fuel range hydrocarbons through the modified CO_2-based FT process. This catalyst was tested in a bench-scale fixed-bed reactor without gas recirculation at our dedicated high-pressure testing facility at the Institute of Sustainability for Chemicals, Energy and Environment (ISCE²) in Singapore. The observed catalytic performance and inventory flow data obtained were utilized to perform an LCA, with process simulation aids and intelligent scale-up assumptions being made. The LCA steps are undertaken as per the guidelines stipulated by the International Organization for Standardization (ISO) [58, 59]. More details are provided in the subsequent sections.

4.2. *LCA model parameters*

As mentioned above, the power of the LCA can be applied to expand the debate on the single-stage of CO_2 reduction/utilization to factor in additional emissions that might arise from the use of energy and other resources [47, 60]. Notably, the choice of the CO_2 source is highly important as the energy requirements of the capture process can have significant impacts on the true viability of the CCU pathway [61].

In this case study, the functional unit is of 1 kg of aviation-standard kerosene jet fuel, and it is assumed that fossil and CO_2-based kerosene both have the same properties (typically known as Jet A-1) for use in aviation. The cradle-to-gate system boundary for fossil-based kerosene production (Fig. 7(a)) is defined from cradle-to-gate:

i) Crude oil production from the Middle East region,
ii) Delivery of crude oil feedstock is done via an ocean tanker from Saudi Arabia's Terminal to Singapore's Jurong Island port in Singapore (marine travel distance of 6,847 km), and
iii) Crude oil is processed in a refinery in Singapore to produce 1 kg of fossil-based kerosene.

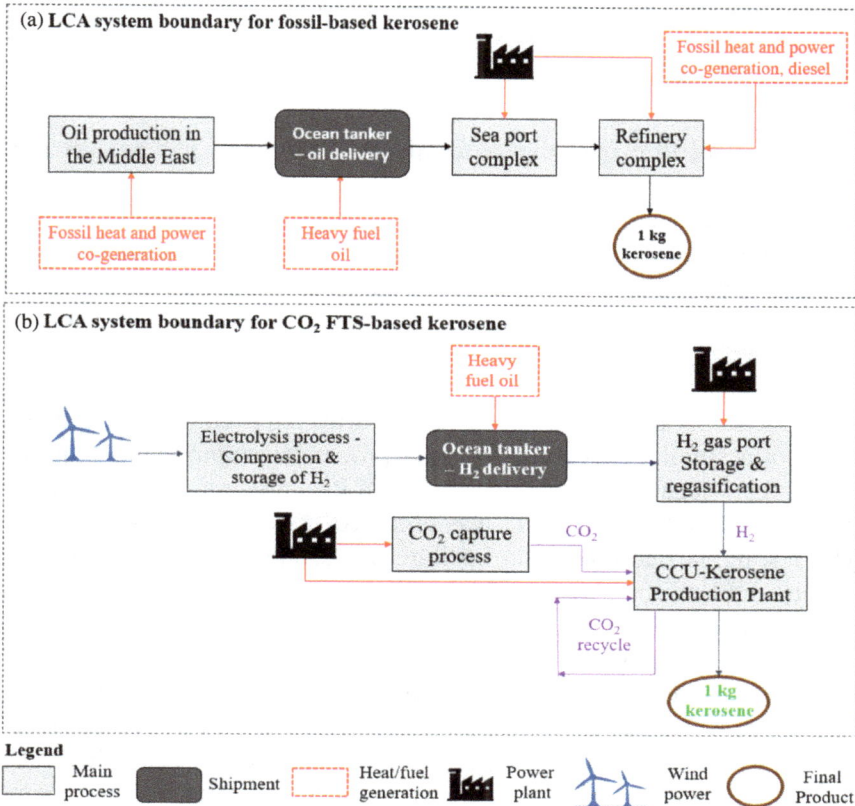

Fig. 7. LCA system boundaries for: (a) Fossil-based kerosene, and (b) CO_2 FT-based kerosene.

Correspondingly, the cradle-to-gate for CCU kerosene production (Fig. 7(b)) is defined as follows:

i) CO_2 feedstock is captured from NCCC (natural gas combined cycle) via amine absorption technology [62] from the flue gas emissions of a power plant.
ii) H_2 feedstock is supplied from renewable sources in Australia; delivery of H_2 to Singapore's port from Australia's West terminal is carried out via an LNG-type ocean carrier (marine travel distance of 3,051 km).
iii) H_2 regasification is carried out at the receiving port [63].
iv) CO_2 and H_2 feed gas are converted to CO_2 FT-based kerosene via a thermocatalytic process in a production plant; with reasonable assumptions made for separation and recycle.

4.3. *LCI data*

It is important to clarify that as catalyst development has not yet progressed to a pilot-plant scale, the data obtained at the laboratory level was scaled-up intelligently with reasonable assumptions for the purposes of simulating the output of an industrial-scale production plant. Accordingly, it was assumed that a CO_2 conversion of 80% was effectively obtained for this simulated plant operating at typical process conditions with recycle in a potential real-life design. Correspondingly, the CO_2 utilization was 80% of 3.14 kg of CO_2 (2.51 kg in total) being converted to 1 kg of kerosene (as per Eqs. (1) and (2)). Utility requirements were determined via simulation of a simplified process, encompassing required downstream separation units, in an ASPEN Plus process simulation suite. Inventory flow data (Table 1) was obtained from the bench-scale catalytic testing carried out at the Institute of Sustainability for Chemicals, Energy and Environment (ISCE[2]) in Singapore. To perform quantitative mass and energy balances over the processing stages [62, 63] depicted in the LCA system boundary, global LCI databases (ecoinvent [64]) and software (Gabi Life Cycle Engineering) were used to provide data detailing the emissions and environmental burdens associated with the production of raw materials and resources required for the respective fuel/feed production pathway [65–67]. LCI data sourced and referenced are compiled in Table 2.

Table 1. Single-pass laboratory-scale mass balance ($ISCE^2$ catalyst), adjusted to 1 kg of kerosene yield.

Species	In (kg)	Out (kg)
CO_2	41.56	27.44
H_2	5.71	3.77
C_1–C_4	—	1.68
C_5–C_8	—	0.92
C_9–C_{16}	—	1.00
CO	—	0.84
H_2O	—	11.56
Total	47.27	47.21

Table 2. Additional LCA data applied.

LCA Stages for Fossil-based Kerosene	Type of Data and Source(s)
Crude oil extraction in the Middle East region	Energy use and emissions [64]
Shipment — crude oil delivery	Energy use and emissions [64]
Sea port complex	Port emissions [65]
Refinery complex	Energy use and emissions [64]

LCA Stages for CO_2 FT-based Kerosene	Type of Data and Source(s)
Renewable H_2 production, compression and storage	Energy use and emissions [66, 67]
Shipment — H_2 delivery	Energy use and emissions [64]
Sea port station and regasification of H_2	Port emissions and H_2 regasification [63, 65]
CO_2 capture — Natural Gas Combined Cycle (NGCC) power plant	Energy penalty [62]
CO_2-FT kerosene production plant	Plant utility supplied by NGCC power [68]

4.4. Lifecycle Impact Assessment

Life Cycle Impact Assessment (LCIA) is the final step in LCA studies, where inventory information (Table 1) is converted into a set of environmental impact categories. Various types of scientific environmental models exist to evaluate global (e.g., climate change) or regional-specific impacts (e.g., acidification). Measurements of various pollutants resulting in the set of environmental impact categories are vital to ensure the LCA system (described in Sec. 4.2) meets the objective to mitigate climate change and ensure the sustainability of the planet.

As one of the most widely applied and updated *impact* assessment methods, ReCiPe 2016 [69] was used to generate the set of environmental impacts that provides characterization factors representative on a global scale while maintaining the possibility for a number of impact categories. The four environmental interventions selected are: (i) Climate Change (CC), (ii) Terrestrial Acidification (TA), (iii) Photochemical Ozone Formation — Human Health (POF-HH), and (iv) Fuel Depletion (FD). The results of CC comparing both conventional (fossil-based) and CO_2-based kerosene production stages are displayed in Fig. 8(a) and 8(b), respectively, and the corresponding comparative results for FD are displayed in Fig. 9(a) and 9(b), respectively. Other environmental parameters are presented in Table 3.

The cradle-to-gate LCA investigation elucidated that the production process was generally environmentally favourable (Fig. 8(a)). Notably, the CC results demonstrate that GHG emissions dropped from 3.57 kg CO_2 (fossil-based) to −0.45 kg (CO_2-based). This implies that climate change impacts mitigation can be achieved by the FCZA catalyst developed for the production of jet fuel through the modified CO_2 FT process designed by ISCE[2]. TA results for CCU-based kerosene were reduced by *ca.* 9×10^{-3} kg SO_2-eq; and POF-HH impacts decreased by 99%. In addition, FD was reduced by a factor of ~2.5 oil-eq, indicating that the new production process is effective in reducing negative environmental and health impacts.

Furthermore, it is clearly demonstrated that the majority of the CC (80%) and FD (>50%) impacts for fossil-based kerosene were generated during the crude oil extraction stage. Additionally, pollution from shipment and refinery complex operation, along with the associated use of fuels/heating systems, contributed significantly to TA and POF-HH impacts, which are not present for the CCU process. Apart from CC, the results for CO_2-based kerosene show that most of the environmental impacts largely arose from the transportation (shipping) of renewable H_2 from Australia to Singapore, and 35% and 40% of the total FD impacts were due to energy use for the CO_2 capture process and H_2 regasification, respectively; both of which are highly energy intensive.

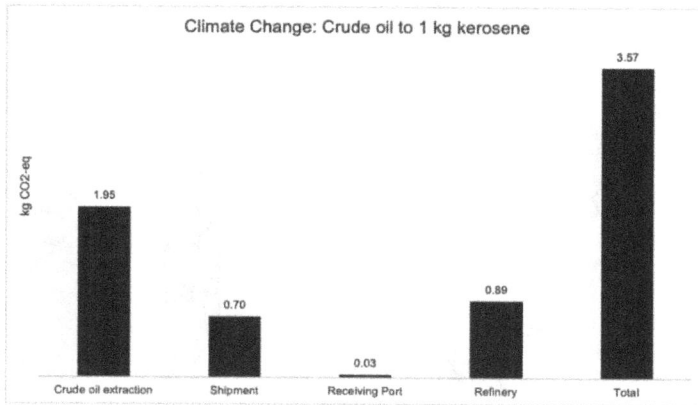

(a)

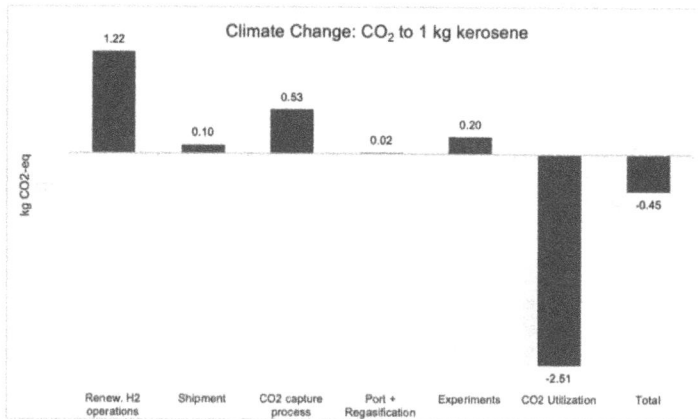

(b)

Fig. 8.   Climate Change results: (a) Fossil-based kerosene production, and (b) CO_2-based kerosene production.

To summarize, the investigation done via an LCA perspective demonstrates that apart from reducing climate change impacts via utilization of CO_2, other sets of environmental factors should be carefully considered. Significantly, specific stages in the production pathway (e.g., transfer of H_2 via ocean tankers and regasification) were observed to contribute notably to the environmental burden.

(a)

(b)

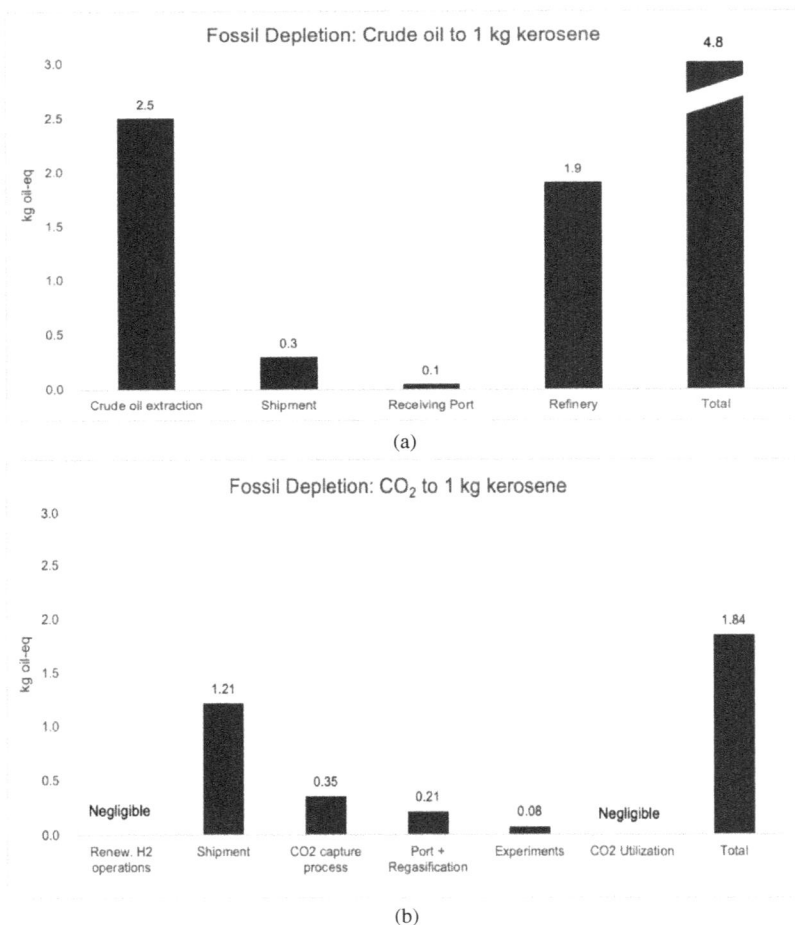

Fig. 9.   Fossil Fuel Depletion results: (a) Fossil-based kerosene production, and (b) CO_2-based kerosene production.

5. Further Discussion: Developments in Sustainable Aviation

The preliminary LCA study conducted demonstrates that CO_2-based kerosene, produced via a simplified process design enhanced with a suitable catalyst, has great potential. However, it should be highlighted that the LCA results generated from the current lab-scale study conducted do

Table 3. Potential environmental profile (1 kg of kerosene) — fossil-based vs. CO_2-based.

Unit Operation	Terrestrial Acidification (kg SO_2-eq)	Photochemical Ozone Formation (kg NOx-eq)
Production of Fossil-based Kerosene		
Crude oil extraction	9.1×10^{-3}	3.9×10^{-3}
Shipment	2.0×10^{-5}	1.0×10^{-6}
Receiving Port	1.0×10^{-6}	1.2×10^{-8}
Refinery	3.2×10^{-4}	2.5×10^{-4}
Total	9.4×10^{-3}	4.2×10^{-3}
Production of CO_2 FT-based Kerosene		
Renewable H_2 operations	1.50×10^{-6}	2.00×10^{-8}
Shipment	2.00×10^{-5}	4.10×10^{-6}
CO_2 capture process	5.10×10^{-6}	4.30×10^{-7}
Receiving port & regasification	0.00	2.30×10^{-7}
CO_2-based kerosene production	1.00×10^{-8}	0.00
CO_2 utilization	N.A.	N.A.
Total	2.66×10^{-5}	4.78×10^{-6}

not practically represent that of an industrial-scale CCU plant integrated with optimized process. Further larger-scale case studies, with associated process-based information for the LCA, are recommended to enhance the evaluation of environmental benefits for industrial-scale production of CO_2-FT-based kerosene. Due to the inherent constraints faced by the aviation industry, the significance of producing a direct, drop-in replacement fuel in a sustainable manner cannot be understated. The development of efficient catalyst design would contribute significantly to the environmental viability and future outlook of this direct conversion process [28]. Process economics will also certainly be improved as a result, leading to the envisaged production pathway becoming more economically viable and environmentally beneficial [34, 35].

Research interest in the evaluation of sustainable aviation fuels with the LCA is also on the rise. In a similar study, Li and Mupondwa evaluated the environmental impact of hydro-processed renewable jet fuel derived from Camelina oil [70]. In their LCA method, the impacts of global warming potential, human health, ecosystem quality, and energy resource consumption are generated and evaluated. Their results highlight that using biomass feedstock for fuel production accounted for a significant contribution to overall environmental performance, hence demonstrating the critical importance of minimizing environmentally damaging agriculture activities. Further investigations focusing on low-carbon jet-fuel oil are also expected to follow.

Recently, Nasriddinov and colleagues [71] investigated the effect of several metallic promoters on FeAlK, a known CO_2 hydrogenation catalyst. The addition of metallic promoters facilitated the formation of ferrite and carbide phases, both of which are key to the promotion of C-C coupling and hence, enhancing modified CO_2 FT activity. The production of kerosene-based aviation fuel from CO_2/CO was also studied by Arslan and colleagues [72]. This investigation (Fig. 10) demonstrated the ultra-high selectivity of single-ring aromatics (precursors for the kerosene-based aviation fuel) over the desired range (C_8–C_{12}) via the catalytic hydrogenation of CO_2/CO over a bifunctional $ZnCr_2O_4$/Sbx-H-ZSM-5 nano-catalyst.

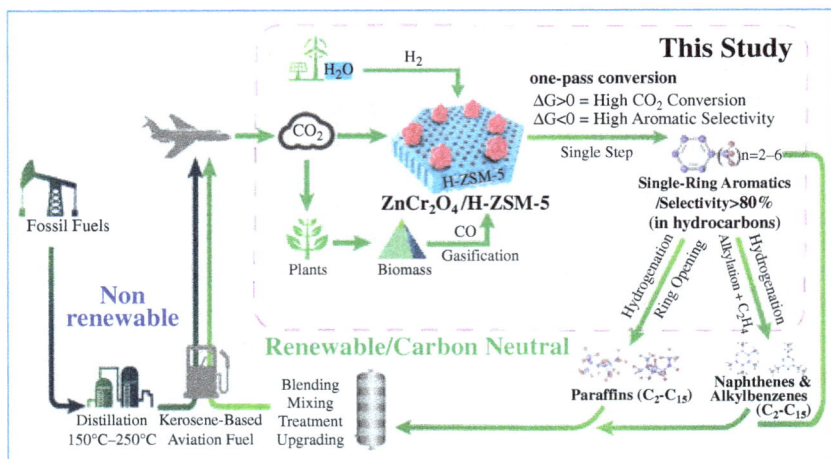

Fig. 10.   Conversion of CO_2/CO into precursors for kerosene-based aviation fuel. (Reproduced from Ref. [72] with permission from the American Chemical Society.)

6. Concluding Remarks

In conclusion, we have demonstrated that the valorization of CO_2 to sustainable aviation kerosene via CO_2-FT technology is highly promising. However, further development is required for economic viability and industrial adoption as a direct replacement for currently adopted sustainable aviation fuel (SAF). As FT fuel production is a complex process with multiple downstream processing steps, more accurate environmental and costing evaluations need to be conducted in the future when the catalytic technology has attained a sufficient technological-readiness level.

As the industrial focus moves towards sustainable processes and away from the traditional reliance on crude feedstocks, the push for the efficient production of renewable fuels will gain increased industrial significance. Catalysis has a key role to play in the production of sustainable fuels and commodity chemicals, as it allows for the transformation of waste carbon emissions to essential chemicals that are crucial for everyday life, whether it be the demand for energy sources or commercial products. As hydrogen technology improves and catalyst development ramps up in the near future, the potential for CO_2 valorization technology will only increase. It has been successfully shown that CCU provides significant environmental benefits — the race for sustainable technological development is only just beginning.

References

1. N. Gruber, D. Clement, B.R. Carter, R.A. Feely, S. van Heuven, M. Hoppema, *et al.*, The Oceanic sink for anthropogenic CO_2 from 1994 to 2007, *Science* **363**, 1193 (2019).
2. P.-C. Chiang and S.-Y. Pan, Post-combustion carbon capture, storage, and utilization, in *Carbon Dioxide Mineralization and Utilization*, P.-C. Chiang and S.-Y. Pan (eds.), Springer, Singapore, pp. 9–34 (2017).
3. M. Aresta, A. Dibenedetto and A. Angelini, Catalysis for the valorization of exhaust carbon: From CO_2 to chemicals, materials, and fuels. Technological use of CO_2, *Chem. Rev.* **114**(3), 1709–1742 (2014).
4. J. Artz, T.E. Muller, K. Thenert, J. Kleinekorte, R. Meys, A. Sternberg, *et al.*, Sustainable conversion of carbon dioxide: An integrated review of catalysis and life cycle assessment, *Chem. Rev.* **118**(2), 434–504 (2018).
5. Q.-W. Song, Z.-H. Zhou and L.-N. He, Efficient, selective and sustainable catalysis of carbon dioxide, *Green Chem.* **19**(16), 3707–3728 (2017).

6. H. Lamberts-Van Assche and T. Compernolle, Economic feasibility studies for Carbon Capture and Utilization technologies: A tutorial review, *Clean Technol. Environ. Policy* **24**(2), 467–491 (2022).
7. W. Zhou, K. Cheng, J. Kang, C. Zhou, V. Subramanian, Q. Zhang, *et al.*, New horizon in C1 chemistry: Breaking the selectivity limitation in transformation of syngas and hydrogenation of CO_2 into hydrocarbon chemicals and fuels, *Chem. Soc. Rev.* **48**(12), 3193–3228 (2019).
8. A. Alcasabas, P. Ellis, G. Williams, C. Zalitis and I. Malone, A comparison of different approaches to the conversion of carbon dioxide into useful products, *Johnson Matthey Technol. Rev.* **65**(2), 180 (2020).
9. Y. Li, X. Cui, K. Dong, K. Junge and M. Beller, Utilization of CO_2 as a C1 building block for catalytic methylation reactions, *ACS Catal.* **7**(2), 1077–1086 (2017).
10. G. Ertl, H. Knozinger, F. Schuth and J. Weitkamp, *Handbook of Heterogeneous Catalysis,* 2nd Revised Edition, Wiley-VCH Verlag GmbH & Co. KGaA (2008).
11. J. Ashok, S. Pati, P. Hongmanorom, Z. Tianxi, C. Junmei and S. Kawi, A review of recent catalyst advances in CO_2 methanation processes, *Catal. Today* **356**, 471–489 (2020).
12. C.H. Tan, S. Nomanbhay, A.H. Shamsuddin, Y.-K. Park, H. Hernández-Cocoletzi and P.L. Show, Current developments in catalytic methanation of carbon dioxide — a review, *Frontiers Energ. Res.* **9** (2022).
13. M. Götz, J. Lefebvre, F. Mörs, A. McDaniel Koch, F. Graf, S. Bajohr, *et al.*, Renewable power-to-gas: A technological and economic review, *Renew. Energ.* **85**, 1371–1390 (2016).
14. R.K. Srivastava, P.K. Sarangi, L. Bhatia, A.K. Singh and K.P. Shadangi, Conversion of methane to methanol: Technologies and future challenges, *Biomass Convers. Biorefin.* **12**(5), 1851–1875 (2022).
15. H. Tomita, K. Takano, T. Endo, Y. Yamanaka, K. Nariai, H. Kamata, *et al.* (2021). Development of catalytic methanation as a solution of CO_2 utilization, *15th Greenhouse Gas Control Technologies Conference*, GHGT-15, Abu Dhabi, UAE.
16. T. Chwoła, T. Spietz, L. Więcław-Solny, A. Tatarczuk, A. Krótki, S. Dobras, *et al.*, Pilot plant initial results for the methanation process using CO_2 from amine scrubbing at the Łaziska power plant in Poland, *Fuel* **263**, 116804 (2020).
17. A. Navajas, T. Mendiara, L.M. Gandía, A. Abad, F. García-Labiano and L. F. de Diego, Life cycle assessment of power-to-methane systems with CO_2 supplied by the chemical looping combustion of biomass, *Energy Convers. Manag.* **267**, 115866–115877 (2022).
18. M. Pérez-Fortes, J.C. Schöneberger, A. Boulamanti and E. Tzimas, Methanol synthesis using captured CO_2 as raw material: Techno-economic and environmental assessment, *Appl. Energy* **161**, 718–732 (2016).

19. C.-Y. Chou and R.F. Lobo, Direct conversion of CO_2 into methanol over promoted indium oxide-based catalysts, *Appl. Catal. A-Gen.* **583**, 117144–117152 (2019).

20. K.A. Ali, A.Z. Abdullah and A.R. Mohamed, Recent development in catalytic technologies for methanol synthesis from renewable sources: A critical review, *Renew. Sust. Energ. Rev.* **44**, 508–518 (2015).

21. C.R. International. *Carbon Recycling International* (2016). https://www.carbonrecycling.is

22. D. Xu, Y. Wang, M. Ding, X. Hong, G. Liu and S.C.E. Tsang, Advances in higher alcohol synthesis from CO_2 hydrogenation, *Chem.* **7**(4), 849–881 (2021).

23. B.G. Harvey and H.A. Meylemans, The role of butanol in the development of sustainable fuel technologies, *J. Chem. Technol. Biotechnol.* **86**(1), 2–9 (2011).

24. M. Sauer, Industrial production of acetone and butanol by fermentation — 100 years later, *FEMS Microbio. Lett.* **363**(13), (2016).

25. T. Takashi, M. Atsushi and T. Hiro-o, Alcohol synthesis from CO_2/H_2 on silica-supported molybdenum catalysts, *Chem. Lett.* **14**(5), 593–594 (1985).

26. W. Gao, Y. Zhao, J. Liu, Q. Huang, S. He, C. Li, *et al.*, Catalytic conversion of syngas to mixed alcohols over CuFe-based catalysts derived from layered double hydroxides, *Catal. Sci. Technol.* **3**(5), 1324–1332 (2013).

27. A.M. Bahmanpour, M. Signorile and O. Kröcher, Recent progress in syngas production via catalytic CO_2 hydrogenation reaction, *Appl. Catal. B: Environ.* **295**, 120319–120329 (2021).

28. M. Tavares, G. Westphalen, J.M. Araujo Ribeiro de Almeida, P.N. Romano and E.F. Sousa-Aguiar, Modified Fischer–Tropsch synthesis: A review of highly selective catalysts for yielding olefins and higher hydrocarbons: A review, *Front. Nanotechnol.* **4** (2022).

29. L. Lindenthal, J. Popovic, R. Rameshan, J. Huber, F. Schrenk, T. Ruh, *et al.*, Novel perovskite catalysts for CO_2 utilization — Exsolution enhanced reverse water-gas shift activity, *Appl. Catal. B: Environ.* **292**, 120183–120194 (2021).

30. L.J. France, P.P. Edwards, V.L. Kuznetsov and H. Almegren. The indirect and direct conversion of CO_2 into higher carbon fuels, in *Carbon Dioxide Utilisation*, P. Styring, E.A. Quadrelli and K. Armstrong (eds.), Elsevier, pp. 161–182 (2015).

31. H. Mahmoudi, M. Mahmoudi, O. Doustdar, H. Jahangiri, A. Tsolakis, S. Gu, *et al.*, A review of Fischer–Tropsch synthesis process, mechanism, surface chemistry and catalyst formulation, *Biof. Eng.* **2**(1), 11–31 (2017).

32. D. Marxer, P. Furler, J. Scheffe, H. Geerlings, C. Falter, V. Batteiger, *et al.*, Demonstration of the entire production chain to renewable kerosene via solar thermochemical splitting of H_2O and CO_2, *Energy Fuels.* **29**(5), 3241–3250 (2015).

33. M. Martinelli, C.G. Visconti, L. Lietti, P. Forzatti, C. Bassano and P. Deiana, CO_2 reactivity on Fe–Zn–Cu–K Fischer–Tropsch synthesis catalysts with different K-loadings, *Catal. Today.* **228**, 77–88 (2014).

34. B. Yao, T. Xiao, O. A. Makgae, X. Jie, S. Gonzalez-Cortes, S. Guan, *et al.*, Transforming carbon dioxide into jet fuel using an organic combustion-synthesized Fe-Mn-K catalyst, *Nat. Commun.* **11**(1), 6395–6406 (2020).

35. J. Wei, Q. Ge, R. Yao, Z. Wen, C. Fang, L. Guo, *et al.*, Directly converting CO_2 into a gasoline fuel, *Nat. Commun.* **8**(1), 15174–15181 (2017).

36. P. Gao, S. Li, X. Bu, S. Dang, Z. Liu, H. Wang, *et al.*, Direct conversion of CO(2) into liquid fuels with high selectivity over a bifunctional catalyst, *Nat. Commun.* **9**(10), 1019–1024 (2017).

37. O.A. Ojelade and S.F. Zaman, A review on CO_2 hydrogenation to lower olefins: Understanding the structure-property relationships in heterogeneous catalytic systems, *J. CO_2 Util.* **47**, 101506 (2021).

38. D. Wang, Z. Xie, M.D. Porosoff and J.G. Chen, Recent advances in carbon dioxide hydrogenation to produce olefins and aromatics, *Chem.* **7**(9), 2277–2311 (2021).

39. M.D. Porosoff, B. Yan and J.G. Chen, Catalytic reduction of CO_2 by H_2 for synthesis of CO, methanol and hydrocarbons: Challenges and opportunities, *Energy Environ. Sci.* **9**(1), 62–73 (2016).

40. F. Jiao, J. Li, X. Pan, J. Xiao, H. Li, H. Ma, *et al.*, Selective conversion of syngas to light olefins, *Science* **351**(6277), 1065–1068 (2016).

41. Z. Li, J. Wang, Y. Qu, H. Liu, C. Tang, S. Miao, *et al.*, Highly selective conversion of carbon dioxide to lower olefins, *ACS Catal.* **7**(12), 8544–8548 (2017).

42. S. Wang, L. Zhang, P. Wang, X. Liu, Y. Chen, Z. Qin, *et al.*, Highly effective conversion of CO_2 into light olefins abundant in ethene, *Chem.* **8**(5), 1376–1394 (2022).

43. S. Wang, L. Zhang, W. Zhang, P. Wang, Z. Qin, W. Yan, *et al.*, Selective conversion of CO(2) into propene and butene, *Chem.* **6**(12), 3344–3363 (2020).

44. L. Tan, F. Wang, P. Zhang, Y. Suzuki, Y. Wu, J. Chen, *et al.*, Design of a core–shell catalyst: An effective strategy for suppressing side reactions in syngas for direct selective conversion to light olefins, *Chem. Sci.* **11**(16), 4097–4105 (2020).

45. K.M. Kwok, L. Chen and H.C. Zeng, Design of hollow spherical Co@ hsZSM5@metal dual-layer nanocatalysts for tandem CO_2 hydrogenation to increase C2+ hydrocarbon selectivity, *J. Mater. Chem. A.* **8**(25), 12757–12766 (2020).

46. M.R. Gogate, Methanol-to-olefins process technology: Current status and future prospects, *Pet. Sci. Technol.* **37**(5), 559–565 (2019).

47. L.J. Müller, A. Kätelhön, M. Bachmann, A. Zimmermann, A. Sternberg and A. Bardow, A guideline for life cycle assessment of carbon capture and utilization, *Front. Energy Res.* **8** (2020).
48. G. Zhao, X. Huang, X. Wang and X. Wang, Progress in catalyst exploration for heterogeneous CO_2 reduction and utilization: A critical review, *J. Mater. Chem. A.* **5**(41), 21625–21649 (2017).
49. D.J. Drury, Formic acid, in *Kirk-Othmer Encyclopedia of Chemical Technology*, C. Ley (ed.), pp. 1–9, John Wiley & Sons Inc. (2001).
50. C.M. Jens, K. Nowakowski, J. Scheffczyk, K. Leonhard and A. Bardow, CO from CO_2 and fluctuating renewable energy via formic-acid derivatives, *Green Chem.* **18**(20), 5621–5629 (2016).
51. M. Pérez-Fortes, J.C. Schöneberger, A. Boulamanti, G. Harrison and E. Tzimas, Formic acid synthesis using CO_2 as raw material: Techno-economic and environmental evaluation and market potential, *Int. J. Hydrog. Energy* **41**(37), 16444–16462 (2016).
52. G. Zang, P. Sun, A. Elgowainy and M. Wang, Technoeconomic and life cycle analysis of synthetic methanol production from hydrogen and industrial byproduct CO_2, *Environ. Sci. Technol.* **55**(8), 5248–5257 (2021).
53. M. Rosental, T. Fröhlich and A. Liebich, Life cycle assessment of carbon capture and utilization for the production of large volume organic chemicals, *Frontiers Clim.* **2** (2020).
54. C. Falter, V. Batteiger and A. Sizmann, Climate impact and economic feasibility of solar thermochemical jet fuel production, *Environ. Sci. Technol.* **50**(1), 470–477 (2016).
55. C. van der Giesen, R. Kleijn and G.J. Kramer, Energy and climate impacts of producing synthetic hydrocarbon fuels from CO_2, *Environ. Sci. Technol.* **48**(12), 7111–7121 (2014).
56. A. Dichter, K. Henderson, R. Riedel and D. Riefer, How airlines can chart a path to zero-carbon flying, McKinsey & Co. (2020). https://www.mckinsey.com/industries/travel-logistics-and-infrastructure/our-insights/how-airlines-can-chart-a-path-to-zero-carbon-flying (accessed 16 Nov 2022).
57. P. Schmidt, V. Batteiger, A. Roth, W. Weindorf and T. Raksha, Power-to-liquids as renewable fuel option for aviation: A review, *Chemie Ingenieur Technik.* **90**(1–2), 127–140 (2018).
58. ISO 14040, Environmental management: Life cycle assessment — principles and framework, International Organization for Standardization, Geneva (2006).
59. ISO 14044, Environmental management: Life cycle assessment — requirements and guidelines, International Organization for Standardization, Geneva (2006).
60. H.H. Khoo, P.N. Sharratt, J. Bu, T.Y. Yeo, A. Borgna, J.G. Highfield, *et al.*, Carbon capture and mineralization in Singapore: Preliminary environmental

impacts and costs via LCA, *Ind. Eng. Chem. Res.* **50**(19), 11350–11357 (2011).

61. L.J. Müller, A. Kätelhön, S. Bringezu, S. McCoy, S. Suh, R. Edwards, *et al.*, The carbon footprint of the carbon feedstock CO_2, *Energy Environ. Sci.* **13**(9), 2979–2992 (2020).

62. S. Jackson and E. Brodal, Optimization of the energy consumption of a carbon capture and sequestration related carbon dioxide compression processes, *Energs.* **12**(9), 1603 (2019).

63. M. Aziz, A.T. Wijayanta and A.B.D. Nandiyanto, Ammonia as effective hydrogen storage: A review on production, storage and utilization, *Energies* **13**(12), 3062 (2020).

64. EcoInvent, EcoInvent Database: Life cycle inventory for chemicals and fuels, Zurich, Switzerland (2017/8). https://ecoinvent.org/the-ecoinvent-database/

65. V. Paulauskas, L. Filina-Dawidowicz and D. Paulauskas, The method to decrease emissions from ships in port areas. *Sustainability.* **12**(11), 4374–4388 (2020).

66. R. Bhandari, C.A. Trudewind and P. Zapp, Life cycle assessment of hydrogen production via electrolysis — a review, *J. Clean. Prod.* **85**, 151–163 (2014).

67. P.L. Spath and M.K. Mann, Life cycle assessment of renewable hydrogen production via wind/electrolysis, Milestone Completion Report, National Renewable Energy Laboratory (NREL), United States of America (2004).

68. R.B.H. Tan, D. Wijaya and H.H. Khoo, LCI (Life cycle inventory) analysis of fuels and electricity generation in Singapore, *Energ.* **35**(12), 4910–4916 (2010).

69. M.A.J. Huijbregts, Z.J.N. Steinmann, P.M.F. Elshout, G. Stam, F. Verones, M. Vieira, *et al.*, ReCiPe2016: A harmonised life cycle impact assessment method at midpoint and endpoint level, *Int. J. Life Cycle Assess.* **22**(2), 138–147 (2017).

70. X. Li and E. Mupondwa, Life cycle assessment of camelina oil derived biodiesel and jet fuel in the Canadian Prairies, *Sci. Total Environ.* **481**, 17–26 (2014).

71. K. Nasriddinov, J.-E. Min, H.-G. Park, S.J. Han, J. Chen, K.-W. Jun, *et al.*, Effect of Co, Cu, and Zn on FeAlK catalysts in CO_2 hydrogenation to C5+ hydrocarbons, *Catal. Sci. Technol.* **12**(3), 906–915 (2022).

72. M.T. Arslan, G. Tian, B. Ali, C. Zhang, H. Xiong, Z. Li, *et al.*, Highly selective conversion of CO_2 or CO into precursors for kerosene-based aviation fuel via an aldol–aromatic mechanism, *ACS Catal.* **12**(3), 2023–2033 (2022).

Chapter 3

Life Cycle Costing and Carbon Dioxide Emissions of Hydrogen Supply Chains using Different Energy Carriers: A Case Study of Japan for 2030

Akito OZAWA* and Yuki KUDOH

Global Zero Emission Research Center,
National Institute of Advanced Industrial Science and Technology,
16-1 Onogawa, Tsukuba, Ibaraki, 305-8569, Japan

*akito.ozawa@aist.go.jp

Hydrogen supply chains (HSCs) are placed among significant development studies for Japan's national aim of achieving carbon neutrality. The comprehensive assessment of both economic and environmental values of HSCs, in parallel with technological challenges involved, is identified as a crucial step forward to effectively provide low-cost clean hydrogen. In this study, we analyze the potential life cycle cost (LCC) and life cycle CO_2 ($LCCO_2$) emissions of HSCs in Japan for 2030, when HSCs will be commercialized in the country. This chapter presents how a process was defined for the configuration of global and local HSCs data for the purpose of appropriate data compilation for suitable process set up. Life cycle inventory database was used to calculate the LCC and $LCCO_2$ of HSCs using different energy carriers. The overall effect of LCC and $LCCO_2$ results generated by the different power inputs involved in HSCs was also investigated.

1. Introduction

Hydrogen is expected to play an important role as an energy source in the future where decarbonization efforts take place. Since hydrogen does not emit CO_2 when combusted, CO_2 emissions from energy consumption can be reduced by using hydrogen instead of fossil fuels. Hydrogen will also be expected to diversify energy supply chains because it can be produced from a variety of primary energy resources. According to the net-zero scenario of the International Energy Agency (IEA), the worldwide consumption of hydrogen energy (including ammonia and synthetic fuels), which was less than 90 Mt in 2020, will increase to 200 Mt by 2030, and 530 Mt by 2050 [1].

Accelerating the use of hydrogen energy on a global scale requires international cooperation. In October 2018, more than 300 representatives from 21 countries, regions and organizations around the world participated in the 1st Hydrogen Energy Ministerial Meeting held in Japan, where they confirmed the importance of international coordination and shared their policy positions towards the global use of hydrogen. The results were published in the *Tokyo Statement* [2]. In 2019, the Clean Energy Ministerial Hydrogen Initiative (CEM H2I) was inaugurated [3]. A total of 22 governments gathered to promote policies, programs and projects that will accelerate the commercialization and spread of hydrogen and fuel cell technologies.

Japan is seen as one of the leaders in the technological development of hydrogen energy and has made hydrogen the cornerstone of its energy and environmental strategy. In 2017, the government of Japan became the first in the world to formulate a national strategy on hydrogen [4], in which it presented its vision of realizing a hydrogen society. This strategy aims to supply 3 Mt of hydrogen by 2030 and 20 Mt by 2050 across Japan. To transport hydrogen produced at low cost abroad, the strategy proposes building a commercial-scale hydrogen supply chain (HSC) by 2030 that utilizes energy carriers, such as liquid hydrogen, methylcyclohexane, ammonia, and synthetic methane, which will supply hydrogen at approximately 30 JPY/Nm³ (0.23 USD/Nm³ [a]). In June 2021, the government formulated a "green growth strategy" [5], an industrial policy for creating a virtuous economic and environmental cycle. This strategy mentions key

---

[a] Converted at a rate of 1 USD = 130 JPY (as of 31 January 2023).

industrial fields, including hydrogen, ammonia and synthetic methane, which are expected to grow with progress towards carbon neutrality.

Disruptive reform of Japan's energy and environmental technologies is required for the country to achieve its greenhouse gas (GHG) reduction targets. The government of Japan established a 2 trillion JPY (USD 15.3 billion) innovation fund [6] to accelerate the change necessary for the country to achieve carbon neutrality. This fund promotes the building of large-scale HSCs in Japan, as well as research and development projects such as producing hydrogen through renewables-powered electrolysis. We classified these projects into two types based on the type of hydrogen used: global hydrogen or local hydrogen (Table 1). With global hydrogen, hydrogen produced from a variety of primary sources is distributed internationally and used instead of fossil fuels, thereby achieving GHG emissions reductions on a global scale. Since substituting hydrogen for fossil fuels requires low-cost distribution of hydrogen, it increases the scale of the supply chain. Conversely, in local hydrogen supply chains, hydrogen produced using domestic renewable energy resources and unused energy resources is utilized within a region, which improves its energy resilience and helps to revitalize its local economy. As the amount of hydrogen

Table 1.   Differences between global and local hydrogen.

	Global Hydrogen	Local Hydrogen
Is considered to be...	Utilizes a variety of primary energies beyond borders to reduce GHGs on a global scale	Uses various domestic renewable energy resources and unused local resources
Can substitute for/compete with...	Substitutes for imported fossil fuels (LNG, crude oil, etc.)	Competes with other energy storage technologies (storage batteries, pump hydro, etc.)
Candidates for carriers	Liquid hydrogen, methylcyclohexane, ammonia, synthetic methane	High-pressure hydrogen, liquid hydrogen, methylcyclohexane, ammonia, synthetic methane
Example of uses	**Energy sector**: Gas turbines **Transportation sector**: Hydrogen stations (for fuel-cell vehicles) **Industrial sector**: Thermal usage and feedstock **Residential and commercial sectors**: Fuel cells	**Industrial, residential and commercial sectors**: Fuel-cell and internal combustion engine cogeneration **Transportation sector**: Intra-fleet usage (public transportation, industrial machinery)

transported in local HSCs is determined by the potential of and demand for local energy resources, these HSCs have a smaller scale than global HSCs.

The comprehensive assessment of both the economic and environmental values of HSCs, in parallel with technological challenges involved, is identified as a crucial step forward to effectively provide low-cost clean hydrogen. Therefore, we analyzed the life cycle cost (LCC) and life cycle CO_2 ($LCCO_2$) emissions of HSCs in Japan for 2030, when HSCs will be commercialized in the country. In this study, a process was defined for the configuration of global and local HSCs data for the purpose of appropriate data compilation for the process set up of HSCs in Japan. We also assessed how differences in the power input into the supply chain affect the LCC and $LCCO_2$.

2. Method and Assumptions

2.1. *Hydrogen supply chains examined in this study*

We configured global and local HSCs based on the Institute of Applied Energy [7]. The system boundary of this study is shown in Fig. 1. In the global HSC, green hydrogen is produced in the UAE by solar

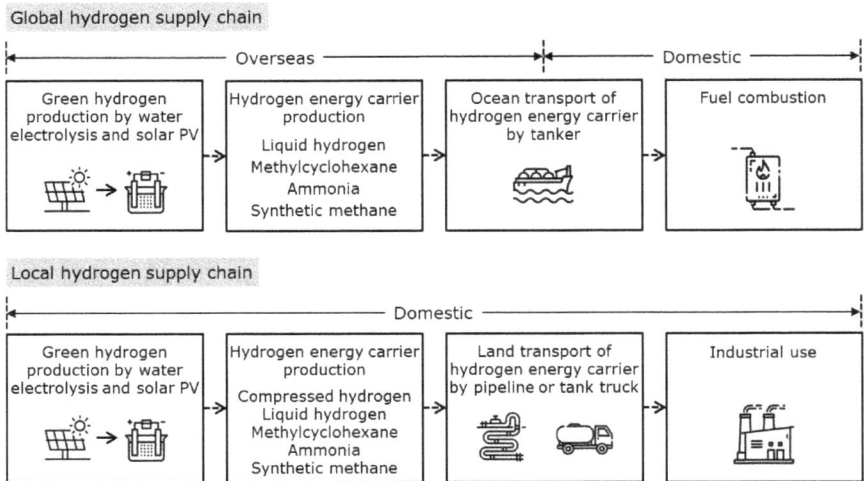

Fig. 1.   The system boundary of this study.

photovoltaic (PV) electrolysis, converted into a hydrogen energy carrier, and then transported by sea on a tanker to Japan, where it is used as fuel. We assumed an annual hydrogen production of 225,000 t/year — hydrogen energy carriers of liquid hydrogen (LH_2), methylcyclohexane (MCH), ammonia (NH_3), and synthetic methane (CH_4), and a transport distance of 10,000 km from the UAE to Japan. For the case of local HSCs, green hydrogen is produced through solar PV electrolysis, converted into a hydrogen energy carrier, and then transported over land by a pipeline or tanker truck to the consumers in the industrial sector. We assumed an annual hydrogen production of 2,838 t/year, the four aforementioned hydrogen energy carriers plus compressed hydrogen (CH_2), and a transport distance of 50 km. Table 2 presents the five hydrogen energy carriers and their characteristics.

Table 2.   Hydrogen carriers and their characteristics.

Hydrogen Carrier	Characteristics
Compressed hydrogen	• Can be directly used as fuel as high-purity hydrogen • Low-energy density per unit volume, not suitable for long-distance transport
Liquid hydrogen	• Can be directly used as fuel as high-purity hydrogen • Liquefaction consumes a large amount of energy
Methylcyclohexane	• Can be used by existing oil transportation infrastructure • Must be dehydrogenated and refined to be used as fuel • Dehydrogenation consumes a large amount of energy
Ammonia	• Can be used by existing ammonia transportation infrastructure • Can be directly used as fuel as ammonia
Synthetic methane	• Can be used by existing gas transportation infrastructure • Can be directly used as fuel as methane • Feedstock CO_2 can be considered as a waste product and does not account for CO_2 inventory [8]

In determination of the set of specifications of the processes for the configuration of HSCs, present technologies adequate for the scale of the supply chain were carefully selected. Any technological selections that were not adequate in terms of scale or efficiency will be omitted from the study. And in their place, technologies suitable and usable by 2030 will be selected. If the prospects of technological developments by 2030 could be found in the literature, we used those results; otherwise, we used estimates based on present/similar technologies and information obtained from interviews.

To calculate the LCC and LCCO$_2$ of the solar PV electrolysis HSCs, we assumed two different cases (a base case and a low-carbon case) of power input into the supply chain. In the base case, we assumed that solar PV power is used only in the process of hydrogen production (renewable electrolysis) and that grid power is used for all other processes. In the low-carbon case, we assumed that solar PV power is used for all power input across the entire supply chain, except for the CO$_2$ capture process necessary for synthetic methane production.

2.2. Calculation of life cycle cost and life cycle CO$_2$ emissions

The life cycle cost (LCC) in the supply chain was calculated based on Eq. (1). The costs consist of fixed costs and variable costs. The fixed costs are costs incurred regardless of the amount produced by the processes and include construction costs, property taxes, demolition costs, labor, maintenance costs, insurance costs, and management costs. The variable costs are costs that change in proportion to the amount produced by the processes and include raw fuel costs, electricity costs, and CO$_2$ transportation/storage costs. The lifetime of the supply chain was assumed to be 30 years, with a discount rate of 5% and a currency exchange rate of 130 JPY = 1 USD (as of 31 January 2023).

$$LCC = \frac{\sum_{t=1}^{n} \frac{F_t + V_t}{(1+r)^t}}{\sum_{t=1}^{n} \frac{H_t}{(1+r)^t}}. \tag{1}$$

In Eq. (1), F_t is the fixed expense in year t, V_t is the variable expense in year t, r is the discount rate, and n is the expected lifetime of the supply chain.

The life cycle CO$_2$ (LCCO$_2$) in the supply chain was calculated based on Eq. (2):

$$LCCO_2 = \sum_{i,j} (x_{i,j} \times e_j), \tag{2}$$

where $x_{i,j}$ is the amount of input material j into the process i of configuring the supply chain and e_j is the CO$_2$ intensity of the input.

For the CO_2 intensities e_j, we used the values in the Inventory Database for Environmental Analysis (IDEA) [9], an LCA database developed by the National Institute of Advanced Industrial Science and Technology (AIST). Because the data in the IDEA is for Japan, the CO_2 intensities can be used as they are for the domestic processes in the supply chains. However, since we could not obtain foreign CO_2 intensities for the inputs into the processes outside of Japan, we determined them as follows. When the input was a non-energy input, we used the IDEA value of the input into the domestic process; when the input was electricity, we used IEA statistics [10], which indicate the emissions per unit of electricity generated from fossil fuels in the energy sector; when the input was fuel, we used the direct CO_2 emission coefficient of that fuel.

Table 3 shows the unit prices of electricity and the CO_2 intensities from different sources. For this analysis, we looked at the CO_2 emissions associated with the amount of raw material input into the energy carrier system/utility amount and ignored emissions caused by capital goods. However, for solar PV, rather than setting the capital goods emissions as zero, we calculated the emissions associated with the production of component materials, construction and transportation. For solar PV intensities, we used the CO_2 intensity per unit of electricity generated divided by the total amount of electricity generated during the expected durable lifetime (years).

Table 3.   Electricity price and CO_2 intensity.

	Japan		UAE	
	Grid	Solar PV	Grid	Solar PV
Electricity price [US cents/kWh]	9.6	5.4	4.3	5.4
CO_2 intensity [g-CO_2/kWh]	573	51.8	407	39.8

2.3. Calculation of life cycle cost and life cycle CO_2 emissions

2.3.1. Hydrogen production

Figure 2 illustrates the process of producing hydrogen using renewables-powered electrolysis. Solar PV electricity is used to produce hydrogen through alkaline water electrolysis. We assumed the use of solar PV

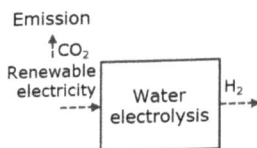

Fig. 2.   Hydrogen production using renewables-powered electrolysis.

power in the UAE when hydrogen was to be produced overseas. When hydrogen was to be produced in Japan, we assumed the purchase of green power certificates.

- **Annual hydrogen production**: 2.5 billion Nm³ H_2/year
- **Equipment utilization rate**: Overseas 20%; domestic 90%
- **Electricity intensity**: 4 kWh/Nm³ H_2
- **Construction expenses:** 0.76 USD/(Nm³ H_2/year)

2.3.2. *Global hydrogen supply chains*

(1) Liquid hydrogen

Figure 3 shows a global HSC with LH_2.

- Hydrogen is liquefied using a liquefier with a capacity of 36,500 t/year. The electricity intensity is 0.55 kWh/Nm³.
- The LH_2 is stored at loading/unloading ports in spherical vacuum double-shell tanks with a capacity of 50,000 m³. The boil-off rate is set to 0.1%/day.
- The LH_2 is transported by ship with a capacity of 160,000 m³, which is equivalent to that of an LNG carrier. The boil-off rate is set to 0.4%/day.
- The LH_2 is vaporized at the unloading port and then supplied to consumers.

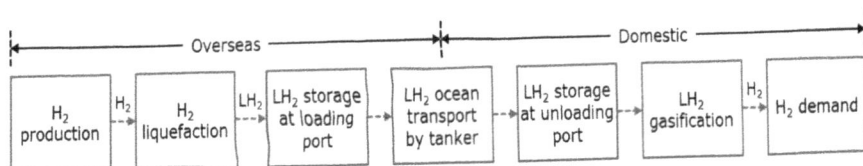

Fig. 3.   Global hydrogen supply chain using liquid hydrogen.

(2) Methylcyclohexane

Figure 4 shows the global HSC with MCH.

- MCH is produced by adding hydrogen to toluene (TOL) at a plant that produces 1.1 Mt toluene/year. The unreacted hydrogen is captured and used at the plant for heating.
- The MCH and TOL are stored at the loading/unloading ports in two 67,200 m³ tanks and one 98,100 m³ tank.
- The MCH and TOL are transported by sea on a tanker with a capacity of 81,000 m³. LNG is assumed as the tanker fuel.
- TOL and hydrogen are obtained by dehydrogenating the MCH at a plant that processes 2.15 Mt MCH/year. The hydrogen is supplied to consumers, and the TOL is reused to produce MCH.

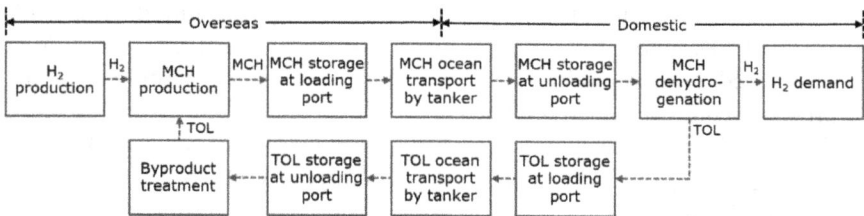

Fig. 4. Global hydrogen supply chain using methylcyclohexane.

(3) Ammonia

Figure 5 shows the global HSC with NH$_3$.

- NH$_3$ is produced from nitrogen and hydrogen at atmospheric pressure using the Haber–Bosch process at a plant with a capacity of 584,000 t NH$_3$/year.

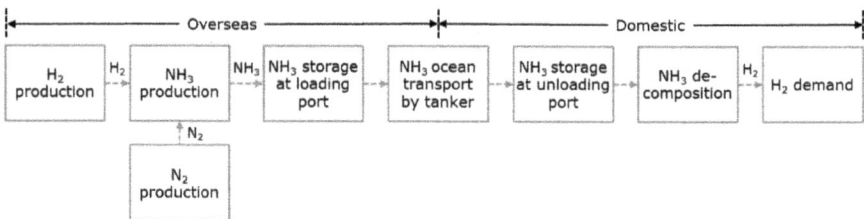

Fig. 5. Global hydrogen supply chain using ammonia.

- The NH_3 is stored at the loading/unloading ports in tanks with a capacity of 57,000–81,000 m³.
- The NH_3 is transported by sea on a ship with a capacity of 780,000 m³, which is equivalent to that of an LPG carrier. LNG is assumed as the fuel for the ship.
- The NH_3 is decomposed at a plant that processes 876,000 t/year, and the hydrogen obtained from it is supplied to consumers.

(4) Synthetic methane

Figure 6 shows the global HSC with CH_4.

- CH_4 is synthesized from hydrogen and CO_2 through methanation at a plant with a capacity of 550,000 t/year. The CO_2 used here is captured by separating CO_2 collected through chemical absorption with an amine solution. The unit price of captured CO_2 is assumed to be 38.5 USD/t CO_2. The CO_2 intensity accounts for indirect CO_2 emissions associated with inputting the necessary utilities for capturing, compressing and transporting CO_2.
- The CH_4 is liquefied using a liquefier with a capacity of 36,500 t/year. The electricity intensity is set to 0.37 kWh/Nm³ CH_4.
- The liquid methane (LCH_4) is stored at the loading/unloading ports in two 90,000 m³ tanks. The boil-off rate is set to 0.1%/day.
- The LCH_4 is transported by sea on an LNG carrier with a capacity of 160,000 m³. Boil-off gas is used to fuel the carrier; if there is not enough, LNG is used.
- The LCH_4 is vaporized and used as fuel at the demand site.
- Feedstock CO_2 is considered as a waste product and does not account for CO_2 inventory.

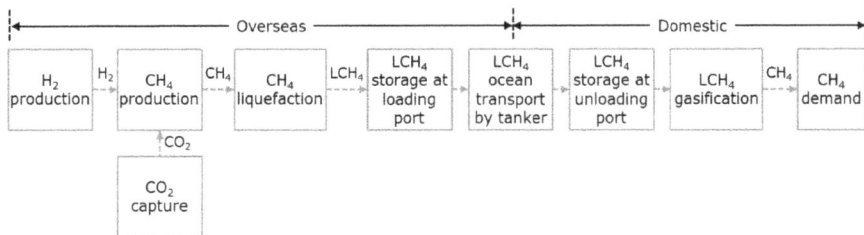

Fig. 6.   Global hydrogen supply chain using synthetic methane.

2.3.3. *Local hydrogen supply chains*

(1) Compressed hydrogen

Figure 7 shows the local HSC with CH_3.

- Hydrogen is stored at the supply site in three 13,600 m³ tanks and then compressed and delivered. The electricity intensity of the storage compressor (1 MPa) is set to 0.11 kWh/Nm³ and the delivery compressor (20 MPa) is set to 1.2 kWh/Nm³.
- The CH_2 is transported over land on a hydrogen tube trailer and delivered to consumers. Sixteen steel gas cylinders with a pressure of 20 MPa and a capacity of 3,000 Nm³ are used to transport the CH_2. The trailer is kept at the demand site, where the used cylinders are collected. The fuel used for transport is diesel fuel.

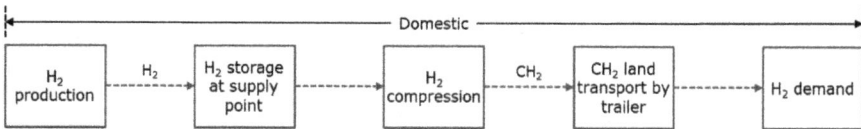

Fig. 7.   Local hydrogen supply chain using compressed hydrogen.

(2) Liquid hydrogen

Figure 8 shows the local HSC with LH_2.

- Hydrogen is liquefied using a liquefier with a capacity of 4,380 t/year. The electricity intensity is set to 0.89 kWh/Nm³.
- The LH_2 is stored at the supply and demand sites in tanks with a capacity of 500 m³. The boil-off rate is set to 0.1%/day.
- The liquid hydrogen is transported over land on three 23 kL tank trucks. The trucks use diesel fuel.

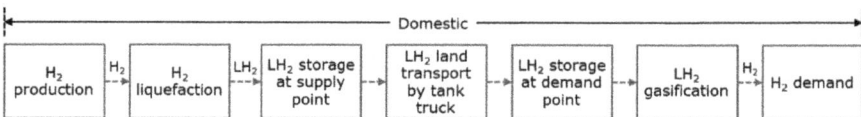

Fig. 8.   Local hydrogen supply chain using liquid hydrogen.

(3) Methylcyclohexane

Figure 9 shows the local HSC with MCH.

- MCH is produced by adding hydrogen to TOL at a plant that produces 56,000 t TOL/year. The heat that results from the addition of hydrogen to the TOL is captured and used at the plant.
- The MCH and TOL are stored at the supply and demand sites in one 250 m³ tank and one 740 m³ tank, respectively.
- The MCH and TOL are transported over land on four 23 kL tank trucks. The trucks use diesel fuel.
- TOL and hydrogen are obtained by dehydrogenating the MCH at the demand site using a device capable of dehydrogenating 56,000 t MCH/year. Off-gas from pressure swing adsorption is used to produce the heat necessary for the dehydrogenation reaction; if this is not enough, natural gas is used.

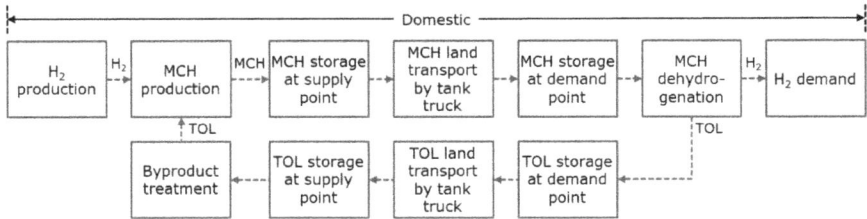

Fig. 9.   Local hydrogen supply chain using methylcyclohexane.

(4) Ammonia

Figure 10 shows the local HSC with NH_3.

- NH_3 is produced from nitrogen and hydrogen at atmospheric pressure using the Haber–Bosch process at a plant with a capacity of 18,000 t NH_3/year.

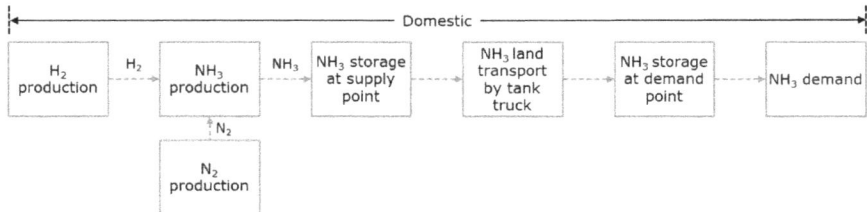

Fig. 10.   Local hydrogen supply chain using ammonia.

- The NH$_3$ is stored in 285 m^3 tanks at the supply and demand sites.
- The NH$_3$ is transported over land on two 23 kL tank trucks. The tank trucks use diesel fuel.
- The NH$_3$ is directly used as fuel at the demand site.

(5) Synthetic methane

Figure 11 shows the local HSC with CH$_4$.

- CH$_4$ is synthesized from hydrogen and CO$_2$ through methanation at a plant with a capacity of 7,000 t/year. The raw CO$_2$ is captured by separating CO$_2$ collected through chemical absorption with an amine solution. The unit price of raw CO$_2$ is 38.5 USD/t CO$_2$. The CO$_2$ intensity accounts for indirect CO$_2$ emissions associated with inputting the necessary utilities for capturing, compressing, and transporting CO$_2$.
- The CH$_4$ is stored at the supply site in one gas container with a capacity of 10,200 m^3.
- The CH$_4$ is supplied to consumers through existing city gas pipelines. The consignment fee is set to 4.8 US cents/Nm^3 CH$_4$.
- The CH$_4$ is directly used as fuel at the demand site. Feedstock CO$_2$ is considered as a waste product and does not account for the CO$_2$ inventory.

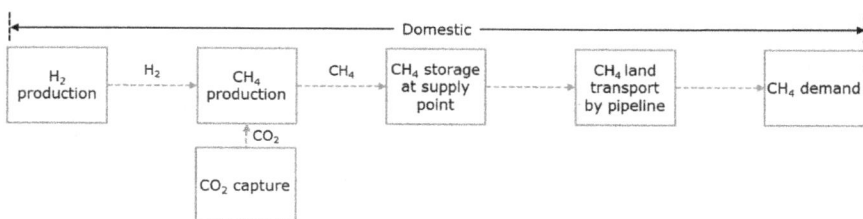

Fig. 11.   Local hydrogen supply chain using synthetic methane.

3. Results

3.1. *Global hydrogen supply chains*

Figure 12 shows the LCC and LCCO$_2$ of the global HSCs. In the base case (grid power is used for all electricity input into the HSC except for the hydrogen production process), the LCC of the global HSC varies according to the hydrogen energy carrier within the range of 3.5–4.5 US cents/MJ, and

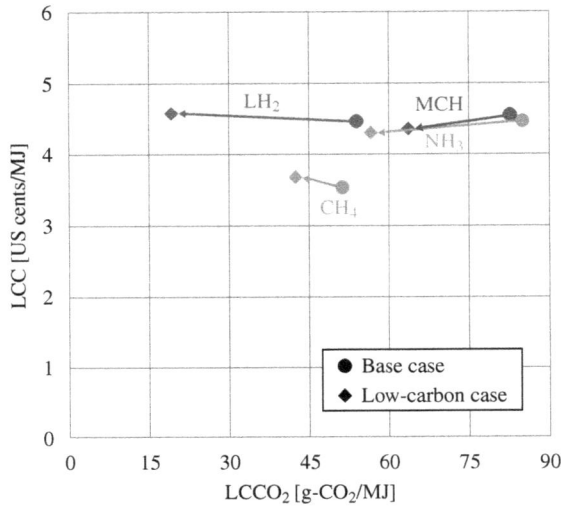

Fig. 12.   LCC and $LCCO_2$ of the global supply chains.

the $LCCO_2$ varies within the range of 51–85 g CO_2/MJ. These results indicate that the choice of energy carrier causes greater variability in $LCCO_2$ than in LCC. Both the LCC and $LCCO_2$ are lowest when CH_4 is used as the hydrogen energy carrier.

In the low-carbon case (PV power is used for all electricity input across the entire supply chain except for the CO_2 processing), the LCC of the global HSC varies according to the hydrogen energy carrier within the range of 3.7–4.6 US cents/MJ, and the $LCCO_2$ varies within the range of 19–64 g CO_2/MJ. The LCC in the low-carbon case increases in comparison to the base case by 0.11–0.13 US cents/MJ (2.5–3.8%) when LH_2 or CH_4 is used as the energy carrier but decreases by 0.17–0.20 US cents/MJ (3.8–4.4%) when MCH or NH_3 is used. In comparison to the base case, the low-carbon case incurs higher electricity costs for the overseas processes and lower electricity cost for the domestic processes. When liquid hydrogen or synthetic methane is used, the LCC increases because the overseas processes consume a large amount of electricity. When MCH or NH_3 is chosen, the LCC decreases because the domestic processes consume a larger proportion of the electricity. The $LCCO_2$ is 8.8–34.5 g CO_2/MJ (17–64%) lower in the low-carbon case than in the base case. Among the hydrogen energy carriers, the $LCCO_2$ is the lowest with LH_2, six times less than the base case. The supply chain using LH_2 consumes a large amount of electricity in the hydrogen liquefaction process; using solar PV power for this process can reduce CO_2 emissions considerably.

3.2. *Local hydrogen supply chains*

Figure 13 shows the LCC and LCCO$_2$ of the local HSCs. In the base case, the LCC of the local HSC varies according to the hydrogen energy carrier within the range of 4.2–5.5 US cents/MJ, and the LCCO$_2$ varies within the range of 37–85 g CO$_2$/MJ. The LCC is lowest when CH$_4$ is used as the hydrogen energy carrier, but LCCO$_2$ is lowest when CH$_4$ is used.

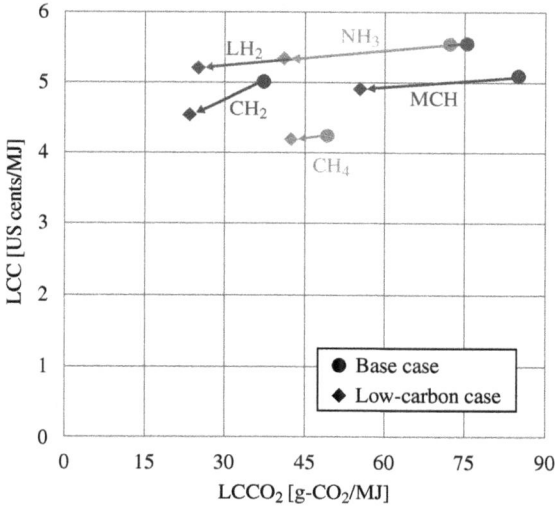

Fig. 13.   LCC and LCCO$_2$ of the local supply chains.

In the case of low-carbon HSC, the LCC of the local HSC varies according to the hydrogen energy carrier within the range of 4.2–5.3 US cents/MJ, and the LCCO$_2$ varies within the range of 24–55 g CO$_2$/MJ. The low-carbon case results in lower electricity costs for domestic processes and lower CO_2 intensity, which causes both the LCC and LCCO$_2$ to decrease. The LCC decreases by 0.05–0.47 US cents/MJ (1.2–9.4%) and the LCCO$_2$ decreases by 6.9–50.2 g CO$_2$/MJ (14–67%) in comparison to the base case. These decreases are considerable when LH$_2$, MCH or NH$_3$ is chosen as the energy carrier.

4. Conclusion

In this study, we analyzed the LCC and $LCCO_2$ emissions of HSCs in Japan in 2030, when HSCs will be commercialized in the nation. We defined a process for configuring global and local HSCs and collected process data. Then, we used a life-cycle inventory database to calculate the LCC and $LCCO_2$ of HSCs using different energy carriers. We also assessed how differences in the power input into the supply chain affect LCC and $LCCO_2$.

The results of the analysis indicated that the LCC and $LCCO_2$ of the HSCs vary according to the choice of energy carrier and are also affected by the electricity input into the supply chain. In terms of environmental factors, the $LCCO_2$ of HSCs with LH_2, MCH and NH_3 is considerably affected by the CO_2 intensity of the input electricity. This suggests that using low-carbon electricity across the entire supply chain (and not just for the hydrogen production process) is an effective way to create a clean HSC. On the other hand, using low-carbon electricity across the entire supply chain increases electricity cost and could also increase the LCC. In the analysis in this study, LCC increased in the low-carbon case of the global HSC when LH_2 or CH_4 was chosen as the energy carrier. This suggests that attention must be paid to the cost and CO_2 intensity of the electricity used across the entire supply chain to supply low-cost clean hydrogen.

Because this analysis was based on certain assumptions, changing the prerequisites will directly affect the results. There is a trade-off between cost and CO_2 emissions [11], which will likely make different conditions desirable according to external factors such as when the supply chain is introduced and the extent of hydrogen power generation. When the uncertainty of the future is considered, the relative merits of the carriers analyzed are not definitive, and future technological developments and circumstances in producing countries could cause them to change in relation to each other.

Analyzing the LCC and CO_2 emissions of technologies at the research and development stage is an effective method for understanding the potential for profit when those technologies are implemented. In this study, we looked at HSCs, which are garnering attention as a strategy for achieving carbon neutrality, and analyzed the LCC and $LCCO_2$ of green hydrogen produced through solar PV electrolysis. However, blue hydrogen (for instance, natural gas reforming or coal gasification with carbon capture and storage) is also expected to play an important role in building large-scale global HSCs and procuring low-cost clean hydrogen. This study also assumed that Japan would import global hydrogen from the UAE; however, there are other prospective producers of green or blue hydrogen, such as Australia and Norway. In future studies, the analyses of the LCC and CO$_2$ emissions of HSCs other than those examined in this chapter should be carried out.

References

1. International Energy Agency, *Net Zero by 2050*, IEA Publications, Paris (2021).
2. Hydrogen Energy Ministerial Meeting, Tokyo statement (2018). https://www.nedo.go.jp/content/100885424.pdf
3. Japanese Ministry of Economy, Trade, and Industry, Outline of strategic energy plan (2021). https://www.enecho.meti.go.jp/en/category/others/basic_plan/pdf/6th_outline.pdf
4. The Ministerial Council on Renewable Energy Hydrogen and Related Issues, The basic hydrogen strategy (2017). https://policy.asiapacificenergy.org/sites/default/files/Basic%20Hydrogen%20Strategy%20%28EN%29.pdf
5. Japanese Ministry of Economy, Trade, and Industry, Green growth strategy through achieving carbon neutrality in 2050 (2021). https://www.meti.go.jp/english/press/2020/pdf/1225_001b.pdf
6. New Energy and Industrial Technology Development Organization, Overview of the green innovation fund projects (2021). https://green-innovation.nedo.go.jp/en/about/
7. The Institute of Applied Energy. Report on the FY 2008 project for cooperation with international organizations in global warming countermeasures (survey on efforts for international cooperation through mission innovation) [in Japanese] (2019). https://www.meti.go.jp/meti_lib/report/H30FY/000477.pdf
8. G. Reiter and J. Lindorfer, Global warming potential of hydrogen and methane production from renewable electricity via power-to-gas technology, *Int. J. Life Cycle Assess.* **20**, 477–489 (2015).
9. National Institute of Advanced Industrial Science and Technology, Summary of the IDEA (2021). https://riss.aist.go.jp/en-idealab/idea/
10. International Energy Agency, *CO$_2$ Emissions from Fuel Combustion*, OECD Publishing, Paris (2016).
11. B. Hurtubia and E. Sauma, Economic and environmental analysis of hydrogen production when complementing renewable energy generation with grid electricity, *Appl. Energy* **304**(15), 117739 (2021).

Chapter 4

Connection of the Methanol Economy to Net-Zero Emissions Supported by LCA-Based Environmental Performance Information

Róbert MAGDA[a,b], Judit TÓTH[c]* and Sarolta IGAZ[c]

[a]*John von Neumann University,*
Izsáki út 10, 6000 Kecskemét, Hungary
[b]*Vanderbijlpark Campus,*
North-West University,
Vanderbijlpark 1900, South Africa
[c]*Atomos Educational Ltd,*
Mária utca 107, 1161 Budapest, Hungary

*toth.judit@atomos.hu

As global energy consumption is constantly increasing, reducing both the energy and carbon intensity of the energy supply is a key priority in tackling climate change. These factors have inevitably led to the need for a sustainable energy transition and the emergence of new technologies. With the large and dynamic growth in the use of renewable energy sources, energy storage has become a key factor in the security of energy supply. Exploiting the technological link between renewable energy and carbon dioxide sequestration will make a positive contribution to reducing atmospheric carbon emissions and increasing independence from fossil fuels. The methanol economy is essentially the creation of a Circular Economy (CE) in which pollutants and waste from energy production and other industrial processes can be captured and integrated into the system using carbon-neutral energy sources. Life Cycle Assessment (LCA) investigations confirm that the concept of the methanol economy, defined as the production and use of renewable methanol, demonstrates the potential to achieve sustainable development and climate policy goals.

1. Introduction

Energy has always been one of the most important resources, as it contributes to the development and prosperity of human societies. Changes in the use of energy resources have always been driven by two main factors: the evolution of energy prices and technological progress [1]. The question is whether today's energy transition will be as smooth as the previous ones. This will depend crucially on the amount of oil and gas available, the impact of greenhouse gases (GHGs) on the Earth's climate, and whether access to alternative energy sources will be economically competitive. The goals are clear, but the way forward to achieve low-carbon energy resources is still faced with multiple challenges. We will be able to replace a significant part of the areas using fossil fuels with energy from alternative sources, but will still need carbon-based products in some areas — just think of the objects in our immediate environment. The use of carbon-based raw materials in both energy and plastics production will continue for a long time, although new technological pathways will emerge. Research into new raw materials and technologies is being driven by the growing demand for energy and environmental sustainability.

Assessing and understanding the risks and environmental impacts of new technologies will facilitate their uptake. Life Cycle Impact Assessment (LCIA) seeks to identify as accurately as possible the direct and indirect impacts of interventions on the environment. There are a number of complementary approaches to reducing atmospheric carbon dioxide concentrations and emissions. To remove large amounts of carbon dioxide from the atmosphere and prevent emissions, Carbon Capture and Sequestration (CCS) technology has a crucial role to play, but it must be complemented by Carbon Capture and Utilization (CCU) technology, which provides added value and can become a key tool for climate change mitigation [2].

2. The Methanol Economy

The quantitative need for the direct use of carbon dioxide is outweighed by the amount produced by the combustion of fossil fuels, and the conversion of captured carbon dioxide into useful products is, therefore, of strategic importance for sustainable development [3].

In 1990, Nobel Laureate György Oláh proposed the concept of a methanol economy to replace fossil fuels, a new approach to reduce dependence on depleting supplies of oil and natural gas (and eventually coal), and a solution to the problem of global warming caused by excessive carbon emissions. His research provides a comprehensive and sustainable solution for the long-term replacement of fossil fuels through the chemical conversion of carbon dioxide into renewable methanol using alternative energy sources such as solar, wind, hydro, geothermal and nuclear power. Methanol has a wide range of applications: it can be used as an energy storage medium, as a fuel or fuel additive, and is a widely used feedstock for the chemical industry [4].

The concept of the methanol economy and its implementation is a promising way to solve global problems [5]. Adopting a Circular Economy (CE) approach, the implementation of a circular methanol economy could be the path to carbon neutrality. The CE philosophy is based on the principle of natural processes. Ecosystems are interconnected to form networks of networks, circulating energy and nutrients in cycles so that nothing is wasted [6]. Figure 1 illustrates the closing of "the carbon loop" by capturing carbon dioxide from power plants to produce renewable and sustainable methanol, which can be used to efficiently recycle the released carbon dioxide [7].

2.1. *Methanol as a feedstock for the chemical industry*

Methanol is an integral part of everyday life as a feedstock for plastics, synthetic fibers, fuels, resins, paints, adhesives, solvents, carpets, insulators, refrigerants, and chipboard. In 2021, nearly 48% of all methanol consumption was used as a feedstock for the chemical industry. Methanol use is increasing year after year, largely due to the spread of the

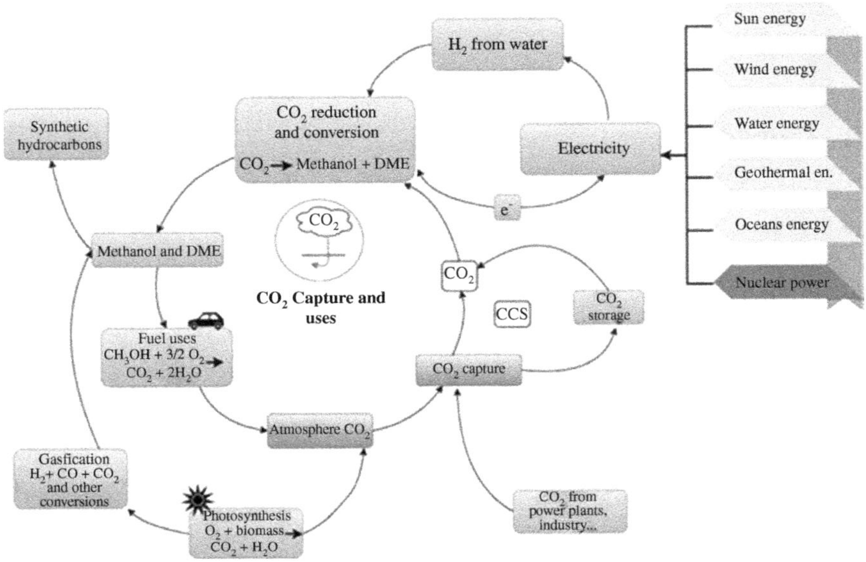

Fig. 1.   The carbon loop.

technology to produce olefins from methanol. In 2014, 9.1 million tons of methanol were used for this purpose, rising to 31.202 million tons in 2020 and 33.5 million tons in 2021. In olefin production, methanol is used to produce ethylene and propylene, which are important monomers for plastics [8]. Methanol-to-olefins usage accounts for more than 31% of methanol consumption in 2021, while alternative fuels have a share of nearly 17% [9].

2.2. *Role of methanol in the energy system*

An important area of carbon dioxide use is the production of synthetic fuels, which would capture at least 23% of carbon dioxide emissions per year [10]. In the energy sector, carbon dioxide emissions are concentrated, so good results can be achieved in reducing emissions through the use of CCS technologies [11]. In recent years, there has been a paradigm shift in the use of methanol, with declining use as a chemical feedstock and increasing use in the energy sector.

2.2.1. *Fuel*

Methanol can be used directly as a fuel with properties very similar to petrol. Methanol has a high octane rating and a high auto-ignition temperature, making it a suitable motor fuel, but it has the disadvantage of having a lower energy density than petrol and ethanol.

Methanol has a low cetane number, so it cannot be used in diesel engines, but dimethyl ether produced from methanol is a very good substitute for gas oil. Methanol is often used to produce biodiesel [12]. It is envisaged that methanol cars of the future are not expected to use internal combustion engines but will be powered by an electric motor and a fuel cell. The fuel cell is the primary power source, fed by the methanol, and an additional battery provides support, e.g., for dynamic loads or cold starts. When energy demand is low, the electricity from the fuel cell recharges the battery. This hybrid arrangement allows the fuel cell to operate at high efficiency while improving the lifetime of the fuel cell [13].

2.2.2. *Energy storage*

One of the main difficulties in deploying renewable energy sources is the intermittent and variable nature of the power supply, depending on the weather and time of day. A continuous supply requires temporary energy storage until the time of use. At present, large-scale energy storage is achieved by mechanical means, such as reservoirs, and electrochemical energy storage, using rechargeable batteries. Each type of energy storage has its own advantages and disadvantages. Those using mechanical energy have an established technology but are characterized by high initial investment costs and geographical limitations. Electrochemical storage is highly efficient but offers short storage times. Energy storage in chemical materials is characterized by long storage times but lower efficiency. The mass use of batteries raises questions about the security of metal supplies, as mining causes significant environmental damage, water pollution and other problems, such as the distance between the extraction and use sites and waste management [14].

Power-to-X technologies are essential to the solution of energy storage. These technologies offer convenient ways to store and transport energy in regions where grid connections are weak, such as rural and remote areas. In the Power-to-X concept, X is most often replaced by hydrogen, methanol or possibly methane. Accordingly, the concept of a hydrogen economy and the concept of a methanol economy have evolved together in recent decades. In the hydrogen economy, the intermediate energy carrier is elementary hydrogen. The hydrogen economy is very attractive in theory because it relies on the process of splitting and forming water; however, questions relating to techno-economic feasibility, safety and other sustainability issues have to be addressed [15]. This material has a number of properties that makes the adoption of the hydrogen economy less attractive. Hydrogen is a gas that is difficult to liquefy, while methanol is a liquid, and handling large quantities of hydrogen poses a major technical challenge: it is challenging to store, transport, and distribute safely.

The energy density of hydrogen per unit volume (0.0097 GJ/m^3) is a fraction of the energy density of methanol [16]. Methanol is a better energy storage than current best hydrogen storage technologies, namely, high-pressure storage in composite vessels (up to 300 bar) and metal hydrides. The challenge of hydrogen storage may be met by storage in hydrogen compounds. Such substances are methanol, ammonia, and formic acid, which have established production technologies and existing infrastructures. Methanol can be considered not only as an alternative to hydrogen but also as a hydrogen storage medium [17].

3. The Relationship Between the Methanol Economy and Life Cycle Analysis

A systematic examination of global methanol supply chains and environmental impacts is needed to ensure that the methanol economy is truly sustainable. Life Cycle Assessment (LCA) is an essential part of research into fossil fuel substitution and the introduction of new technologies. The use of carbon dioxide as a feedstock can be a very effective tool to reduce global carbon dioxide concentrations and dependence on fossil fuels, but an assessment of the environmental impacts of the technologies developed is needed to show whether the particular technological pathway is actually helping to achieve sustainability goals.

The LCA is the most appropriate tool for assessing the environmental impact of products. This analysis allows the quantification and comparability of the environmental impacts of products, processes, and models based on metrics [18]. The strength of the LCA is that it collects and summarizes data on environmental impacts at all stages of a product's life cycle. Then it interprets the aggregated results and weights and evaluates them in terms of the significance of the environmental impacts. It presents the results in simplified indicators in an easy-to-understand format.

Analysis of a set of LCA results can be used to make previous LCAs easier to interpret. The analysis allows comparisons between studies, helps to identify the main drivers of environmental impacts, and reduces the uncertainty of estimates [19, 20].

4. Environmental Impacts of Methanol Fuel Use

Efforts to transform road transport focus on the electrification of the transport sector. The main drawback of this strategy is that it requires the replacement of the existing vehicle fleet and a completely new infrastructure. Liquid fuels from renewable sources, such as methanol, represent an interesting alternative that can also be used in existing vehicles [21].

The use of methanol as a vehicle fuel is expected to contribute to the decarbonization of the transport sector, which is of crucial importance, as the transport sector will be the main determinant of the scale and composition of global energy consumption in the future; the energy demand of this sector is growing much more dynamically than that of other sectors.

4.1. *Greenhouse gas emission relations between biomethanol and other fuels*

Biomethanol is produced from biomass, of which forestry and agricultural waste, municipal waste, biogas from wastewater, waste, and byproducts from the paper industry are all significant. In support of a longer-term strategy for fuels, the Danish Technological Institute has investigated the use of three energy sources in road transport: E5 (petrol with 5% ethanol), M85 (petrol with 85% methanol), and battery. The ethanol used in the study is produced from the fermentation of maize, and the methanol is produced from biogas derived from wet manure. It should be emphasized that the technological pathway for the production of biomethanol contributes to reducing both the waste problem and methane emissions associated with agricultural activities. The results of the study are shown in Fig. 2.

The use of methanol produced from biogas results in significantly lower carbon dioxide emissions in road transport, despite the fact that

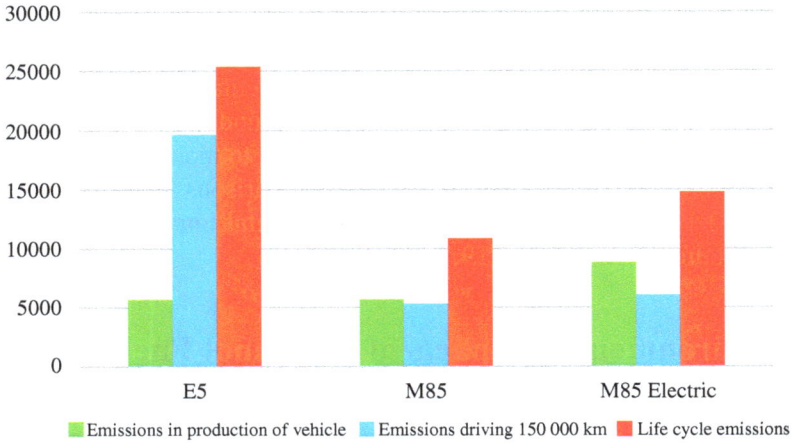

Fig. 2.   Comparison of Danish electric, methanol, and petrol cars.

electricity generation in Denmark has a very low environmental impact of only 56 g CO_2eq/MJ, and if we consider the European Union's electricity mix, which is 129 g CO_2eq/MJ, then the use of methanol shows even more environmental benefits [22].

Table 1 shows the default values for renewable fuels under the European Renewable Energy Directive (RED II) from 2021 [23], which is the total value for production, generation, transport, and distribution. As can be seen, methanol is potentially a relatively low GHG emitter compared to other fuels. An additional advantage would be the production of biomethanol from biogas derived from wet manure (biomethane), as the use of raw manure would lead to a reduction in carbon dioxide emissions and could even result in negative GHG emissions.

Table 1.   Default values of methanol and other renewable fuels according to RED II.

Renewable Fuel	Default Value According to RED II gCO_2eq/MJ
Methanol	10.4–16.2
Bioethanol	15.7–71.7
Biodiesel	14.9–75.7
Biomethane	−100.0–73.0

Shipping is a sector where the use of methanol is particularly important, as maritime transport accounts for 80–90% of commercial transport and, therefore, the level of GHG emissions and the environmental impact of the high sulfur content of the diesel fuel traditionally used are very significant. In the maritime sector, significant changes are also needed in terms of sulfur dioxide, nitrogen oxides, and GHG emissions. In April 2018, the International Maritime Organization (IMO) adopted a strategy to reduce GHG emissions from ships, with the aim of reducing the total annual GHG emissions by at least 50% by 2050 compared to 2008. Newly built ships have to meet increasingly stringent environmental standards, which are tracked using an Energy Efficiency Design Index (EEDI) measurement.

In order to comply with the regulation, shipping companies will have to change the fuel they use. Liquefied natural gas (LNG) and methanol could be an alternative to the dominant diesel fuel, although there are attempts to use all-electric power. The GHG emissions of Marine Gas Oil (MGO) and methanol produced from different feedstocks are shown in Fig. 3.

An investigation carried out via an LCA has shown that methanol produced from natural gas is no more beneficial than diesel fuel in terms

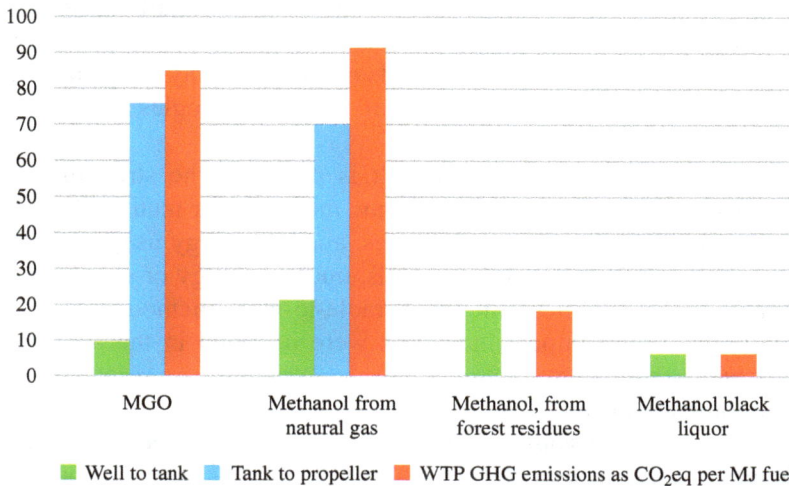

Fig. 3.   The greenhouse gas emissions of MGO and methanol produced from different feedstocks.

of preventing climate change; only the environmental impact during transport is reduced. Methanol produced from renewable feedstocks such as wood residues and black liquor from pulp mills can reduce GHG emissions by 75–90% [24, 25].

5. Greenhouse Gas Emission Relations of E-methanol

The E-methanol made from renewable hydrogen is synthesized with a carbon feedstock captured with renewable electricity. The largest industrial carbon dioxide emitters are the cement, iron and steel industries, partly due to the energy-intensive nature of their technology (using carbon-based fuels) and partly due to the use of coal as a reducing agent and the carbon dioxide produced in the manufacturing process. The "capture", management and use of the carbon dioxide released is facilitated by its local presence, which can be considered a point source. The use of Emissions-to-Liquids (ETL) technology helps to recover carbon dioxide generated as waste in various industries. The process involves reacting carbon dioxide from an industrial source with hydrogen, which is predominantly obtained from the electrolysis of water (refer to Fig. 1).

The use of carbon dioxide from the atmosphere or the flue gas of power plants as a feedstock may suggest that the process is clearly environmentally friendly because it reduces the amount of carbon dioxide in the atmosphere. In order to identify the most sustainable solutions, it is essential that the environmental impacts of each pathway are comparable, quantified, and reflects the benefits of new technologies compared to existing processes [26].

What makes the analysis difficult is precisely the strength of the methanol economy, which is that the raw materials for methanol production can come from a variety of sources, and the energy used in the process can also originate from fossil fuels, nuclear energy or some form of renewable energy. Studies using the "cradle-to-gate" method and containing global warming impact indicators were included in the analysis of LCA results (Table 2).

By comparing each ETL technology pathway with a conventional fossil technology, it is possible to determine whether there is an environmental benefit to using a particular technology in terms of its impact on global warming. Due to the different technology pathways using different

Table 2. Parameters of the analyzed studies.

Authors	Electricity Source
Kim *et al.* [27]	Photovoltaic; fossil
Al-Kalbani *et al.* [28]	Photovoltaic; fossil, wind
Sternberg *et al.* [29]	Electricity (EU-27 in 2020)
Meunier *et al.* [30]	Electricity (ENTSO-E)
Thonemann and Maga [31]	Electricity (ESDP 2030); wind
Kaiser *et al.* [32]	Wind

feedstock sources, the global warming potential compared to the conventional methanol production pathway is shown in bars in Fig. 4.

The analysis shows that the carbon dioxide emissions of the energy source used to produce the hydrogen needed to synthesize methanol determine whether the use of the methanol produced is beneficial in terms of the global warming impact. The use of pure photovoltaic or wind energy is clearly positive, while the use of fossil sources (coal, natural gas) has a negative impact. In the case of the electricity mix, the carbon intensity of the electricity mix used is the determining factor.

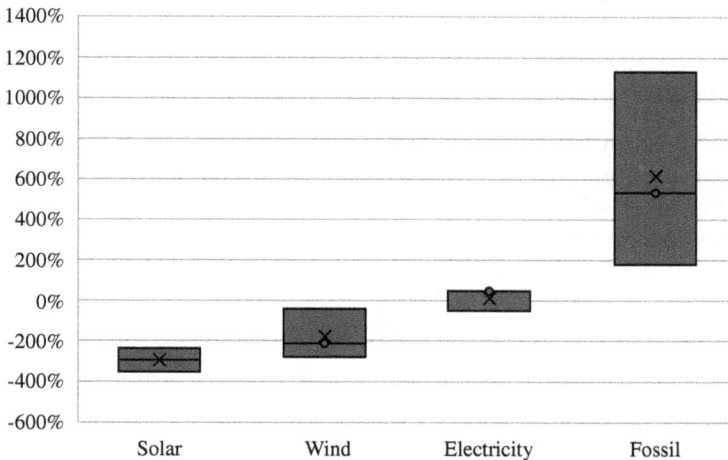

Fig. 4. Impact on global warming for energy sources.

In order to reduce GHG emissions, the use of renewable energy sources is an essential requirement for the design and selection of technologies, but according to González-Garay and colleagues [33], the extent of energy use, freshwater, and land use must be also taken into account to express the real environmental impact.

6. Conclusions

Exploiting the technological link between renewable energy and carbon dioxide capture will lead to a positive change in reducing atmospheric carbon dioxide emissions and increasing independence from fossil fuels. In this chapter, we demonstrated the approach of the formation of a carbon dioxide loop as the material released during combustion is recycled and reused. There is a growing interest in Power-to-X technologies because they are actually able to convert renewable energy into chemicals and fuels that are easy to store and transport. Ideally, the use of fossil fuels can be exiled from all areas of life by using atmospheric carbon dioxide.

In essence, the methanol economy has the potential to incorporate the creation of a CE, through which it is possible to capture and connect pollutants and waste generated during energy production and other industrial processes to the system with the help of carbon-neutral energy sources.

References

1. R. Fouquet, A brief history of energy, in *International Handbook on the Economics of Energy*, J. Evans and L.C. Hunt (eds.), Edward Elgar Publishing, Cheltenham, UK (2011).
2. P. Styring, Carbon dioxide utilization as a mitigation tool, in *Managing Global Warming*, Trevor M. Letcher (ed.), Academic Press (2019).
3. D. Milani, R. Khalilpour, G. Zahedi and A. Abbas, A model-based analysis of CO_2 utilization in methanol synthesis plant, *J. CO_2 Util.* **10**, 12–22 (2015).
4. Gy. Oláh, A. Goeppert and G.K. Surya Prakash, *Kőolaj és a Földgáz Után: a Metanolgazdaság*, Better, Budapest (2007).
5. J. Kothandaraman, S. Kar, R. Sen, A. Goeppert, G.A. Olah and G.K.S. Prakash, Efficient reversible hydrogen carrier system based on amine reforming of methanol, *J. Am. Chem. Soc.* **139**(7), 2549–2552 (2017).
6. G. Pauli, *A Kék Gazdaság — 10 év — 100 innováció — 100 millió munkahely*, PTE KTK, Pécs (2010).

7. G.A. Olah, A. Goeppert and G.K.S. Prakash, *Beyond Oil and Gas: The Methanol Economy*, Wiley-VCH, Weinheim (2018).
8. M.R. Gogate, Methanol-to-olefins process technology: Current status and future prospects, *Pet. Sci. Technol.* **37**(5), 559–565 (2019).
9. Methanol Institute, https://www.methanol.org/methanol-price-supply-demand/
10. G.R.M. Dowson and P. Styring, Demonstration of CO_2 conversion to synthetic transport fuel at flue gas concentrations, *Front. Energy Res.* **5**(26), 1–11 (2017).
11. H.J. Herzog, Scaling up carbon dioxide capture and storage: From megatons to gigatons, *Energy Econ.* **33**(4), 597–604 (2011).
12. P. Eichler, F. Santos, M. Toledo, P. Zerbin, G. Schmitz, C. Alves, *et al.*, Produção do biometanol via gaseificação de biomassa lignocelulósica. *Quimica Nova.* **38**(6), 828–835 (2015).
13. C.F. Shih, T. Zhang, J. Li and C. Bai, Powering the future with liquid sunshine, *Joule.* **2**(10), 1925–1949 (2018).
14. I. Bársony, Fenntarthatóság — fenntartásokkal, *Magyar Tudomány.* **181**(7), 948–967 (2020).
15. U. Bossel, Does a hydrogen economy make sense? *Proc. IEEE.* **94**(10), 1826–1837 (2006).
16. Z. Mayer and Á. Kriston, *Hidrogén és Metanol Gazdaság*, EDUTUS Főiskola, Budapest (2011).
17. J. Andersson and S. Grönkvist, Large-scale storage of hydrogen, *Int. J. Hydrog. Energy* **44**(23), 11901–11919 (2019).
18. Integrált Termékpolitika, Integrated product policy. http://ec.europa.eu/environment/ipp/.
19. G.A. Heath and M.K. Mann, Background and reflections on the life cycle assessment harmonization project, *J. Ind. Ecol.* **16**(s1), S8–S11 (2012).
20. R. Magda and J. Tóth, The connection of the methanol economy to the concept of the circular economy and its impact on sustainability, *Visegr. J. Bioecon. Sustain. Dev.* **8**(2), 58–62 (2019).
21. J. Tóth and R. Magda, The green methanol — playing a role in sustainable energy management, *Int. J. Manag. Sci.* **4**(4), 249–258 (2019).
22. K. Winther, Methanol as motor fuel, Danish Technological Institute, p. 59 (2019).
23. European Parliament and Council of the European Union, Directive (EU) 2018/2001 of the European Parliament and of the Council of 11 December 2018 on the promotion of the use of energy from renewable sources (recast): RED II (2018).
24. S. Brynolf, E. Fridell and K. Andersson, Environmental assessment of marine fuels: Liquefied natural gas, liquefied biogas, methanol and biomethanol, *J. Clean. Prod.* **74**, 86–95 (2014).

25. J. Ellis and M. Svanberg, SUMMETH — Sustainable marine methanol deliverable D5.1 Expected benefits, strategies, and implementation of methanol as a marine fuel for the smaller vessel fleet. http://summeth.marine-methanol.com/?page=reports (2018)

26. A. Zimmermann, L.J. Müller and A. Marxen, *Techno-Economic Assessment & Life-Cycle Assessment Guidelines for CO_2 Utilization*, Chem Media and Publishing Ltd (2018).

27. J. Kim, C.A. Henao, T.A. Johnson, D.E. Dedrick, J.E. Miller, E.B. Stechel, *et al.*, Methanol production from CO_2 using solar-thermal energy: Process development and techno-economic analysis, *Energy Environ. Sci.* **4**, 3122–3132 (2011).

28. H. Al-Kalbani, J. Xuan, S. García and H. Wang, Comparative energetic assessment of methanol production from CO_2: Chemical versus electrochemical process, *Appl. Energy* **165**, 1–13 (2016).

29. A. Sternberg, C. M. Jens and A. Bardow, Life cycle assessment of CO_2-based C1-chemicals, *Green Chem.* **19**(9), 2244–2259 (2017).

30. N. Meunier, R. Chauvy, S. Mouhoubi, D. Thomas and G. De Weireld, Alternative production of methanol from industrial CO_2, *Renew. Energ.* **146**, 1192–1203 (2020).

31. N. Thonemann and D. Maga, Life cycle assessment of steel mill gas-based methanol production within the Carbon2Chem® project, *Chemie-Ingenieur-Technik.* **92**(10), 1425–1430 (2020).

32. S. Kaiser S, F. Siems, C. Mostert and S. Bringezu, Environmental and economic performance of CO_2-based methanol production using long-distance transport for H_2 in combination with CO_2 point sources: A case study for Germany, *Energies* **15**(7), 2507 (2022).

33. A. González-Garay, M.S. Frei, A. Al-Qahtani, C. Mondelli, G. Guillén-Gosálbez and J. Pérez-Ramírez, Plant-to-planet analysis of CO_2-based methanol processes, *Energy Environ. Sci.* **12**(12), 3425–3436 (2019).

Chapter 5

Life Cycle Assessment of Different Solar to Hydrogen Routes and Hydrogen Carriers

Zhihua WANG*, Jinxu ZHANG and Runfan ZHU

*State Key Laboratory of Clean Energy Utilization,
Zhejiang University,
Hangzhou, 310027, China*

*wangzh@zju.edu.cn

A comprehensive life cycle assessment (LCA) was carried out for three methods of hydrogen production and two types of hydrogen carriers. The LCA work included the comparison of hydrogen (H_2) production by solar energy and two types of H_2 carriers, i.e., methanol and ammonia. The assessment also contains an evaluation of four environmental factors, which are global warming potential, acidification potential, ozone depletion potential, and nutrient enrichment potential. After conducting a quantitative analysis of all scenarios with environmental factors being considered, the results conclude that using solar photothermal (PT) technology coupled with thermochemical water splitting by the sulfur–iodine (SI) cycle for hydrogen production, SI-PT has significant advantages in the environmental impact of the whole ecosystem, whereas the solar-photovoltaic-based ammonia (PV-NH_3) production route has the lowest greenhouse effect in all scenarios.

1. Introduction

With increasing attention to global warming or climate change, along with the need for energy security and ecological environment protection, developing renewable energy systems has become a major strategy for global energy transformation. Hydrogen is a promising energy source that can be produced through renewable energy such as hydropower, wind power, solar power, etc. As an energy carrier, hydrogen has the advantages of high efficiency, and being clean, safe, and sustainable, and is considered as "21st Century Energy". Hydrogen has become a research hotspot to expand the usage and promotion of renewable energy, either by being incorporated into the existing natural gas pipeline network, through fuel cells to generate power, or through methanation reaction with carbon dioxide (CO_2) to provide natural gas.

Due to their high hydrogen density (12.5% (wt) gravimetrically for methanol and 17.7% (wt) gravimetrically for ammonia), methanol and ammonia are regarded as attractive chemical hydrogen carriers, which are more suitable for large-scale storage and long-distance transportation than hydrogen. As bulk chemicals, methanol and ammonia are widely used in the chemical industry, fuel cells, and combustion engines, besides hydrogen storage materials.

Hydrogen energy is secondary energy that needs to consume primary energy sources for preparation. Therefore, the development of green and efficient hydrogen production technology coupled with renewable energy is needed for the large-scale utilization of hydrogen energy. As ideal renewable energy, solar energy has the characteristics of being clean and inexhaustible. Using solar energy to produce hydrogen as well as effective hydrogen carriers can greatly reduce greenhouse gas (GHG) emissions and produce "Green Hydrogen".

2. Goal and Scope Definition

2.1. *Hydrogen production methods*

The goal of this life cycle assessment (LCA) is to find the best matching and environmentally friendly solar energy hydrogen production methods, i.e., proton exchange membrane (PEM) water electrolysis coupled with photovoltaic (PV) power, PEM water electrolysis coupling photothermal (PT) power, and thermochemical water splitting method using sulfur–iodine (SI) cycle coupling solar PT power. The main assessment method was the CML 2001 assessment system developed by the Institute of Environmental Sciences, Leiden University, Netherlands [1].

To better compare the differences of the hydrogen production processes, it is necessary to determine the same function unit to make different hydrogen production processes comparable. This study does not consider the usage of hydrogen energy after hydrogen production. Here, 1 kg H_2 from production was determined as the function unit.

The system boundary of hydrogen production methods compared in this chapter is shown in Fig. 1. Solar power generation methods are mainly divided into PV power generation and PT power generation. PV power generation refers to the semiconductor interface PV effect, where solar cells are used to directly convert solar energy into electricity. PV power generation has the characteristics of less regional impact, safety, and reliability.

2.2. *Hydrogen carriers' supply chain*

Two types of carbon-free or carbon-neutral hydrogen (H_2) carriers were compared in the LCA method. Focus was given to the total amount of GHGs emitted throughout the entire supply chain, from raw materials acquisition and processing, liquefied energy carrier production, and storage to transportation until final product utilization.

According to the different sources of raw materials and final product types, there exist six comparative scenarios in this study: (1) coal-based methanol (Coal-CH_3OH), (2) natural gas-based methanol (NG-CH_3OH), (3) PV and carbon capture and utilization (CCU)-based methanol (PV/CCU-CH_3OH), (4) coal-based ammonia (Coal-NH_3), (5) natural gas-based ammonia (NG-NH_3), and (6) solar PV-based ammonia (PV-NH_3). The detailed flow diagram is shown in Fig. 2 and Fig. 3, respectively.

3. Life Cycle Inventory

3.1. *Proton exchange membrane coupled with photovoltaic power for hydrogen production*

The case study in this chapter is based on the Large Hydrogen Plant in Guyuan, Hebei, China. The PEM electrolyzer equipment studied herein

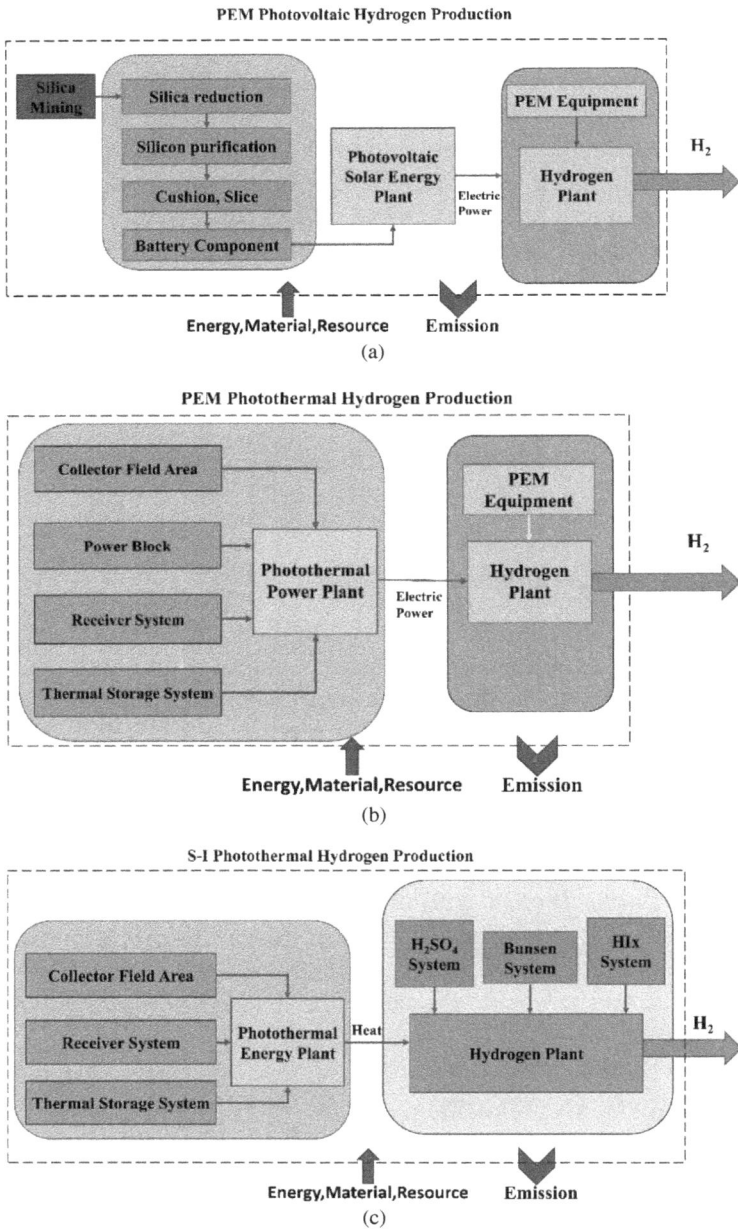

Fig. 1.   System boundary of different hydrogen production methods using solar energy. (a) PEM photovoltaic hydrogen production, (b) PEM photothermal hydrogen production, and (c) SI photothermal hydrogen production [2].

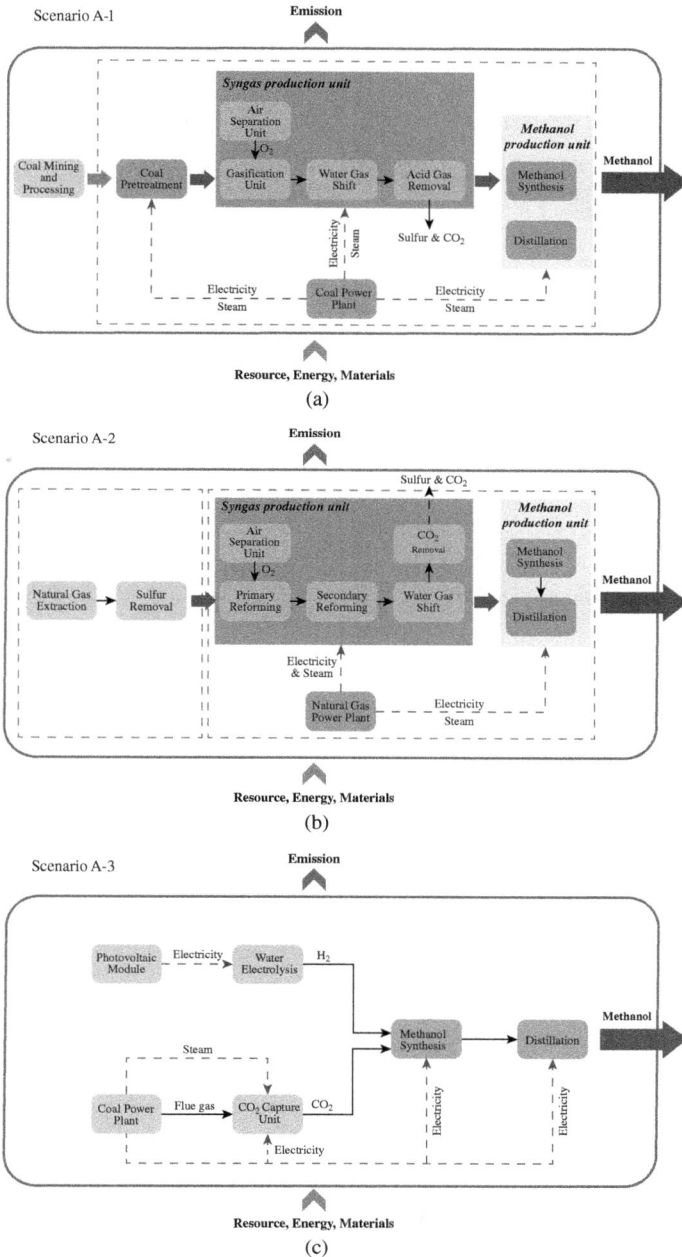

Fig. 2. The schematic system diagram of different methanol production methods. (a) Coal-CH$_3$OH route, (b) NG-CH$_3$OH route, and (c) PV/CCU-CH$_3$OH route [3].

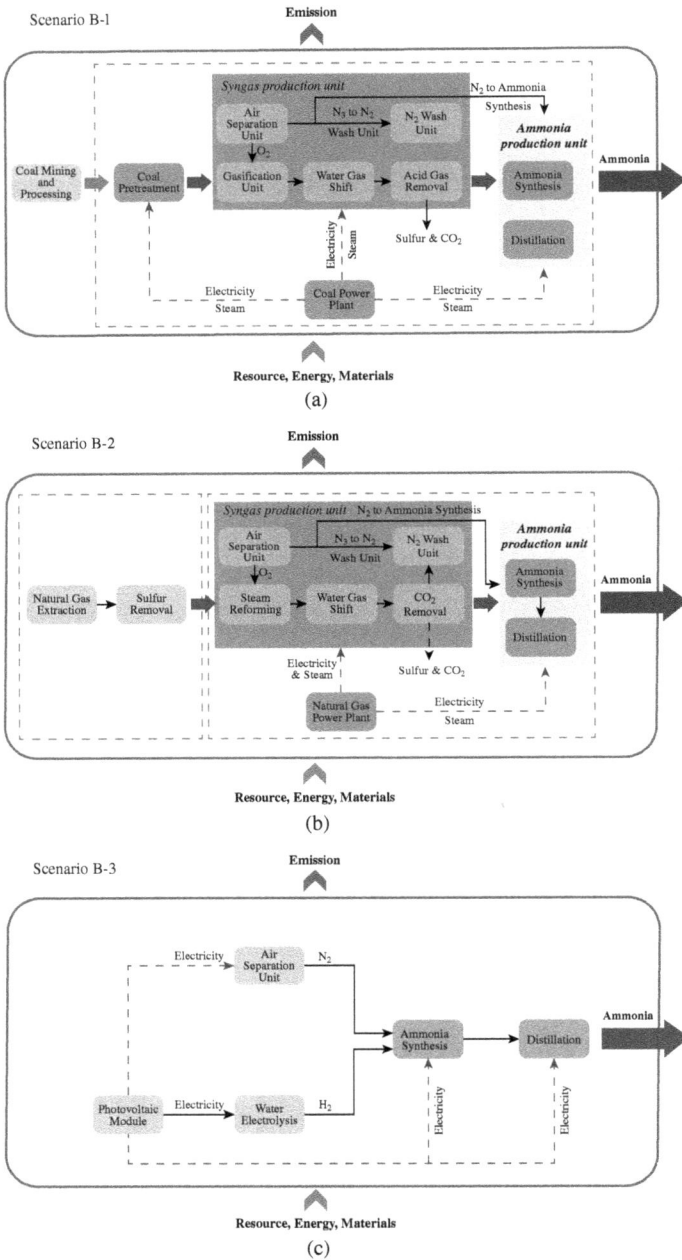

Fig. 3.   The schematic system diagram of different ammonia production methods: (a) Coal-NH$_3$ route, (b) NG-NH$_3$ route, and (c) PV-NH$_3$ route [3].

has the power consumption of 5 MW, capacity of 90 kg/h H_2, and efficiency of 60%. The annual output of the equipment is 17.52 MNm^3 H_2. Considering the coupling between the electrolyzer and solar energy, the equipment is assumed to operate for 10 hours per day, and the life of the entire hydrogen production plant is 30 years. For the full operation state, the energy consumption of the whole system is calculated to be 54.81 kWh/kg H_2.

The cathode of the PEM electrolysis device is sprayed on the carbon brazing plate with platinum-based metal, and the anode is sprayed with iridium oxide [4]. The highly recommended membrane material obtained for PEM, Nafion from DuPont, exhibited the quality of the Teflon material developed by Dr. Walther Grot.[a] The overall structure of the electrolysis device is composed of a chamber structure and a skeleton structure. Apart from the membrane and electrode, most of the chamber structures are constructed by stainless steel. The skeleton part is constructed by resin, stainless steel, and chloroprene rubber [5].

The detailed input list is shown in Tables 1 and 2. Table 1 displays the construction input of the whole PEM electrolytic water hydrogen plant for each 1 kg of hydrogen production. Table 2 shows the input of the PEM electrolysis device during operation for each 1 kg of hydrogen production.

At present, most of the installed capacity of PV power stations in China is around 10 MW. Therefore, the energy supply of the studied PV grid power station used here is 10 MW. The PV power generation system boundary includes polysilicon production, PV cell module production,

Table 1.   Consumption list of the construction of PEM hydrogen production plant (1 kg H_2) [2].

Input	Unit	Amount
Concrete	kg	0.00619
Aluminum	kg	0.000111
Copper	kg	0.000331
Polycarbonate Compound	kg	0.000111
Steel	kg	0.0074

[a] https://www.permapure.com/environmental-scientific/nafion-tubing/nafion-physical-and-chemical-properties/

Table 2.   Consumption list of the operation of PEM electrolysis water (1 kg H_2) [2].

Input	Unit	Amount
Activated carbon	kg	9.94E-6
Aluminum	kg	2.98E-5
Copper	kg	4.97E-6
Electricity	KWh	60
Iridium	kg	8.3E-7
Platinum	kg	8E-8
Stainless steel	kg	0.000111
Titanium	kg	0.000583
Water	kg	9
Nafion™	kg	1.76E-5

and power plant construction. The solar cells are multi-use crystalline silicon solar cells in the current market. Since the cost price of a polycrystalline silicon battery is lower than that of a single-crystal silicon battery, polycrystalline silicon battery components are finally adopted to reduce the project cost. Based on the Dunhuang PV grid-connected power station in Gansu, China, the specific annual average grid-connected power and annual average utilization hours are shown in Table 3 [6]. The life cycle of the PV power station is set to be 30 years, during which a total of 509.7 million kWh can be generated. From Sec. 4 onward, this method is denoted as PEM+PV.

Table 3.   Average annual grid electricity and annual utilization hours [2].

Item	Amount	Unit
Average annual solar radiation	1,782	$kW \cdot h \cdot m^{-2}$
Average annual utilization hours	1,691	H
Installation capacity	10,049.6	kWp
Average annual grid electricity	16.99	Million kWh

As shown in Fig. 1, the production process of PV modules includes silica reduction, silicon purification, and the ingot slicing process, and it is finally assembled into PV cell components. The input and output lists for PV module production are from Ref. [7].

In addition to the production of PV modules, the main materials consumed in the construction of power plants are concrete and steel. The input list of PV power plant construction is shown in Table 4.

Table 4. Consumption list of the construction of a 10-MW Photovoltaic Power Station [2].

Input	Amount	Unit
Aerated Concrete	7.36E7	Kg
Stainless Steel	7.26E3	Kg

3.2. *Proton exchange membrane coupled with photothermal power for hydrogen production*

To keep the contrast parameters consistent, the setting of the PEM electrolysis equipment and hydrogen production plant is the same as Sec. 3.1. The energy supply of the whole system is solar PT power generation. PT power generation uses 20 MW centralized solar tower power generation technology. PT power plants include four modules, namely, collector module, receiver module, heat storage module, and steam power generation module. Because it is difficult to obtain the construction data of a 20 MW solar tower power station, all the construction data used in this paper are from the ecoinvent Database. Due to the hydrogen production from electrolytic water not requiring large-scale electricity, large installed capacity power plants are not considered in our study. The data set is scaled for the modeling of a 440 MW power plant from South Africa [8]. The design of a 20 MW power plant is represented by linear and nonlinear equations with specific parameters [9]. From Sec. 4 onwards, this method is denoted as PEM+PT.

3.3. *Sulfur–iodine cycle coupled with photothermal for hydrogen production*

The hydrogen production capacity of the thermochemical water splitting by the SI cycle hydrogen plant calculated in this chapter is 200 tons of hydrogen per day [10], and the life of the hydrogen plant is set as 30 years. Ozbilen estimated the input materials of thermochemical water splitting by the Cu-Cl cycle hydrogen plant [11]. Since the main difference in thermochemical hydrogen production lies in the operation process, it is assumed that the construction materials of the thermochemical water splitting by the SI cycle hydrogen plant are similar. In addition to the basic plant construction, other main construction materials are from the Bunsen reactor, sulfuric acid and HIx distillation columns, purification towers and reactors, and the construction of various pipelines. For each 1 kg of hydrogen production, the list of all construction materials for the hydrogen plant is shown in Table 5. From Sec. 4 onwards, this method is denoted as SI+PT.

The main reactions for hydrogen production from thermochemical water splitting by the SI cycle and the respective required temperatures are shown below:

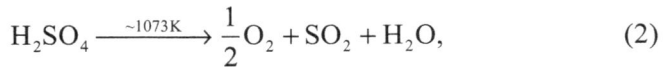

$$I_2 + SO_2 + 2H_2O \xrightarrow{\sim 393K} H_2SO_4 + 2HI, \tag{1}$$

$$H_2SO_4 \xrightarrow{\sim 1073K} \frac{1}{2}O_2 + SO_2 + H_2O, \tag{2}$$

$$2HI \xrightarrow{\sim 723K} H_2 + I_2. \tag{3}$$

The heat required in the reaction process comes from solar heat. To reduce the operation cost, we reduce the step of heat conversion into

Table 5.   Consumption list of the construction of thermochemical water splitting by the SI cycle hydrogen (1 kg H_2) [2].

Input	Unit	Amount
Aluminum	kg	0.0223
Copper	kg	0.312
Lead	kg	0.0505
PVC	kg	0.173
Spruce log	kg	3.8
Steel	kg	10.2
Concrete	kg	40.1
Titanium	kg	0.0148

Table 6. Consumption list of the operation of a hydrogen plant by thermochemical water splitting by SI cycle (1 kg H_2) [2].

Input	Unit	Amount
Aluminum	Kg	8.6E-8
Copper	Kg	7.14E-7
Lead	Kg	1.95E-7
PVC	Kg	6.68E-7
Steel	Kg	2.62E-6
Titanium	Kg	5.74E-9
Water	Kg	0.000714
Iodine	Kg	5.03E-5
Thermal energy	MJ	184

electricity in the PT power station. The heat in the heat storage module is directly used as the energy source of hydrogen production in the SI cycle. The other components of the PT power station are unchanged. The heat value of the whole system comes from the process simulation by Aspen Plus [12].

Only water is consumed during the whole reaction process, but considering that a certain amount of iodine will deposit and adhere to the pipe wall during the actual operation, we consider the iodine loss during the operation of hydrogen production from the thermochemical water splitting by the SI cycle. For every 1 kg of H_2 produced, the consumption list of SI cycle hydrogen production is shown in Table 6.

3.4. *Hydrogen carrier scenarios*

A flow-process diagram for the H_2 carriers supply chain is shown in Fig. 4. The "cradle to gate" is defined as the system boundary of all the six comparison scenarios, which covers the material and energy production chain and all processes from the raw material extraction through production, transportation, and up to the final stage of H_2 utilization. This work assumes that both methanol and ammonia production plants are located in northwest China, where coal, natural gas, and solar energy resources are abundant.

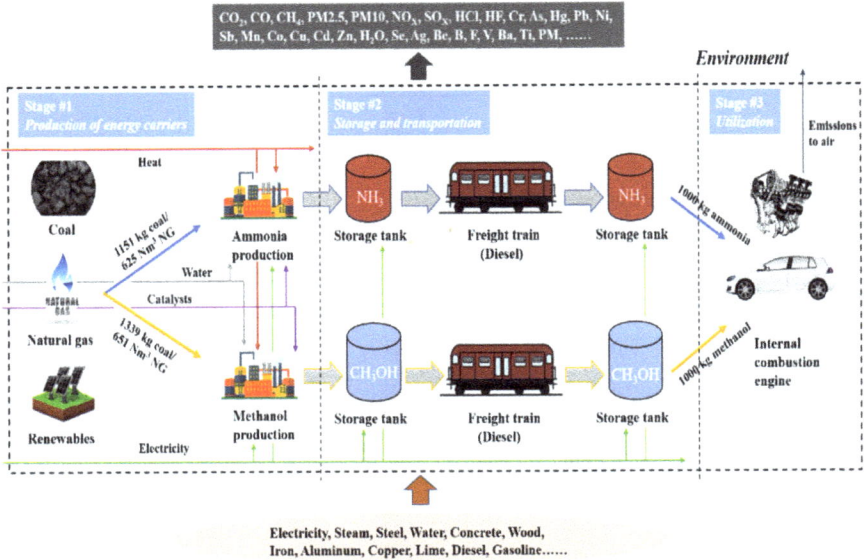

Fig. 4.   Hydrogen carriers supply chain LCA scheme [3].

The transportation and application of methanol and ammonia are considered, as the two H_2 carriers emit different amounts of GHGs into the environment during the transportation and utilization stage. The transportation stage contains three parts: land storage, loading and unloading, and railway transport. We chose freight train as the transportation mode, and the transport distance is selected as 1,500 km. After arriving at the destination, the product is used as fuel in an internal combustion engine in all involved routes. The transportation of feedstock and auxiliary materials is excluded from the scope of this study, since the manufacturing plant is a pithead plant in all the cases. This simplified method has also been reported in other literatures. All the relevant energy (electricity, heat, etc.) and material (water, catalyst, etc.) consumption and treatment at each stage of the supply chain are considered, which adhere to China standards as far as possible. In addition, the GHG emission during infrastructures and facilities construction are contained in the LCA system. However, due to the lack of relevant data, this study does not involve the recycling of materials and the utilization of byproducts.

4. Result and Discussion

4.1. *Environmental impacts of different hydrogen production methods*

The environmental impacts of the three hydrogen production methods are studied, and the results of CML 2001 impact categories (GWP, AP, EP, ODP) are given. To ensure the comparability of the results, all impact categories are calculated based on the 30-year service life of the equipment aimed at producing 1 kg of hydrogen. The calculation results are shown in Fig. 5.

As a whole, the environmental impacts of PEM+PV and PEM+PT are higher than that of SI+PT. In terms of GWP, PEM+PV and PEM+PT are 9.37 kg CO_2-eq and 8.67 kg CO_2-eq, respectively, which are 9 times and 8 times higher than that of SI+PT. Unlike GWP, the AP and EP of PEM+PT are slightly higher than that of PEM+PV, and they are both 10 times higher than SI+PT. For ODP, PEM+PV is 3.19E-11 kg R11-eq, much higher than PEM+PT and SI+PT, which are 100 times higher than

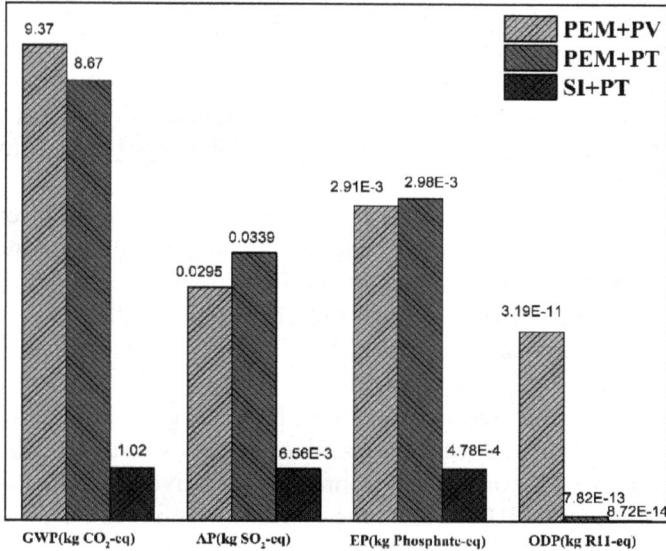

Fig. 5.   Environmental impact of different hydrogen production methods [2].

both PEM+PT and SI+PT. On one hand, the energy consumption of hydrogen production from electrolytic water is higher than that from thermochemical water splitting by the SI cycle. The results also concluded that when the use of PV or PT is applied, the process of light to electricity or heat to electricity should be considered. It is well noted that SI+PT can directly use the energy of heat (thus avoiding steam turbine power generation) and lead to significant reduction of impact on the environment.

In terms of hydrogen production from PEM electrolysis water, the hydrogen production from electrolysis water using PT power generation is lower than that of using PV power generation in GWP while being slightly higher than that of using PV power generation in AP and EP. It is worth noting that the ODP of hydrogen production from electrolysis water using PT power generation is two orders of magnitude lower than that of using PV power generation.

4.2. *Proportion of environmental impact in each hydrogen production methods*

To better understand the contribution of each step to the environmental impact in different hydrogen production methods, the environmental impact weights of the four main steps in the hydrogen production process, i.e., the construction of solar power plants, the operation of solar power plants, the construction of hydrogen plants, and the operation of hydrogen plants, are analyzed.

Figure 6 shows the proportion of environmental impact in each step of three hydrogen production methods. As shown in the figure, no matter the type of environmental impact, the construction process of solar power plants always contributes the most. In terms of GWP, the construction of solar energy power plants in PEM+PV, PEM+PT, and SI+PT accounts for 78%, 73%, and 77%. The second-largest impact on the environment comes from the construction process of the hydrogen plant, and whether it is the operation of the solar power plant or the operation of the hydrogen plant, the impact on the environment is relatively small. It is worth noting that in the SI+PT process, the construction of the PT power plant accounts for 48% in AP, while the proportion of the construction of the S-I hydrogen production plant is slightly higher, accounting for 51% in AP, which is different from the other two hydrogen production methods.

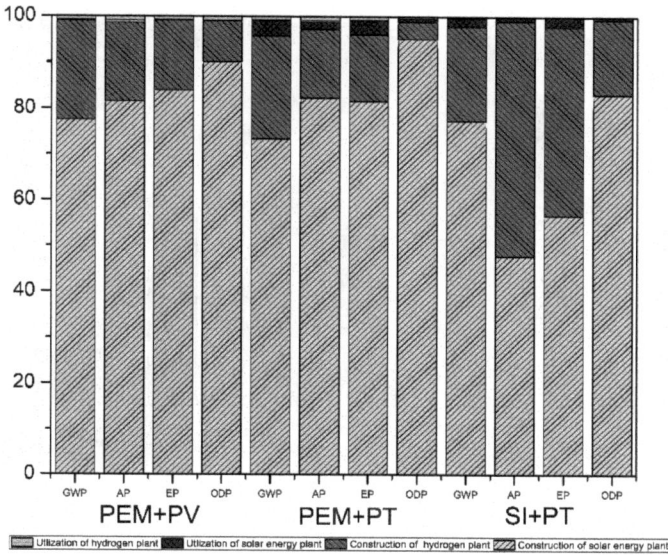

Fig. 6. Contribution of each step to the environmental impact in three hydrogen production methods.

In summary, the construction process is the main source of environmental pollution. Therefore, the selection of low-pollution construction materials is the key to reducing the overall solar power hydrogen production.

4.3. *Environmental impacts of different hydrogen carrier scenarios*

Figure 7(a) and 7(b) illustrates the GWP balance of the entire supply chain of fuels per kg and MJ, including production, transportation, and the utilization stage of methanol and ammonia. For ammonia, GHG emissions are mainly concentrated in the fuel production phase, while transportation and utilization phases contribute little to the greenhouse effect. Methanol, on the other hand, emits a considerable amount of CO_2 when burned in the internal combustion engine, resulting in a large GWP in the utilization phase. As shown in Fig. 7(a), in NG-CH_3OH and PV/CCU-CH_3OH, GHG emissions in the fuel utilization stage account for 61.71% and 56.55% of

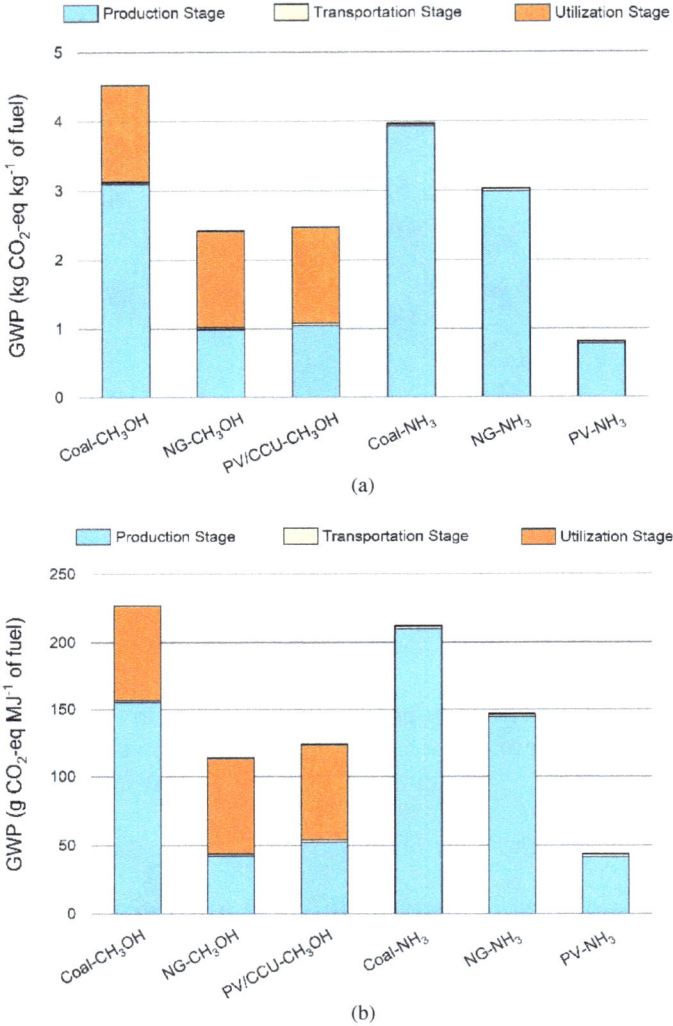

Fig. 7. Total GHG emissions from the entire supply chain of energy carrier under various scenarios: (a) per unit mass, and (b) per unit energy [3].

the total GHG emissions through the whole life cycle, respectively. Although methanol is considered as a clean fuel as compared with conventional fossil fuels such as gasoline and diesel, it still contains a certain amount of carbon, which will be converted into CO_2 during the combustion process and released into the atmosphere. On the other hand, ammonia is a carbon-free energy carrier with almost no direct GHG effect. This is the main reason for the above differences.

Considering a complete life cycle, Coal-CH$_3$OH and Coal-NH$_3$, which use coal as raw material to manufacture H$_2$ carriers, have the highest GHG emissions. When 1 kg or 1 MJ of methanol is produced from hard coal and transported 1,500 km away by rail and finally used in the internal combustion engine, it emits around 4.52 kg or 227.05 g of CO$_2$-eq, respectively. For 1 kg or 1 MJ of ammonia, this quantity of emissions slightly reduces to 3.97 kg or 212.36 g of CO$_2$-eq, respectively. Using natural gas to produce H$_2$ carriers has a significant effect on reducing GHG emissions, especially for methanol. In NG-CH$_3$OH, producing 1 kg methanol emits about 0.84 kg CO$_2$-eq, which is approximately 73% less than in Coal-CH$_3$OH.

In all the investigated scenarios, the route of ammonia production via electrolysis of PV solar has the lowest GHG emissions, which are 80% and 71% less than the coal-to-ammonia route and natural gas-to-ammonia during the production phase, respectively. Regardless of whether it is calculated by unit mass (kg) or unit energy (MJ), electrolysis-based ammonia production has the lowest GWP throughout the full supply chain, at 0.82 kg of CO$_2$-eq/kg of ammonia and 43.87 g of CO$_2$-Equation MJ^{-1} of ammonia.

5. Conclusion

In this chapter, the LCA is carried out for hydrogen production by solar PV power generation coupled with PEM electrolysis water, solar PT power generation coupled with PEM electrolysis water, and solar PT coupling with thermochemical water splitting by an SI cycle. Four environmental impacts (GWP, AP, EP, ODP) of each method are quantified. The proportion of the environmental impact in each step of each method and the change of environmental impact with lifetimes are studied and analyzed. The research conclusions of this paper are as follows:

(1) From the aspect of the environmental impact of the whole system, solar PT coupling thermochemical water splitting by an SI cycle for hydrogen production (SI+PT) has the lowest GWP, AP, EP, ODP. The environmental impact of solar PT power generation coupled with PEM water electrolysis (PEM+PT) is smaller than that of solar PV power generation coupled with PEM water electrolysis (PEM+PV).

(2) During solar PV power generation and solar thermal power generation coupled PEM water electrolysis process (PEM+PV/PEM+PT), the main environmental impact comes from the construction of the solar power plant and PEM water electrolysis plant. In the process of solar PT coupling thermochemical water splitting by the SI cycle for hydrogen production (SI+PT), in terms of AP, the construction of the hydrogen production plant is the main influence, followed by the construction of a solar PT power plant, while GWP, EP, and ODP mainly come from the construction process of the solar thermal power plant.

(3) The results of life cycle emissions show that the production phase is the largest contributor to GHG emissions in most scenarios. For methanol, the utilization phase has an important contribution to overall emissions, while the utilization phase contributes little to overall emissions in the ammonia supply chain. Considering the completed life cycle, solar-PV-based ammonia production route (PV-NH$_3$) emits 43.9 g of CO$_2$-Equation MJ^{-1} of ammonia, which has the lowest greenhouse effect in all scenarios. In the case of fuel production using a conventional manufacturing method, syngas production process generates the most GHG, followed by the fuel synthesis process. However, in the scenarios of using renewable H$_2$ to produce fuel, the H$_2$ generation process replaces the syngas production process to having the most GHG emissions during the fuel production phase.

(4) With the implementation of a new round of energy industry planning, China is accelerating the transition process of clean energy, and the demand for decarbonization and emission reduction is increasing daily. As mentioned above, there are great expectations on ammonia as a net-zero emission society enabler. At present, ammonia is still dominantly used in the fertilizer industry, and its application for power needs to be further explored. Government agencies and industry investors should consider appropriate capital investment to support follow-up, high-quality research.

References

1. J. Guinee, Handbook on life cycle assessment — operational guide to the ISO standards, *Int. J. Life Cycle Assess.* **6**, 255 (2001).
2. J. Zhang, B. Ling, Y. He, Y. Zhu and W. Wang, Life cycle assessment of three types of hydrogen production methods using solar energy, *Int. J. Hydrog. Energy* **47**(30), 14158–14168 (2022).

3. R. Zhu, Z. Wang, Y. He, Y. Zhu and K. Cen, LCA comparison analysis for two types of H_2 carriers: Methanol and ammonia, *Int. J. Energy Res.* **46**, 11818–11833 (2022).
4. A. Mayyas and M. Mann, Manufacturing competitiveness analysis for hydrogen refueling stations, *Int. J. Hydrog. Energy* **44**, 9121–9142 (2019).
5. M. Miller, A.S. Raju and P.S. Roy, The development of lifecycle data for hydrogen fuel production and delivery, Institute of Transportation Studies, Working Paper Series (2017).
6. A. Stoppato, Life cycle assessment of photovoltaic electricity generation, *Energy* **33**, 224–232 (2008).
7. J. Hong, W. Chen, C. Qi, L. Ye and C. Xu, Life cycle assessment of multi-crystalline silicon photovoltaic cell production in China, *Sol. Energy* **133**, 283–293 (2016).
8. B.D. Kelly, Advanced thermal storage for central receivers with supercritical coolants, Office of Scientific and Technical Information, US Department of Energy (2010).
9. T. Telsnig, Standortabhängige analyse und bewertung solarthermischer kraft-werke am beispiel südafrikas (2015).
10. NETL D, Life-cycle analysis of greenhouse gas emissions for hydrogen fuel production in the United States from LNG and coal, National Energy Technology Laboratory, United States (2006).
11. A. Ozbilen, I. Dincer and M.A. Rosen, Environmental evaluation of hydro-gen production via thermochemical water splitting using the Cu–Cl Cycle: A parametric study, *Int. J. Hydrog. Energy* **36**, 9514–9528 (2011).
12. J. Zhou, Y. Zhang, Z. Wang, W. Yang, Z. Zhou, J. Liu, *et al.* Thermal effi-ciency evaluation of open-loop SI thermochemical cycle for the production of hydrogen, sulfuric acid and electric power, *Int. J. Hydrog. Energy* **32**, 567–575 (2007).

https://doi.org/10.1142/9789811275661_0006

Chapter 6

Hydrogen Production from Biomass and Waste: Life Cycle Assessment Perspective and Deployment Opportunities

Massimiliano MATERAZZI*, Andrea PAULILLO and Paola LETTIERI

Department of Chemical Engineering,
University College London,
Torrington Place, London WC1E 7JE, United Kingdom

*massimiliano.materazzi.09@ucl.ac.uk

Hydrogen is widely recognized to have a key role in decarbonizing various industries as well as the transportation, heating and power sectors. Three main pathways to produce low-carbon hydrogen have been identified in this chapter — (1) green hydrogen (Green-H2), produced from electrolysis combined with power sourced from renewable energy, (2) blue hydrogen (Blue-H2), produced from the reformation of natural gas with Carbon Capture and Storage (CCS), and (3) biohydrogen (BioH2) produced from the gasification of waste feedstocks. Using the United Kingdom (UK) as a case study, this chapter provides a summary of the Life Cycle Assessment (LCA) application for an industrial process producing BioH2 from biomass and waste, and compares this with other low-carbon hydrogen routes. The overall production levels of BioH2 is observed to be limited by the availability of sustainable feedstocks; however, the results of negative carbon dioxide (CO_2) emissions achieved via

BioH2 production shows that its overall potential to reduce greenhouse gas (GHG) emissions is significantly better, as compared to Blue-H2 and Green-H2. In particular, BioH2 application is capable of generating net-negative CO_2 emissions required to make a very important contribution to achieving net-zero commitment in the UK.

1. Introduction

Over the last three decades, the United Kingdom's (UK) greenhouse gas (GHG) emissions have fallen by approximately 47% to the current 405 Mt CO_2-eq/year [1]. The majority of this reduction was achieved largely by the closure of coal-fired power plants and the increasing use of renewable electricity generation. The UK has set a target to reach net-zero GHG emissions by 2050 [1]. This requires the current 522 Mt CO_2-eq emissions per year to reduce to be as close to zero over the next three decades. As one of the methods for carbon emission reduction, the adoption of low-carbon fuels, such as hydrogen, is expected to play a significant role to reach net-zero targets in the UK.

Hydrogen is currently used as an industrial feedstock, mainly for ammonia production and in oil refineries. Hydrogen can also be used for electricity generation in fuel cells or for heat and electricity generation. As hydrogen is carbon-free, no CO_2 emissions are released during combustion, with water vapor as the only byproduct. However, certain types of hydrogen production can release significant GHG emissions, namely, fossil-based hydrogen production, which are produced from Steam Methane Reforming (SMR) and Autothermal Reforming (ATR) of natural gas. There are three main technologies that can produce hydrogen with low carbon impact:

- Electrolysis using renewable electricity to produce "green hydrogen" (Green-H2),
- Reformation of natural gas (ATR and SMR) with Carbon Capture and Storage (CCS) to produce "blue hydrogen" (Blue-H2), and
- Reformation of biogas or gasification of biomass with CCS to produce "biohydrogen" (BioH2).

This chapter looks at the forecast environmental impact emissions associated with BioH2 from waste and biomass using Life Cycle

Assessment (LCA) methodologies, alongside the estimated trend of deployment for hydrogen for the UK to fully decarbonize and reach net zero.

2. The Biohydrogen Process

Biohydrogen (BioH2) is a low-carbon energy vector and commodity made largely from renewable feedstocks, such as biomass resources that are composed of long chain hydrocarbons. During biomass cultivation stages, carbon is captured from the atmosphere through photosynthesis. The BioH2 production process reforms these hydrocarbon chains into BioH2 and carbon dioxide (CO_2). The BioH2 is then used as an energy resource, whilst CO_2 is used as a feedstock for CCS applications. The overall impact is to convert a low-grade resource (such as waste wood or refuse derived fuel (RDF)) into a low-carbon fuel while capturing CO_2 to generate negative carbon emissions. In this sense, the technology, which qualifies as BECCS (Bioenergy with Carbon Capture and Storage), has the potential to remove GHGs from the atmosphere more cost effectively than competing approaches. This means that BioH2 can contribute twice to Net Zero targets because it can sequester CO_2 from the atmosphere whilst reducing carbon emissions from the energy sector.

Figure 1 illustrates the schematic of the BioH2 production process from waste feedstock based on typical two-stage gasification and CO_2 capture processes [2].

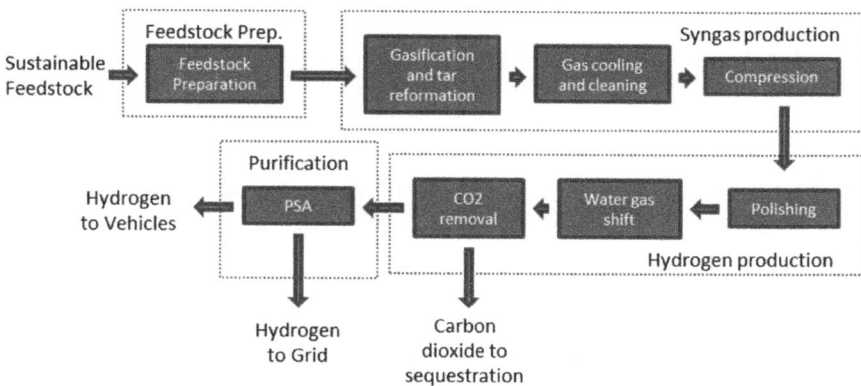

Fig. 1.   Schematic of the biohydrogen production process from waste feedstock.

The process can be summarized as follows:

- Feedstock is made up of waste and biomass residues such as household waste, waste wood, straw, sawdust or sugarcane bagasse. It is prepared for the process through shredding, sorting and drying.
- There are several options for syngas production [3]. They all rely on heating the feedstock in the presence of limited amounts of oxygen (gasification). This generates a crude synthesis gas, which is made up of hydrogen, carbon monoxide, carbon dioxide (CO_2), tars and ash. The tars and ash must be removed for the syngas to be processed further. This is achieved through some combination of further heating, wet scrubbing, and catalytic conversion. The clean syngas is then compressed.
- The carbon monoxide in the syngas reacts with water to produce CO_2 and hydrogen in a process known as the water–gas shift (WGS). Conventionally, this is carried out over an iron or copper catalyst under carefully controlled conditions to optimize the amount of hydrogen produced. In more advanced processes, such as sorption-enhanced water–gas shift (SEWGS), the CO_2 is removed as it is produced to boost hydrogen production.
- The output from WGS is a mixture of CO_2 and BioH2. The CO_2 is removed either through a chemical process, where a compound that reacts with CO_2 is used, or a physical process, where the CO_2 is dissolved or absorbed by a material. In either case, the CO_2 is then released, compressed and passed to a sequestration network.
- The BioH2 that is released from the CO_2 capture system will normally contain significant levels of contaminants, such as carbon monoxide or methane. The gas is cleaned using a pressure swing adsorption (PSA) device. The output from this is a high-quality BioH2 stream.

In advanced separation technologies, such as the SEWGS system, the WGS and CO_2 steps are combined into a single vessel [4]. This has the advantage of producing a high-purity BioH2 stream that may not require a PSA purification step if the target market for the hydrogen is heat. Purification will always be required if the BioH2 is intended for use in fuel cell electric vehicles.

Syngas produced from biomass gasification can have issues with contaminants such as tars and ash from the input waste stream [5]. Two-stage technologies, such as plasma-assisted gasification, have been designed to overcome this by utilizing high temperatures and plasma catalysts to remove tars and vitrify ash to produce a clean syngas [6]. Like Blue-H2, the operation of the BioH2 plant can also provide a consistent supply of hydrogen throughout the year and does not face the same intermittency challenges as Green-H2 from variable renewable energy (wind and solar). The size of BioH2 plants can be scaled at smaller sizes than ATR plants for Blue-H2, with, for example, a typical commercial plant for BioH2 being from a 40-MW hydrogen HHV capacity. Although the chemical engineering processes used in these gasification processes have been proven, the combination of the system, as a whole, still needs to be proven at a close-to-commercial scale. For example, in the UK (Swindon), the company Advanced Biofuel Solutions (ABSL) is commissioning and demonstrating the technology with an output capacity of 3 MW hydrogen HHV. This will be scaled up to produce 44 MW at a commercial scale.

The availability of feedstock (either RDF or biomass) limits the potential production quantity of BioH2. The value of gate fees for waste feedstock will also have a large impact on the economics of the BioH2 plant. An important byproduct of the BioH2 production process is high-quality CO_2, which can be compressed as a feedstock for CCS applications. It is displayed in Sec. 4, where the utilization of biomass feedstock can give negative CO_2 results via the photosynthesis of CO_2 from the atmosphere.

2.1. *Technological aspects*

2.1.1. *Feedstock quality*

Thermochemical treatment of biomass feedstock and gasification, in particular, is gaining strong traction in Europe, giving the numerous opportunities associated to product flexibility and low environmental impact. Recent studies have proven that BioH2 offers the largest potential in terms of GHG removal [7–9]. However, BioH2 production should ideally rely on the use of second or third-generation biomass as primary feedstock to avoid land use competition with food crops and intensification of deforestation, habitat loss, and loss of soil fertility. Municipal Solid Waste (MSW) represents an ideal source because of their large availability and low cost. From a climate change perspective, the use of waste as feedstock not only ensures large and economical availability for consistent hydrogen supply but also avoids pollution due to waste sent to landfills and incineration facilities. However, the use of waste as a feedstock poses technical challenges, which are still under investigation (see the next section).

BioH2 plant performances and environmental attributes are obviously strictly dependent on feedstock composition. Generally, the design point for the waste composition for a thermochemical facility is derived from several datasets for representative residual municipal, commercial and trade waste collected nationally as well as locally. This typically shows a substantial quantity of organic (biomass) content in the waste material, which is typically between 40% and 60% in weight (as received basis) [10]. The waste composition used in this study is indicated in Table 1. This is generated from averaging a number of datasets collected in the UK [11], and is used in this study to run the BioH2 models and determine the environmental attributes.

Waste cannot be thermochemically treated in its original form when collected. The untreated municipal or commercial waste is first mechanically processed in a material recycling facility (MRF). This is done to homogenize the material and remove part of the moisture, recyclables (e.g., metals and dense plastics) and reject materials (e.g., oversize and inert). The material is then shredded using a tearing motion to achieve a rough shred of waste residues, with a homogenous, predetermined particle size between 1 and 50 mm, depending on the gasification reactor requirements. The final feedstock is in the form of floc of RDF, which is then further dried on-site using waste heat from the process. Typically, a 100,000-tonne MSW feed produces an output of ca. 60,000 to 80,000 tonnes of RDF with a moisture content of 10–17%, 10–20% ash content, and 15–25 MJ/kg calorific value (CV), as shown in Table 1.

2.1.2. *Waste gasification development stage*

Compared to pure biomass, RDF introduces a greater concentration and diversity of contaminants due to the high number and variability of sourcing points. This presents a major challenge, which is compounded by the fact that more sophisticated applications (including catalytic processes for BioH2 production and fuel cells for transportations) have very low tolerances.

The state of technology development for biomass or waste gasification is generally seen to be in the TRL (Technology Readiness Level) range of 7 to 8. This has recently been reviewed by the Department for

Table 1. Waste feedstock composition analysis.

Waste Fractions [wt% as received]	MSW	Waste Wood
Paper and cardboard	22.7	0.8
Wood	3.7	93.4
Metals	4.3	1.7
Glass	6.6	—
Textile	2.8	—
WEEE	2.2	—
Plastics	10	0.5
Inert/aggregates/solid	5.3	2.5
Organic fines	35.5	1.1
Miscellaneous	7.1	—
Proximate Analysis [wt%, dry]	**RDF**	
Fixed carbon	8.90	22.8
Volatile matter	64.70	68.0
Ash	11.80	0.5
Moisture	14.60	8.7
Ultimate Analysis [wt%, dry ash free (DAF)]	**RDF**	
Fossil carbon	20.51	0.6
Biogenic carbon	36.23	44.6
Hydrogen	6.86	6.46
Oxygen	31.78	45.38
Nitrogen	4.1	0.26
Sulfur	0.18	0.01
Chlorine	0.34	0.25
Energy Content [MJ/kg DAF]	**RDF**	
Gross calorific value (GCV)	28.99	23.0
Net calorific value (NCV)	27.02	21.8

Business, Energy & Industrial Strategy (BEIS) in the UK [12]. Firstly, most biomass and waste-fueled gasifiers are fundamentally unsuited to the production of syngas as an intermediate to hydrogen or synfuel production, principally because they are air-aspirated rather than oxygen blown. Air-aspirated gasifiers entrain large volumes of nitrogen in the syngas — the removal of nitrogen from the product (hydrogen, biomethane, etc.) being expensive and difficult to accomplish. Gasifiers suited to the production of BioH2 must be indirectly heated or oxygen/steam blown, and ideally, they can operate significantly above ambient pressure. Another class of suitable technologies is that of multi-stage conversion processes, which combine bulk gasification in conventional fluidized bed reactors with advanced reforming steps to deal with tars and ashes, naturally abundant in waste feedstock. Some of these technologies have been tested and demonstrated at pilot scale, but major challenges arose during the scale-up [13]. To address this problem, ABSL embarked on a programme of developments at Swindon some years ago, beginning with a pilot-scale gasifier and a 50 kWth BioSNG demonstration project [14]. The pilot plant experience has enabled ABSL to continue the development of the RDF to the BioSNG and BioH2 concept, with a semi-commercial (1/10th scale) demonstration plant currently being commissioned in Swindon.

2.1.3. *Pre-combustion CO_2 capture*

Whilst post-combustion capture from the flue gas of a biomass power station is not yet a common practice, the technologies used for both power generation and post-combustion capture are mature and at a state of development where they could be classed as commercially proven (TRL 9). Hence, the technology risks associated with applying BECCS to biomass power generation are low. By contrast, the production of BioH2 via gasification of biomass is not a mature technology. Thus the barriers to BioH2 production, despite the conceptual advantages described herein, are not only associated with market immaturity and near-term lack of CCS and hydrogen offtake infrastructure, the sector faces risks connected with technological novelty associated to pre-combustion CO_2 separation.

Pre-combustion capture refers to removing CO_2 from syngas, typically post-WGS stage. Compared to post-combustion technology, which removes dilute CO_2 (~5–15% CO_2 concentration) from flue gas streams and is at a low pressure, the shifted synthesis gas stream is rich in CO_2 and at an ideally higher pressure, which allows for easier removal. Due to the more concentrated CO_2, pre-combustion capture typically is more efficient, but the capital costs of the base waste gasification process and gas cleaning sections are often more expensive than traditional fossil-based power plants.

Today's commercially available pre-combustion carbon capture technologies generally use physical or chemical adsorption processes, and will cost around $60/tonne to capture CO_2 generated by an integrated gasification combined cycle (IGCC) power plant [15]. The current goal of BioH2's research efforts is to reduce this cost to $30/tonne of CO_2. Research focuses on three key separation technologies — advanced solvents, solid sorbents, and membranes — in order to meet this goal. Most of the commercial technologies for pre-combustion CO_2 capture available today share a similar process layout consisting of two successive phases of absorption and desorption of CO_2. The absorption phase uses a solvent to remove CO_2 from the shifted syngas, producing a hydrogen (H_2)-rich stream. These technologies are characterized as physical or chemical, depending on whether the CO_2 is simply physically dissolved or is chemically bound to the solvent. A key difference is that chemical absorption requires increasing temperature for desorbing CO_2; whilst in physical absorption, this can be achieved by solely reducing the pressure [16]. MEA and Benfield, which uses potassium carbonate as a solvent, rely on chemical absorption. Selexol® and Rectisol® use physical solvents, respectively a mixture of dimethyl ethers of polyethylene glycol (DEPG) and refrigerated methanol.

Notably, the Rectisol® technology operates at very low temperatures (ranging between −40°C and −60°C), which requires a refrigeration cycle. The H_2-rich stream is purified via a methanation reactor before PSA to obtain a H_2 stream suitable for gas grid injection (>98% v/v), which is then compressed to 46 bar for storage. The tail gas from PSA contains CO_2 and H_2 primarily, as well as traces of other combustible (e.g., CH_4) and non-combustible (e.g., N_2) substances from syngas and the solvent. The gas is burnt in a gas engine to generate electricity and thermal energy. The former can be sold to the electric grid operator, whilst the latter is recovered in the WGS phase. On the other hand, the CO_2-rich stream from the desorption phase is compressed to 35 bar and stored, pending utilization or storage.

3. LCA Model of BioH2

This work summarizes the application of the LCA to a hydrogen production plant using MSW and waste wood as primary feedstocks, in order to study the overall environmental performance of the process, identify the operational units that have higher environmental impact (hotspot analysis), and quantify GHG removal potential when CCS is applied. For the construction of this LCA model, primary inventory data for a full-scale commercial 50 MW BioH2 plant have been collected from ABSL.

The functional unit is the production of 1 MWHHV of transport-grade H_2 (>99.97% purity) produced at the plant according to ISO 14687 specifications. The system under study is depicted in Fig. 2, where the boundary between the background and the foreground is highlighted. The foreground system comprises all processes that are the main focus of the study and may be directly affected by decisions based on the study's results; the background system is composed of the other processes that provide the foreground system with materials and energy, typically through a homogenous market. The environment is the receptor of emissions into air, water, and soil.

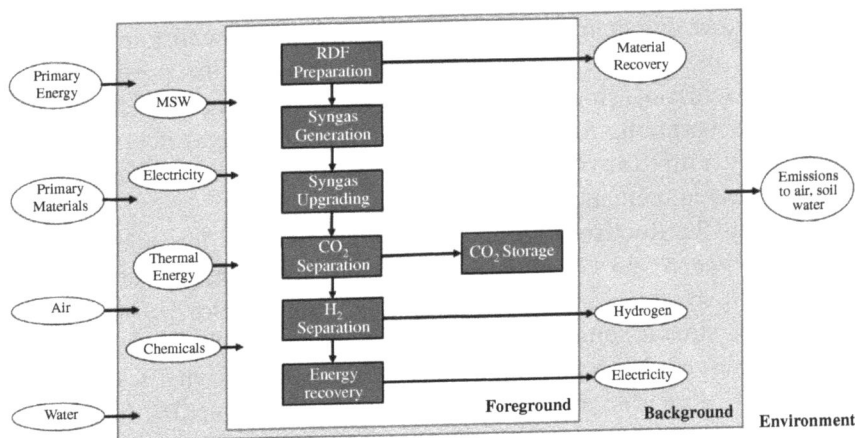

Fig. 2.   System boundary of the analysis where circles identify the flows and squares identify the processes.

The LCA system boundary does not consider the source and generation of waste, its transportation from the recovery facility, and environmental burdens related to the infrastructure of the plant. According to the LCA methodology applied in the present study, the production of BioH2 from waste is considered a multifunctional process, defined as an activity that fulfils more than one function; in this case, besides generating BioH2, the system also generates electricity, recovers materials for recycling, and provides a means to manage specific waste streams. Following the relevant ISO standards [17, 18], the environmental impacts of managing waste streams are estimated by expanding the system boundaries to include the avoided burdens associated with the additional functionalities achieved via conventional processes. Complying with the methodological approach of Clift and colleagues [19], three different burden categories were considered and evaluated: (1) direct burdens, associated with the use phase of the process, (2) indirect burdens due to upstream and downstream processes (e.g., MSW and energy provision), and (3) avoided burdens associated with the additional products or services supplied by the process, e.g., energy produced and material recovered from RDF preparation.

Two additional analyses of the environmental burdens of the BioH2 production process are performed to show differences between the end uses of carbon. One analysis in which all the CO_2 produced at the plant is liquefied and stored for permanent sequestration (CCS). Transportation and underground injections are not indicated; however, BECCS projects are favorable when located close to the sources of emissions, or vice versa, and would take advantage of the UK's existing built and natural assets. A conservative uncertainty analysis on the impact of BioH2 has been carried out to account for the application of different technologies and corresponding energy requirements, as well as the variation due to the waste composition, which turned out to be the main contributor to the overall uncertainty. This approach ignores the issues associated with waste feedstock. In fact, if not treated in thermochemical facilities, waste must be always disposed of, generally either by incineration or landfill, both responsible for substantial GHG emissions. Therefore, a third analysis would include the counterfactuals against which waste-to-hydrogen processes are evaluated. A conservative approach is to take the less polluting disposal method, i.e., incineration with energy and material recovery as the counterfactual in the current UK scenario, and discount all direct (fossil carbon) and indirect CO_2 emissions arising from MSW incineration in modern Waste-to-Energy (WtE) facilities.

3.1. *Allocation*

As explained above, the product system described above provides more than one function. Besides capturing carbon, the system recovers material from the MRF, produces BioH2 and generates electricity, which are three valuable products that qualify for credits. Multi-functional systems present a methodological challenge in the LCA when the objective of the analysis is to compare the environmental performance of different technologies for the provision of one function.

In accordance with ISO standards, we apply the system expansion approach, which entails crediting the system with the avoided environmental impacts associated with the products that are displaced, for example, by electricity and BioH2. We assume that electricity replaces what is available in the UK grid mix, and that BioH2 is used for residential heating, displacing an equivalent amount (in heating potential) of natural gas. As the hydrogen produced by the plant is compliant with transport grade hydrogen, which has a much higher purity than heat-grade hydrogen, the choice of natural gas displacement in this work is highly conservative. It must be noted that the LCA's results are strongly dependent on these assumptions.

3.2. *Life cycle inventory*

It must be noted that the validity and accuracy of LCA results are strongly dependent on that of the underlying inventory; for this reason, having access to high-quality data, e.g., collected on site or extrapolated from design flowsheet, is of the utmost importance. Partly to facilitate compilation of the inventory data, the product system has been distinguished between the foreground and background (Sec. 3). Foreground activities are described by primary data while background activities are described by industry-average data. There are commercially available databases available as sources of background data; notable examples are ecoinvent Database [20] and Sphera [21].

The inventory of the processes analyzed is summarized in Table 2. This reports key inventory data of the three hydrogen production technologies, summarizing the total input and output flows per functional unit (1 MWHHV of transport-grade H2). Further description of the inventory for each process and associated bibliographic references are presented in Ref. [22]. Primary and secondary data used are specifically referred to the UK. The key inventory data for the BioH2 production process considers the input flows for the total energy consumption in terms of thermal energy and electricity and the oxygen required for the process. The output flows quantify the internal electricity production at the plant and the

Table 2. Key inventory data of the three hydrogen production processes. Flow quantity is referred to as a functional unit (1 MWHHV transport-grade H2) and 1 h as a unit of time.

Key Flows	BioH2 (RDF)	BioH2 (Biomass)	Blue-H2		Green-H2
			SMR	ATR	
Input					
Feedstock type	MSW	Waste wood	Natural gas		Water
Feedstock [kg]	566.5	606.6	80.56	79.65	226.8
Oxygen [kg]	134	128	—	77.4	n.a.
Electricity [MJ]	800	740	217.1	296.9	4,974
Thermal energy [MJ]	256	245	—	—	n.a.
Output					
Hydrogen [MJ]	3,600	3,600	3,600		3,600
Materials recovered [kg]	21.8	—	—	—	—
CO_2 released [kg]	16.3	12.7	21.6	13.9	0
Sequestered CO_2 [kg]	516.7	555.0	194.4	265.5	n.a.

material recovery from the RDF preparation, namely ferrous and non-ferrous metals recovered at the feedstock preparation stage.

Throughout the process, several operational units require steam. However, a large quantity of heat is recovered at various points (e.g., waste heat boiler and water–gas shift reactors), and thus steam is re-used within the process, whereby a near-equilibrium between its consumption and production is reached.

4. Results

The environmental impacts of BioH2 are compared to two other alternative routes of hydrogen production, i.e., Blue-H2 and Green-H2. To do this, a cradle-to-gate LCA model has been built for each of them. All models consider the same system boundary, from feedstock acquisition to transport grade hydrogen production. The evaluation of the energetic and environmental impacts of the various hydrogen production routes has been performed using GaBi 10.0.0.71 software. The Life Cycle Impact Assessment (LCIA) stage has been performed using the methodology EF

3.0 [23]. Specific details for the construction of each LCA model are reported in Ref. [22]. The impact categories, which are considered most significant for the purpose of the comparison of the three hydrogen production routes, are Climate Change (kg CO_2 eq.), Acidification (Mol H+ eq.), and Eutrophication freshwater (kg P eq.).

4.1. *Climate change impact*

Different scenarios of the BioH2 production process are presented with regard to the climate change impact (CCI). These scenarios showcase the consequences of capturing point carbon emissions via CCS and considering (thereby crediting) the biogenic carbon fraction of feedstock. A carbon capture rate of 90% is employed in all cases with CCS. The baseline biogenic fraction of RDF used is ~60%, as per Table 1. When accounting for the difference between biogenic and fossil carbon, biogenic carbon emissions to air is considered carbon neutral. Corresponding scenarios produce a carbon negative impact when CCS is applied to the system, as carbon is effectively being removed from the natural carbon cycle. This translates to the total climate change impact of the BioH2 of -217 CO_2-eq for a scenario using RDF as a feedstock and -304 kg CO_2-eq for a scenario using waste wood.

In Fig. 3, an additional analysis is presented in which the avoided emissions associated to the MSW counterfactual are included. If not treated in advanced thermochemical facilities, current waste management practices call for disposal either through incineration or landfill. Incineration with energy recovery (WtE) represents the most common practice around the world and is thus considered a realistic counterfactual. Similarly, in previous cases, only emissions associated to the fossil carbon fraction of feedstock have been accounted for. Although electricity and materials are recovered from the process and thus credited on the final GHG output, the incineration option still shows a substantial climate change contribution of 202 kg CO_2-eq. Therefore, by diverting waste from being incinerated, the benefit of BioH2 on climate change can be further pronounced, with a negative contribution to a climate change of -419 kg CO_2-eq per MWHHV of H2 produced. Even higher benefits can be observed if considering other waste management practices to be counterfactual, such as a landfill or incineration with no energy recovery. There are currently several studies in the literature that report similar results on

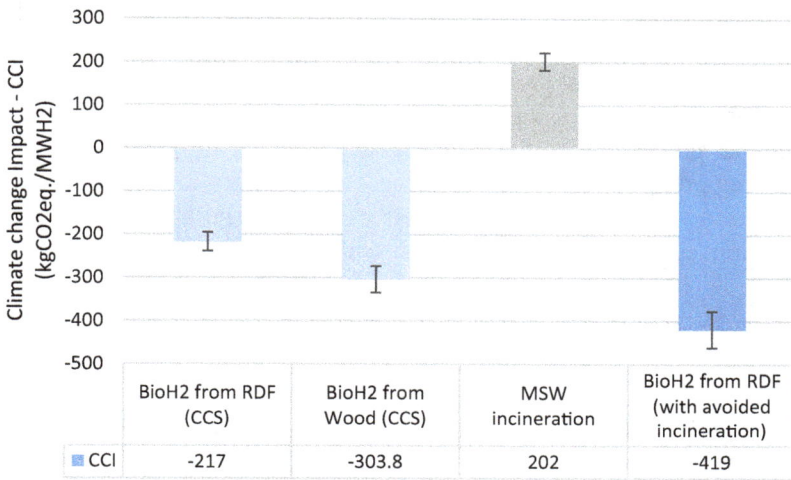

Fig. 3. Climate change impact (CO_2-eq. per functional unit) regarding Carbon Capture and Storage and considering the biogenic fraction of the CO_2 stream. Uncertainties (error bars) calculated based on technical variations in energy usage.

the production of hydrogen from first- and second-generation biomass as an environmentally friendly fuel source [24, 25]. However, the production of BioH2, either from biomass or waste feedstock, would still need to be proven at a commercial scale to validate model assumptions. The future realization of commercial plants, as demonstrated by the authors of Refs. [2] and [22]. and the results of the present work highlight the important advantages of large-scale production of BioH2 from MSW when coupled with CCS.

4.2. *Comparative analysis between BioH2, Blue-H2, and Green-H2*

In the present analysis, the environmental performance of the BioH2 technology is compared to two other competitive low-carbon technologies, Blue-H2 and Green-H2, considering the three most relevant impact categories: Climate Change (kg CO_2-eq), Acidification (mol H+ eq.) and Eutrophication (kg P eq.). The results are expressed per functional unit, 1 MWHHV of transport grade hydrogen produced from all examined processes. The comparison of the environmental performance of the three routes has been performed, taking into account the environmental burdens allocated solely to the production of hydrogen, i.e., excluding the system expansion methodology. The key inventory data for Blue-H2 and Green-H2 production routes are also reported in Table 1, while more details regarding efficiencies and process assumptions are reported in Ref. [22]. These values are largely aligned with industrial and literature data, as different commercial and industrial scale plants are already available in the UK.

4.2.1. *Climate change*

To accurately compare BioH2 with other technologies, credits associated to material recovery and the counterfactual effect of MSW incineration are not considered for analysis to ensure consistency in system boundaries between technologies. The emissions displayed for BioH2 and Blue-H2 are referred to processes that include CCS. The contributions to climate change of the technologies are depicted in Fig. 4. BioH2 production shows the lowest contribution to climate change, equating to -183 kg CO_2-eq.

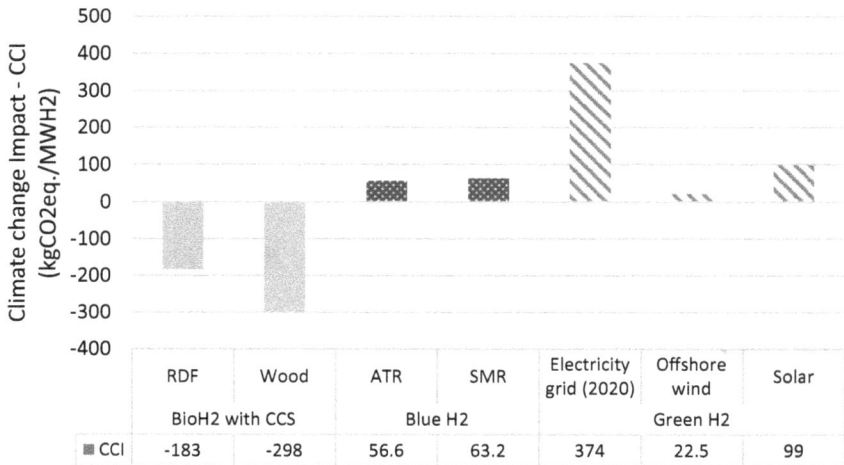

	RDF	Wood	ATR	SMR	Electricity grid (2020)	Offshore wind	Solar
	BioH2 with CCS		Blue H2			Green H2	
■ CCI	-183	-298	56.6	63.2	374	22.5	99

Fig. 4.   Climate change contribution comparison of BioH2, Blue-H2 and Green-H2 production technologies.

These results show that the production of hydrogen from MSW, together with the sequestration of carbon, is not only an effective solution to waste disposal, but it is also appropriate to achieve the objectives proposed by Net Zero 2050; its implementation involves the removal of nearly a quarter ton of CO_2 per MWHHV of H2 produced every hour.

Blue-H2 produced via the SMR process with the CCS process (carbon capture rate of 90%) produces 63.2 kg CO_2-eq. per MWHHV transport grade H2. Approximately 40% of the impact derives from the embodied carbon of natural gas feedstock, rendering the process sensitive to changes in the natural gas source. The upstream emissions are associated to its processing, and for imported natural gas, its liquefaction and shipping. An additional ~40% arises from the SMR process, including 10% of the CO_2 process stream emitted into the air. The remaining CCI is ascribed to the electricity required for CO_2 liquefaction and H_2 compression. The difference between SMR and ATR in favor of ATR is related to the higher CO_2 fraction in the syngas generated by the latter and, therefore, more efficient carbon capture.

A competitive Green-H2 route of production is limited by the high electricity demand of the electrolyzer. This is evident when operating an electrolyzer using the current UK electricity grid mix, leading to marked environmental underperformance, with a climate change impact of 374 kg of CO_2-eq. per MWHHV H2. This limitation can be overcome by exclusively using renewable sources. As shown in Fig. 4, the electricity demand of the electrolyzer and H_2 compression unit, met by electricity produced 100% from solar and 100% from offshore wind, contribute 99 kg CO_2-eq. per MWHHV H2 and 22.5 kg CO_2-eq. per MWHHV H2, respectively. The greater impact from solar compared to offshore wind is primarily from the energy intensive manufacturing of silicon solar cells. The evolving pertinence of these technologies within the energy transition landscape is an important consideration, as LCA results have been reported to be strongly affected by the energy supply, particularly electricity. Thus, any processes with high electricity input will benefit from the decarbonization of the power grid.

The most notable decrease is presented by H_2 production via electrolysis, with electricity supplied from the grid (not a Green-H2 process), with a 33% reduction from 374.78 kg CO_2-eq. by 2030 and a 77% reduction to 87.17 kg CO_2-eq. by 2050. This sensitivity to changes in grid carbon intensity is reflective of the large net electric power necessary for operation and thus constitutes the main burden. The forecasted future efficiencies, as reported by Schmidt and colleagues in consultation with industry and academic experts, is expected to reach a low of ~53.6 kWh/kg of H_2 by 2030 (modeled efficiency is 54.0 kWh/kg of H_2) [26]. Thus, for standard alkaline water electrolysis, impacts in the future are unlikely to change drastically on account of improved efficiencies.

4.2.2. *Other impact categories*

The results for acidification — terrestrial and freshwater (Mole of H+ eq.) — per functional unit are shown in Fig. 5.

The main source of acidification is air-borne emissions of SO_x and NO_x gases from combustion processes that release hydrogen when they are either degraded in the atmosphere or deposited into the soil, vegetation or water. Results show that direct emissions dominate this environmental category for the three routes compared. The higher contribution is presented by Green-H2 with electricity from the 2020 UK grid. Electrolysis with solar reveals an impact of 0.44 mol of H+ eq., relatively higher compared to the other Green-H2 production via renewable energy sources. This result is in line with those reviewed by Bhandari and colleagues, where a number of studies show that solar PV presents a higher

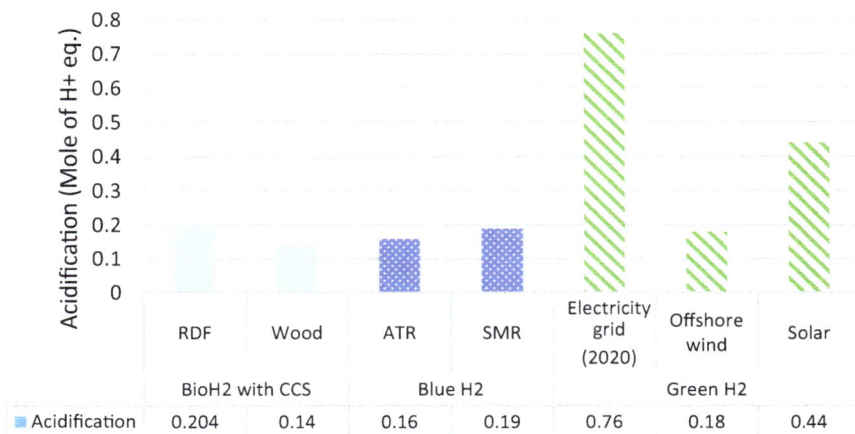

	BioH2 with CCS		Blue H2		Green H2		
	RDF	Wood	ATR	SMR	Electricity grid (2020)	Offshore wind	Solar
Acidification	0.204	0.14	0.16	0.19	0.76	0.18	0.44

Fig. 5. Acidification contribution comparison of BioH2, Blue-H2 and Green-H2 production technologies. The uncertainties were calculated based on the technical variations in energy usage.

contribution to acidification, mainly due to emissions during the manufacturing of mono- and multi-crystalline silicon solar cells [27]. In parallel, electrolysis with offshore wind and Blue-H2 via SMR present the same impact of ~0.18 mol H+ eq. BioH2 from RDF shows an impact of 0.21 mol H+ eq. Although from the hotspot analysis, it is evident that though there are phases that contribute negatively to this category, the absolute impact of BioH2 is positive. This is due to the NO_x and SO_x emissions that take place in certain parts of the production chain, especially those related to the syngas cleaning and bulk H_2 production stage.

Impacts of the eutrophication (kg P eq.) contribution are shown in Fig. 6.

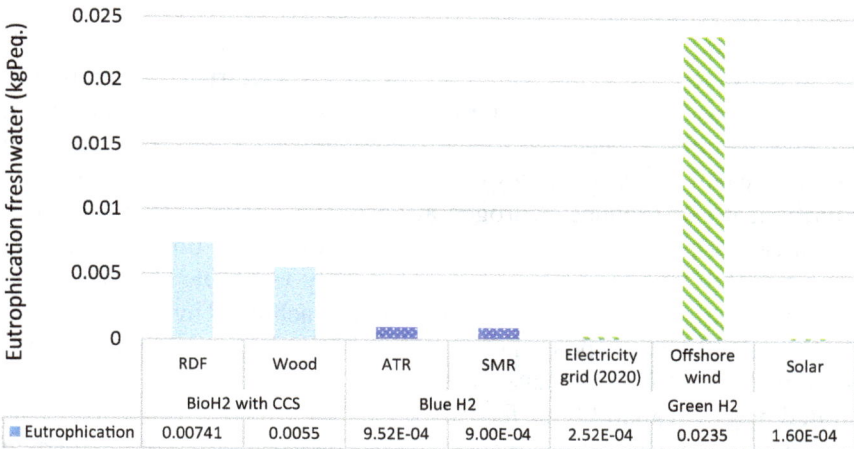

	RDF	Wood	ATR	SMR	Electricity grid (2020)	Offshore wind	Solar
	BioH2 with CCS		Blue H2			Green H2	
■ Eutrophication	0.00741	0.0055	9.52E-04	9.00E-04	2.52E-04	0.0235	1.60E-04

Fig. 6. Eutrophication contribution comparison of BioH2, Blue-H2 and Green-H2 production technologies. The uncertainties were calculated based on the technical variations in energy usage.

Green-H2 offshore wind presents the highest contribution to eutrophication (0.02 kg P eq.), followed by solar (0.009 kg P eq.). The BioH2 contribution to this impact category is 0.007 kg P eq. for the RDF case, while Blue-H2 production shows an impact of 0.0009 kg P eq. The lowest values to the eutrophication contribution are presented by Green-H2 with electricity and 100% solar (0.0002 kg P eq. and 0.00016 kg P eq., respectively).

Results for Green-H2 show that the main contributor to the environmental impact is the energy source. According to the literature, for an equal capacity factor, the environmental performance of wind power hydrogen production appears to be worse when compared to other renewable sources. The size of the turbine, the maintenance requirements, or even the type of generation system is responsible for this impact [28].

5. Interaction Between Low-Carbon Hydrogen Production Pathways

It will be extremely challenging for any one of the low-carbon hydrogen technologies to meet the expected level of hydrogen demand set out by the Net Zero ambition. It seems likely that all options will play a role in the transition to hydrogen.

Green-H2 has the potential to be produced sustainably in large volumes. However, it will take time for low-carbon electricity generation to grow to the scale that meets current electricity demand, plus the additional demand required to decarbonize heat and transport with hydrogen. In addition, electrolyzer technology requires several years to develop to the point that it can produce hydrogen at costs that compete with Blue-H2, which can be produced on a large scale in a few years' time at relatively low cost. However, it is a less environmentally favorable solution in the long term and cannot match the carbon savings achieved by Green-H2 and BioH2, which has the potential to generate negative carbon emissions if combined with CCS. However, the overall production of BioH2 can be limited by the availability of feedstock.

Therefore, there are important synergies between different low-carbon hydrogen production pathways. For example, Blue-H2 might establish the hydrogen market that Green-H2 will meet in future or build the carbon sequestration network required for BioH2 to deliver negative emissions. These negative emissions can offset the residual emissions from Blue-H2 and Green-H2 production. Green-H2 might supply hydrogen to consumers that are remote from Blue-H2 production centers.

Notably, the different hydrogen production options all have different infrastructure requirements, with Blue-H2 and BioH2 reliant on CCS infrastructure. The large-scale ATR plants required for Blue-H2 production also suit the large industrial clusters, where infrastructure such as a supply of natural gas and potentially byproduct oxygen are available. The industrial clusters across the UK would therefore suit Blue-H2 production, with locations along the East Coast and North West of England developing plans for CO_2 pipelines for offshore CO_2 storage (see Fig. 7). The industrial clusters in South Wales and Southampton would require shipping CO_2 to offshore storage sites.

Green-H2 production can be developed at a smaller scale than Blue-H2, and although Green-H2 does not require CCS infrastructure, there are benefits to installing electrolyzers alongside renewables or close to hydrogen demands/infrastructure. The map of industrial clusters in Fig. 7 also highlights regions where there are large energy demands from industrial processes, which could become early adopters of hydrogen. At a smaller scale, where hydrogen can be transported via road tankers to serve transport demands, Green-H2 production plants could be located where renewables are best sited (to access the lowest cost power).

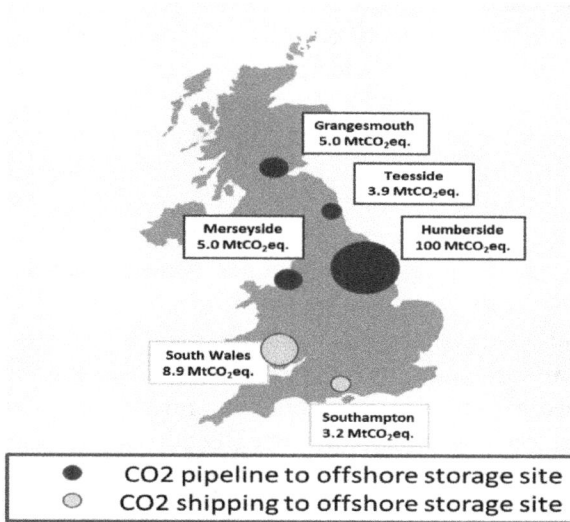

Fig. 7. Location of industrial clusters in the UK and annual GHG emissions from each. (Adapted from BEIS, 2021 [1].)

In the medium term, larger Green-H2 production plants will be developed either in locations close to very large renewable assets (e.g., in coastal locations where offshore wind farm electricity is landed) or locations that are closer to large-scale users, such as industrial clusters, so as to avoid long-range hydrogen transport before a wider conversion of the gas network becomes available to transport 100% hydrogen. A 100%

hydrogen gas network would open up more options for Green-H2 production sites, including the production of hydrogen offshore, connected to offshore wind farms. At a certain scale, the cost of transporting energy in a gaseous form (as hydrogen) can be lower than the costs of transporting energy via electricity. There would be further cost benefits for hydrogen transport if oil and gas pipelines could be repurposed for hydrogen transport.

BioH2 would require CCS infrastructure to deliver very high GHG savings and is therefore suited to the industrial clusters shown in Fig. 7. The use of BioH2 without CCS can still provide GHG emission savings relative to incumbent fuels and converts waste streams into a valuable product, with hydrogen a higher value output than electricity from energy from waste plants. There could therefore be a degree of flexibility with regard to siting some of the plants at locations without CCS infrastructure across the UK, although the full benefits of the technology would require siting around the industrial clusters or locations with CO_2 demand. BioH2 technology can also be deployed at far smaller scales than Blue-H2, allowing it to offer a more distributed approach to hydrogen production.

There is a significant scale-up challenge for low-carbon hydrogen production if the UK is to meet its Net Zero target by 2050. As detailed in the latest UK reports, a range of hydrogen demand forecasts have been published, which vary from 160 TWh/year to almost 600 TWh/year [29]. Given the significant demands for low-carbon hydrogen, it is clear that all of the three low-carbon production routes are needed, and that these need to be developed at pace. The build out rates for all the options presented above will be challenging to meet, and support to develop low-carbon hydrogen markets will be needed to encourage investment in delivering the scale-up of the hydrogen production capacity. The total emissions from the three production pathways to meet the demand for hydrogen to 2050 and beyond were estimated by Element Energy (Fig. 8) [30]. This was based on similar emission factors for Blue-H2 with an average UK natural gas mix, Green-H2 from offshore wind (including embedded emissions of wind), and BioH2 with CCS, to those calculated in this work.

The majority of hydrogen is produced from Blue-H2 and Green-H2 production pathways (>85%), although negative emissions from the BioH2 production route can result in net negative emissions.

It should be noted that even though there are GHG emissions from the production of hydrogen through to the early 2040s, this results in

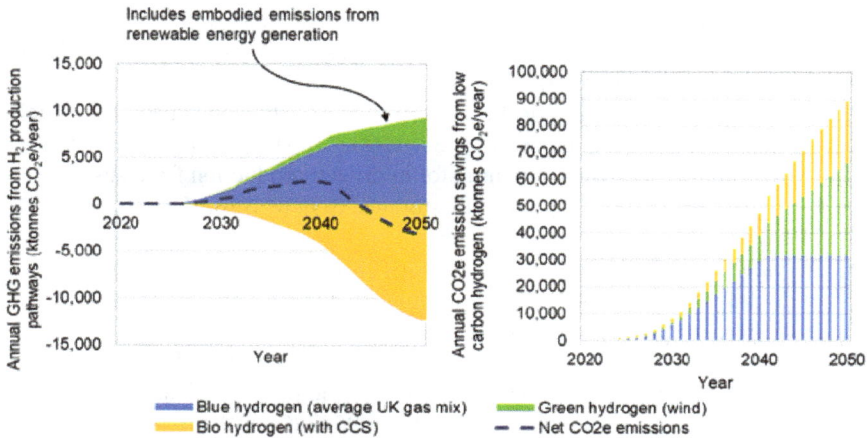

Fig. 8. (left) Estimated annual GHG emissions from the hydrogen production pathways, and (right) CO_2e emission savings relative to business as usual (from natural gas and diesel usage) [30].

significant GHG emissions savings relative to business-as-usual use of natural gas and diesel, with annual GHG emissions savings estimated to be around 90 ktonnes CO_2e/year (which is approximately 18% of the UK's current GHG emissions). This highlights the important role hydrogen can have to decarbonize the hard-to-decarbonize sectors and ensure that the UK's GHG emissions reach Net Zero by 2050.

References

1. BEIS, 2020 UK greenhouse gas emissions, provisional figures, Department for Business, Energy & Industrial Strategy, United Kingdom (2021).
2. M. Materazzi, R. Taylor and M. Cairns-Terry, Production of biohydrogen from gasification of waste fuels: Pilot plant results and deployment prospects, *Waste Manag.* **94**, 95–106 (2019).
3. S. Iannello, S. Morrin and M. Materazzi, Fluidised bed reactors for the thermochemical conversion of biomass and waste, *KONA Powder Part. J.* **37**, 11–131 (2020).
4. T.J. Stadler, P. Barbig, J. Kiehl, R. Schulz, T. Klövekorn and P. Pfeifer, Sorption-enhanced water-gas shift reaction for synthesis gas production from pure CO: Investigation of sorption parameters and reactor configurations, *Energies* **14**, 1–22 (2021).

5. M. Materazzi, Gasification of waste derived fuels in fluidized beds: Fundamental aspects and industrial challenges, in *Clean Energy from Waste*, Springer Theses, Springer, Cham, pp. 19–63 (2017).

6. M. Materazzi and R. Taylor, Plasma-assisted gasification for waste-to-fuels applications, *Ind. Eng. Chem. Res.* **58**, 15902–15913 (2019).

7. L. Rosa and M. Mazzotti, Potential for hydrogen production from sustainable biomass with carbon capture and storage, *Renew. Sust. Energ. Rev.* **157**, 112123 (2022).

8. A. Valente, D. Iribarren and J. Dufour, Life cycle sustainability assessment of hydrogen from biomass gasification: A comparison with conventional hydrogen, *Int. J. Hydrog. Energy.* **44**, 21193–21203 (2019).

9. L. Cao, I.K.M. Yu, X. Xiong, D.C.W. Tsang, S. Zhang, J.H. Clark, *et al.*, Biorenewable hydrogen production through biomass gasification: A review and future prospects, *Environ. Res.* **186**, 109547 (2020).

10. S. Andreasi Bassi, T.H. Christensen and A. Damgaard, Environmental performance of household waste management in Europe — an example of 7 countries, *Waste Manag.* **69**, 545–557 (2017).

11. M. Materazzi, *Clean Energy from Waste*, Springer Theses, Springer (2017).

12. BEIS, Advanced gasification technologies — review and benchmarking (2021). www.gov.uk/government/publications/advanced-gasification-technologies-review-and-benchmarking

13. M. Materazzi and R. Taylor, The GoGreenGas case in the UK, in *Substitute Natural Gas from Waste*, M. Materazzi and P.U. Foscolo (eds.), Elsevier, pp. 475–495 (2019).

14. M. Materazzi, R. Taylor, P. Cozens and C. Manson-Whitton, Production of BioSNG from waste derived syngas: Pilot plant operation and preliminary assessment, *Waste Manag.* **79**, 752–762 (2018).

15. P. Balcombe, J. Speirs, E. Johnson, J. Martin, N. Brandon and A. Hawkes, *et al.*, The carbon credentials of hydrogen gas networks and supply chains, *Renew. Sustain. Energy Rev.* **91**, 1077–1088 (2018).

16. S. L'Orange Seigo, S. Dohle and M. Siegrist, Public perception of carbon capture and storage (CCS): A review, *Renew. Sust. Energ. Rev.* **38**, 848–863 (2014).

17. M. Finkbeiner, A. Inaba, R.B.H. Tan, K. Christiansen and H.-J. Klüppel, The new international standards for life cycle assessment: ISO 14040 and ISO 14044, *Int. J. Life Cycle Assess.* **11**, 80–85 (2006).

18. V.W. Tam, Y. Zhou, C. Illankoon and K.N. Le, A critical review on BIM and LCA integration using the ISO 14040 framework, *Build. Environ.* **213**, 108865 (2022).

19. R. Clift, A. Doig and G. Finnveden, The application of life cycle assessment to integrated solid waste management. Part 1 — Methodology, *Process Saf. Environ. Prot.* **78**, 279–287 (2000).

20. R. Frischknecht and G. Rebitzer, The ecoinvent database system: A comprehensive web-based LCA database, *J. Clean. Prod.* **13**, 1337–1343 (2005).
21. SPHERA, What is Life Cycle Assessment (LCA)? (2020).
22. G. Amaya-Santos, S. Chari, A. Sebastiani, F. Grimaldi, P. Lettieri, and M. Materazzi, Biohydrogen: A life cycle assessment and comparison with alternative low-carbon production routes in UK, *J. Clean. Prod.* **319** 128886 (2021).
23. S. Fazio, V. Castellani, S. Salasa, L. Zampori, S. Sala and E. Diaconu, Supporting information to the characterisation factors of recommended EF Life Cycle Impact Assessment method, JRC Technical Report (2018).
24. C. Antonini, K. Treyer, E. Moioli, C. Bauer, T.J. Schildhauerc, M. Mazzotti, *et al.*, Hydrogen from wood gasification with CCS-a techno-environmental analysis of production and use as transport fuel, *Sustain. Energy Fuels.* **5**, 2602–2621 (2021).
25. A. Susmozas, D. Iribarren, P. Zapp, J. Linssen and J. Dufour, Life-cycle performance of hydrogen production via indirect biomass gasification with CO_2 capture, *Int. J. Hydrog. Energy.* **41**, 19484–19491 (2016).
26. O. Schmidt, A. Gambhir, I. Staffell, A. Hawkes, J. Nelson and S. Few, Future cost and performance of water electrolysis: An expert elicitation study, *Int. J. Hydrog. Energy.* **42**, 30470–30492 (2017).
27. R. Bhandari, C.A. Trudewind and P. Zapp, Life cycle assessment of hydrogen production via electrolysis — A review, *J. Clean. Prod.* **85**, 151–163 (2014).
28. G. Palmer, A. Roberts, A. Hoadley, R. Dargavilled and D. Honnerya, Life-cycle greenhouse gas emissions and net energy assessment of large-scale hydrogen production via electrolysis and solar PV, *Energ. Environ. Sci.* **14**, 5113–5131 (2021).
29. BEIS, UK hydrogen strategy, Department for Business, Energy & Industrial Strategy, United Kingdom (2021). https://assets.publishing.service.gov.uk/government/uploads/system/uploads/attachment_data/file/1011283/UK-Hydrogen-Strategy_web.pdf
30. Element Energy, Review of low carbon hydrogen production (2021)

Chapter 7

Life Cycle Assessment of Chemical Recycling: Recommendations for a Systemic Assessment of its Contribution to the Circular Carbon Economy and Zero Waste Transition

Roh Pin LEE[1]*, Florian KELLER[2] and Raoul VOSS[3]

[1]*Chair of Decarbonization and Transformation of Industry,
Brandenburg University of Technology Cottbus-Senftenberg, Germany*
[2]*Institute of Energy Process Engineering & Chemical Engineering,
TU Bergakademie Freiberg, Germany*
[3]*Professorship Circular Economy,
Technical University of Munich Campus Straubing, Germany*

rohpin.lee@b-tu.de

The challenge of sustainable waste management has reached a new political dimension in recent years. In order to achieve the transition towards a Circular Carbon Economy (CCE) and net-zero carbon emission, the coupling between waste management and chemical sectors is necessary to ensure that carbon-containing waste materials are recirculated as secondary resources to substitute fossil resources in the chemical industry. In this context, Chemical Recycling (CR) is eliciting increasing global interest. However, challenges associated with CR for CCE applications range from uncertainties regarding its environmental impacts, resource efficiency, economic competitiveness to the possibility for easy integration in existing supply chains, and chemical production infrastructure. A Life Cycle Assessment (LCA) could potentially support decision-makers in their evaluation of the environmental impacts associated with CR. However, as it consists of a range of emerging technologies, an LCA of CR technologies poses specific challenges in terms of assessment framework and inventory generation. In this chapter, methodological issues are identified, and suggestions for overcoming challenges associated with LCA application in the CR context are presented. Checklists are also introduced to support practitioners in their LCA evaluation of CR technologies in the form of waste pyrolysis and waste gasification.

1. Introduction

A Life Cycle Assessment (LCA) enables a structured evaluation of direct and indirect environmental impacts of a product/process/service throughout its life cycle. With growing global attention on decarbonization and sustainability, its contribution to the evaluation of opportunities and risks, which are associated with products/technologies/services to support decision processes, is increasing. Today, the process of conducting an LCA and its structural elements are standardized in ISO 14040 and 14044 [1, 2]. Additionally, extensive documents and guidelines are available on specific approaches for conducting LCA applications [3, 4], including methodological guidelines for an LCA in the waste management context [5, 6].

Chemical Recycling (CR) — in enabling the use of carbon-containing waste materials as secondary resources for chemical production — has the potential to contribute to the transition of our carbon intensive society towards decarbonization and zero waste [7–10]. The use of waste to generate a wide range of products ranging from industrial waxes to fuels solvents to plastic building blocks is also referred to as advanced recycling [11]. Via CR, carbon in waste could be bonded in chemical products rather than being emitted 100% into the environment through waste incineration and/or landfilling. Moreover, the use of waste as an alternative carbon feedstock for chemical production could also reduce the demand for fossil feedstock (i.e., oil and natural gas) for production. In addition to contributing to fossil resource conservation, this could also reduce carbon leakages along international fossil resource supply chains [12]. However, an LCA's application for the assessment of CR technologies faces two main issues: a lack of clarity regarding the LCA boundary and insufficient data for LCA evaluations.

In extant literature of LCA applications for CR, a lack of clarity regarding the LCA boundary relates to ambiguity not only in the assessment scope and interfaces for the integration of CR processes in conventional waste treatment and chemical production processes, but also in the reference process for waste treatment. This is in part due to the predominant and limited focus on *"Plastics-to-Plastics"* (i.e., the use of plastic waste for plastic production), which neglects the applicability of CR for recirculating a wide range of carbon-containing waste as secondary carbon resource for chemical production[a] [7–8, 12]. To address this issue, this chapter illustrates the full potential of CR in channelling carbon-containing waste back into the chemical value chain as an alternative feedstock to conventional fossil resources such crude oil and natural gas, i.e., beyond *"Plastics-to-Plastics"* to *"Waste-to-Chemicals"*.

Insufficient data for LCA investigations is the result of limited large-scale implementation of CR technologies on the market. This problem is compounded by the use of confidential information in LCA evaluations, which challenges independent critical evaluation of LCA analyses in the CR context. The resulting lack in transparency, comprehensiveness and comparability of LCA evaluations — reported for different CR technologies [13–14] — thus limits its value in supporting decision processes and socio-political debates about CR's contribution to circularity and sustainability transformation of carbon intensive industries such as the waste management and chemical sectors.

To address the above challenges, the objectives of this chapter are to contribute to clarity in the assessment framework and to address the absence of a well-founded and transparent data basis, with suggestions on how such issues can be overcome. The chapter is structured as follows: First, an overview of alternative CR routes, namely, solvent-based purification, depolymerization, liquefaction and gasification processes, are presented. This is followed by the integration of the CR routes into conventional waste treatment and chemical production routes — based on Best Available Techniques (BATs) in Europe — to generate an integrated reference system for an LCA of CR technologies. Next, recommendations for the LCA to address specific challenges for its application in the CR context are presented. Finally, two checklists to support LCA evaluation of CR technologies in the form of waste pyrolysis and waste gasification are introduced.

2. Alternative Chemical Recycling Routes

This section provides an overview of four technological routes for CR, namely, (1) solvent-based purification, (2) depolymerization, (3) liquefaction, and (4) gasification (see Fig. 1). Note that a number of publications provide general technological classifications/categorizations of associated definitions for CR [15–18]. The classification presented here is related to waste stream input and applicable plastic precursor output to illustrate relevant LCA aspects for inventory balancing of CR processes and pathways [7].

2.1. *Solvent-based purification*

Solvent-based purification — also known as dissolution — includes processes that directly recover high-purity polymers (i.e., without breaking them down into monomers) for direct reintegration in the plastic production process without a polymerization step [19–20]. It targets the removal of additives and the selective separation of one polymer type from other plastics.

Depending on the feedstock type, preconditioning (e.g., shredding, washing, drying) may be necessary before selective dissolution of the target polymer and residue removal by filtration. This is followed by either vaporization or precipitation of the target polymer by heating or the addition of a non-solvent. Finally, the wet target polymer is recovered and dried, with the mixture of solvent/non-solvent/recovered additives thermally separated. Figure 2 illustrates the main process steps for solvent-based purification. Note that as a physical solution process, it is debatable whether solvent-based purification can be considered a CR process. In most classifications, it is not considered as CR, as the structure of the polymer is mostly maintained during the process. Rather, it is considered as a mechanical recycling process or as a separate process category. Nevertheless, due to similarities to other CR routes (application of a chemical as a solvent, innovative/pre-commercial development stage), it is considered in this chapter [21].

2.2. *Depolymerization*

Depolymerization includes all processes in which specific plastic monomers are directly recovered as feedstock for conventional polymerization

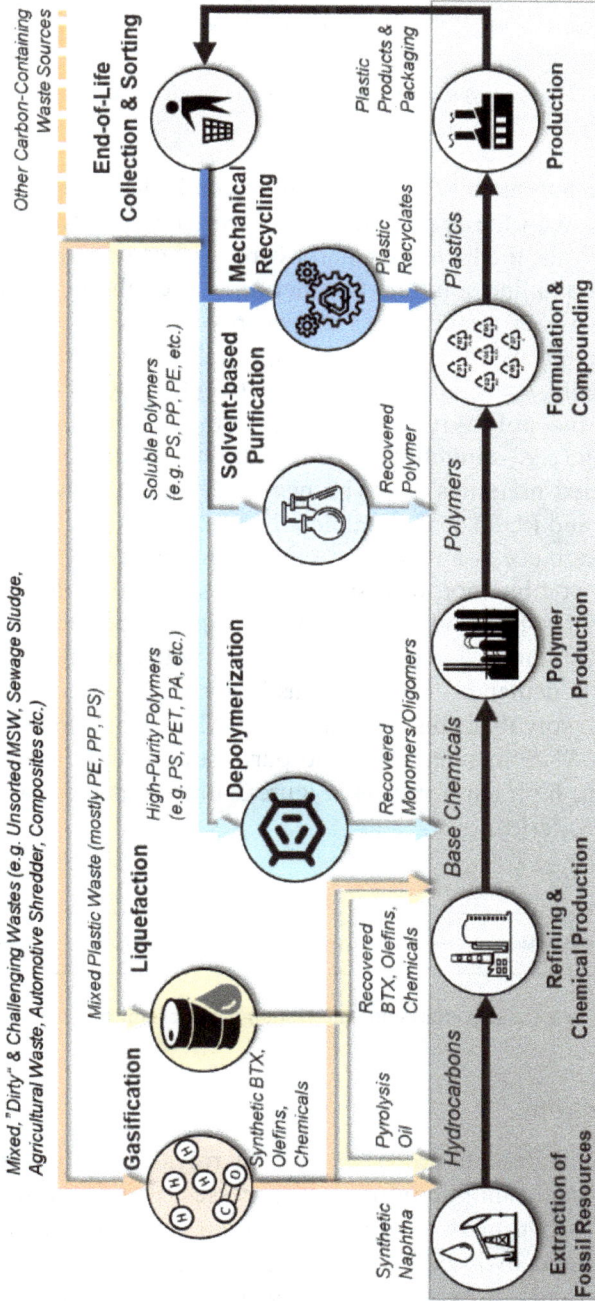

Fig. 1. Chemical recycling routes and typical carbon feedstock [18].

Waste input → Preconditioning → Dissolution Filtration → Precipitation → Recovery Drying → Purified polymer

Fig. 2.   Process steps for solvent-based chemical recycling.

of virgin-grade polymers [22–26]. It can be conducted either chemically (i.e., solvolysis with the application of solvent such as glycolysis, alcoholysis, aminolysis, methanolysis, and hydrolysis, or with the application of catalyst and alkaline solutions) or thermally (i.e., thermal depolymerization, where higher temperatures and energy demand avoid the use of solvents and catalysts). It is primarily applicable for condensation and addition polymers, especially polyethylene terephthalate (PET), polyurethane (PUR), and polystyrene (PS). Depending on the feedstock type, preconditioning (e.g., shredding, washing, drying, melting, and compacting for expanded materials) may be necessary before solvolysis (especially for PET and PUR) or thermal depolymerization (especially for PS). Solvolysis takes place at a moderate temperature and pressure. Its recovery includes a combination of distillation, crystallization, and filtration, and treatment of solid residues from the process is required before disposal. In contrast, thermal depolymerization — with a higher temperature level and energy demand — not only has lower target monomer selectivity compared to solvolysis but gas and heavy oil fractions are also formed in the process. Monomer recovery and purification can be energy intensive and require high temperatures. Figure 3 illustrates the main process steps for depolymerization.

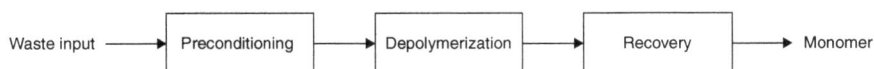

Waste input → Preconditioning → Depolymerization → Recovery → Monomer

Fig. 3.   Process steps for depolymerization-based chemical recycling.

2.3. *Liquefaction*

Liquefaction includes all processes (i.e., pyrolysis, catalytic cracking, hydrocracking, and catalytic reforming processes) that primarily produce a liquid hydrocarbon mixture directly from the waste feedstock [17–18, 27–29]. The liquid hydrocarbon produced can subsequently be further processed (internally or externally) to steam cracker feedstock and/or for the recovery of BTEX (benzene, toluene, ethylbenzene, and xylene) aromatics. Depending on the feedstock type, preconditioning may be required (e.g., presorting, dichlorination) to remove undesirable fractions such as other plastic fractions (especially PET and polyvinyl chloride, i.e., PVC, which could lead to clogging and corrosion issues, respectively) as well as non-plastic combustible fractions and inert fractions (especially metals), which could generate problems for feedstock feeding.

An external heat supply via combustion or electric is generally required. Note that direct hydrocracking/hydropyrolysis-based processes can be net exothermic. However, in such cases, hydrogen demand should be considered. Product oil quality from liquefaction (i.e., contents of unsaturated hydrocarbons, boiling range, hydrogen-carbon ration, content of hetero atoms, heating value) is dependent on both the feedstock type and technology utilized. This, in turn, determines its applicability and upgrading demand, e.g., as a naphtha substitute for steam cracker feedstock or BTX (benzene, toluene, and xylene) aromatic recovery. Figure 4 illustrates the main process steps for liquefaction.

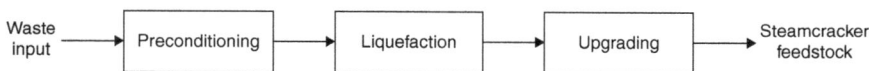

Fig. 4. Process steps for liquefaction-based chemical recycling.

2.4. *Gasification*

Gasification covers all processes (i.e., fixed-bed, fluidized-bed, entrained-flow to a mixture/combination) that primarily produce a synthesis gas (i.e., syngas) with CO and H_2 as the main components [18, 27, 30–32]. The gases produced can be utilized individually or converted downstream to a precursor for plastics or other chemicals/fuels (e.g., methanol, olefins, ammonia) via chemical synthesis. Gasification process can either be autothermal (i.e., heat supply via partial oxidation of utilized feedstock), allothermal (i.e., external heat supply via the combustion of fuel gas/natural gas/side products or via electric heating) or a combination of both. Depending on the type of technology applied, feedstock specifications (e.g., grain size, heating value, fixed-carbon content), limitations (e.g., ash and chlorine context), and treatment requirements will apply.

Syngas composition also varies widely depending on feedstock and gasification technologies/conditions. This, in turn, affects the requirements for downstream gas purification, its applicability, and the resulting product yield for chemical production. Syngas treatment (i.e., from cooling/scrubbing to remove dust/tars/oils and water-soluble components, especially NH_3 and HCl, a CO shift to adjust syngas composition, especially the CO-H_2 ratio, acid gas removal to remove contaminant gas components such as CO_2, H_2S and NH_3, to sulfur recovery and tail gas post-treatment) is associated with utility demand and generation of process CO_2. Figure 5 illustrates the main process steps for gasification.

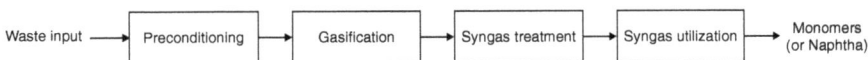

Fig. 5.   Process steps for gasification-based chemical recycling.

Compared to other CR processes, gasification has a more complicated process chain (see Fig. 6). However, syngas from gasification-based CR also opens up a broad range of potential downstream utilization (see Fig. 7).

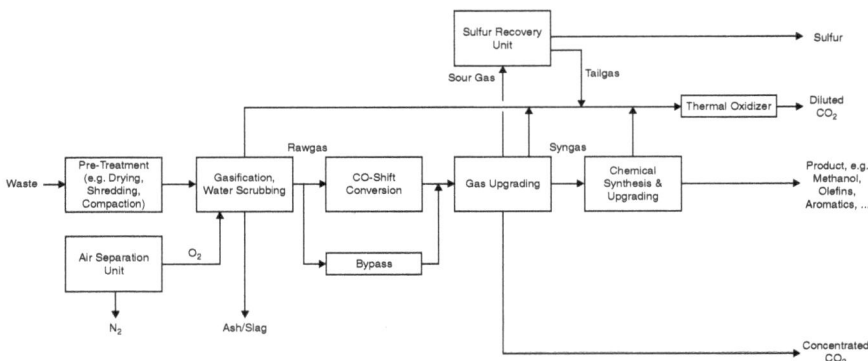

Fig. 6.   Generalized process flowsheet of gasification-based chemical recycling [18].

3. Integration of Chemical Recycling in Conventional Waste Treatment and Chemical Production Systems

To the best of our knowledge, there is currently no conclusive definition of CR technologies available. However, a number of qualifying criteria have been repeatedly stated [33–35], including:

- Application of chemical agents or thermochemical processes,
- Changing of the chemical structure of plastic-containing waste,

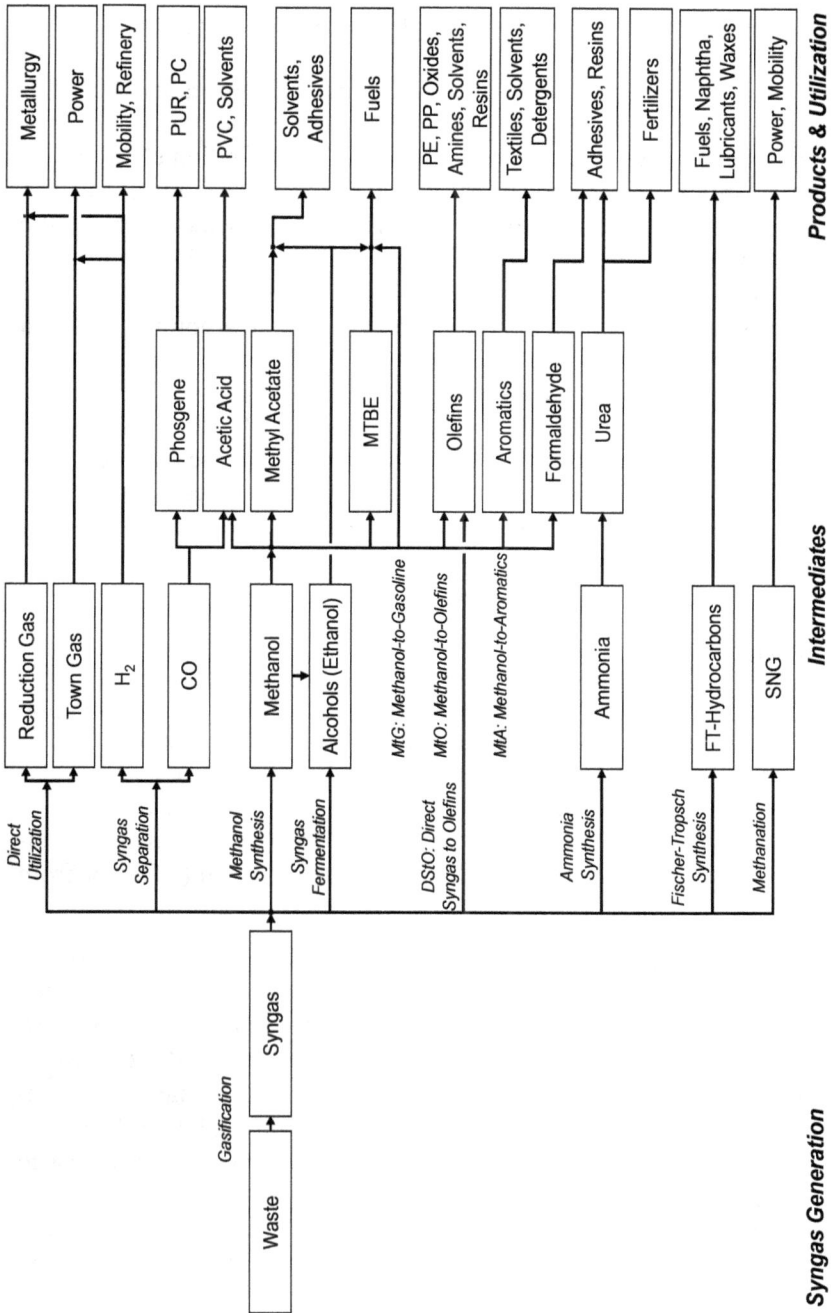

Fig. 7. Overview of syngas utilization pathways [18].

- Production of base chemicals or feedstock for the chemical industry, and
- Exclusion of energy recovery and fuel production processes.

A reference system is developed to specify the definition elements for application in CR assessment (see Fig. 8). Its purpose is to assist LCA practitioners in selecting appropriate assessment frameworks to ensure the transparency, comprehensiveness, and comparability of LCA data and results for different CR pathways. Furthermore, it supports a distinction of CR pathways from other technology pathways (e.g., Carbon Capture and Utilization (CCU) or waste-to-fuels).

Integration of CR pathways in the reference system affects two aspects that are relevant for the assessment, namely, (1) the interfaces to conventional waste treatment (i.e., waste fractions) and the chemical production system (i.e., product chemicals and intermediates), and (2) the associated conventional reference processes that should be considered for substitution or benchmarking [7].

3.1. *Waste treatment*

Waste feedstock: Possible feedstock fractions for CR include primary and secondary plastic-containing waste fractions. Secondary waste fractions refer to product/residual fractions from waste treatment processes. These include waste fractions following mechanical sorting for the recovery of plastics and other value materials, from mechanical (biological and physical) pretreatment for the production of refuse-derived fuel (RDF) or from variations of these processes [36].

Waste treatment process: To the best of our knowledge, there is currently no consistent best practice definition available to determine the reference waste treatment process for specific waste origins and compositions. An orientation is provided by the waste hierarchy defined in the EU Waste Framework Directive, which prioritizes prevention and preparation for re-use before recycling, (thermal) recovery, and disposal [37]. Landfilling — as a disposal technology — should not be considered as a reference waste treatment process, as thermal recovery via incineration is a widely applicable alternative for the treatment of carbon-containing waste. BAT documentations, which are available for waste treatment [36] and incineration processes [38], should be consulted during inventory balancing.

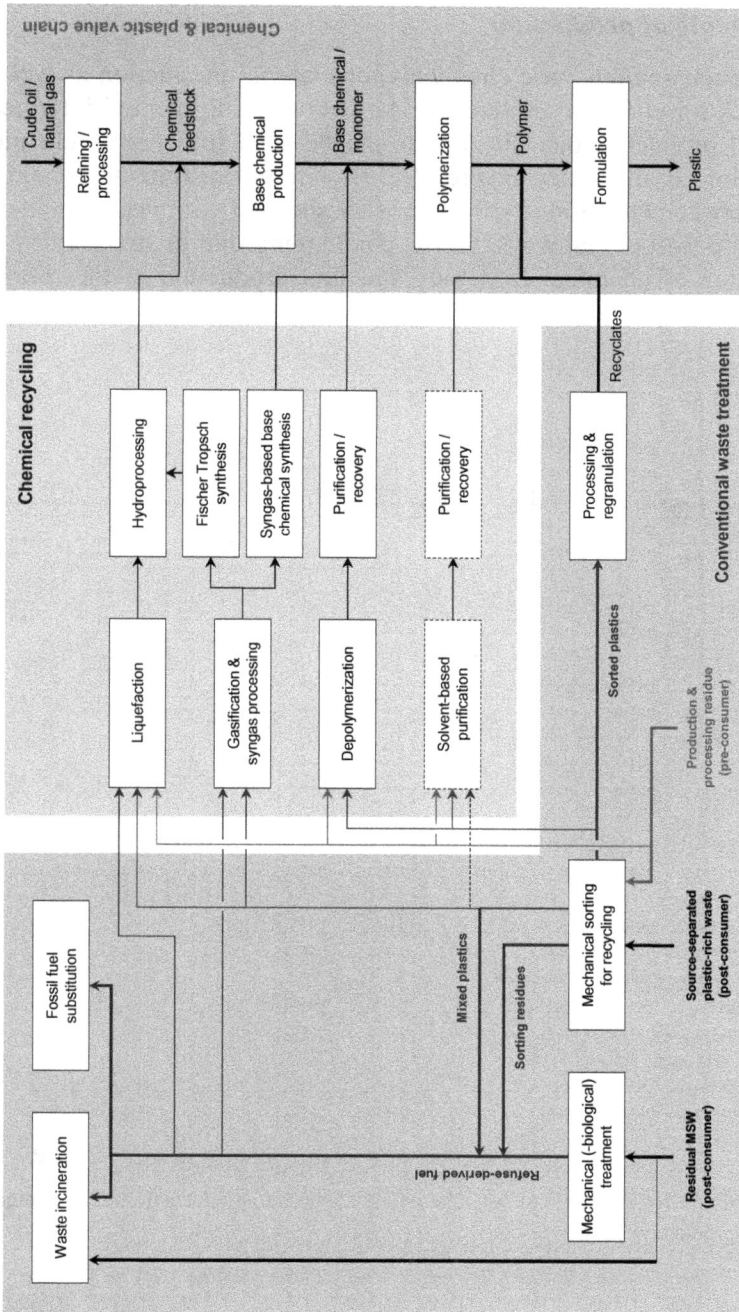

Fig. 8. Integrated reference system.

3.2. *Chemical production*

CR products include base chemicals for material production as well as feedstock for chemical production. An overview of potential intermediates and products from CR, which can be used for organic chemical production, is illustrated in Fig. 9 [39]. While a distinction should be made between CR products from those associated with waste-to-energy and waste-to-fuel pathways, this is challenging due to structural interconnection of chemical feedstock and fuel production in the refinery

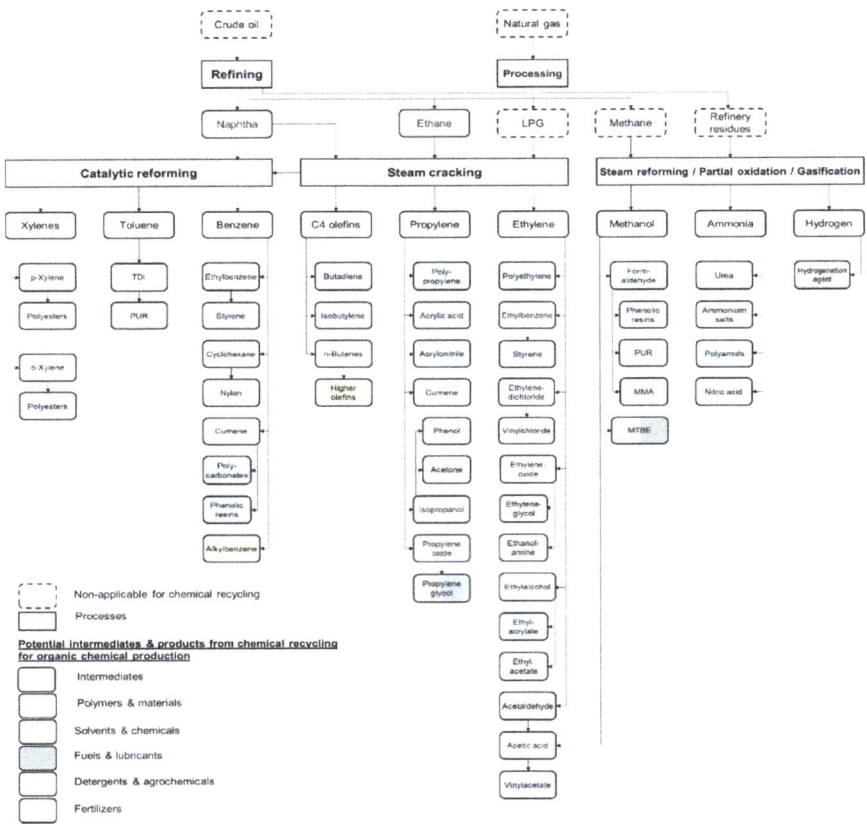

Fig. 9.   Simplified flowchart of potential intermediates and products from CR for organic chemical production [39].

context. The following criteria for addressed CR products are therefore proposed:

- The addressed product is a product/intermediate in the chemical production value chain,
- The quantitative demand for chemical production warrants a large-scale CR production in the targeted scale,
- The addressed product should currently be primarily applied for chemical production (i.e., no substitution of crude oil, refinery fuel products, SNG/fuel gas and coal), and
- No substitution of associated refinery low value or waste products that are currently applied for chemical production, including heavy distillation residues and petcoke.

Chemical production process: To define the reference production pathway or mixtures, the BAT documentation by the European Commission can be applied [40–41]. Note that national deviations in supply and production should be considered (e.g., significant fractions of lower olefin production from natural gas liquids in the United States and from coal in China) [42–43].

3.3. Chemical recycling as "Waste-to-Chemicals" and not "Plastics-to-Plastics"

In view of the wide spectrum of waste feedstock that is applicable for CR as well as the range of chemical intermediates and downstream chemical products that is possible, the recommended scope for CR assessment is *"Waste-to-Chemicals"* i.e., beyond the narrow focus on *"Plastics-to-Plastics"*. The reasons are:

- CR does not inherently target to maintain the functionality of the applied waste (plastic) feedstock. A restriction to plastic production is thus not meaningful.
- Even in countries, such as Germany, with developed source separation systems [44–46], a significant portion of plastics remains in mixed/residual waste fractions [47–48]. While plastics may still form the main

waste fraction, the applicability and characteristics of mixed/residual waste fractions are predominantly determined by the waste mixture characteristics. Therefore, a focus solely on plastic fractions is misleading.

- A predominant focus on plastic products neglects possible target chemicals that are not direct feedstock of major polymers but are still associated with significant environmental impacts in the conventional chemical production process (e.g., C4 olefins, methanol, ammonia).
- Note that it remains debatable whether the production of non-carbonaceous chemicals such as ammonia and hydrogen from waste can or should be considered as CR. While such products do not recirculate waste-based carbon into chemical products [21], their production can exhibit similar/better environmental impacts compared to the production of carbonaceous products. Furthermore, high purity CO_2 can also be recovered during the production process (instead of diluted and dissipated CO_2 as in waste-to-fuel routes) [1–2].
- A focus on plastic products adds unnecessary steps in inventory balancing (e.g., from naphtha to plastics), which complicate and inflate the assessment scope without impacting LCA results.

Finally, it should also be noted that the proposed reference system for integration of CR into conventional waste treatment and chemical production processes are based on an attributional consideration of the current production situation. A more extensive framework to assess the potential impacts of CR application would require a consequential assessment, which includes the projected developments in waste treatment and chemical production (e.g., developments in waste sorting, decreasing refinery fuel demand due to electrification of mobility, etc.) to identify sustainable current and future integration options for CR technologies.

4. Recommendations for LCA for Chemical Recycling

In this section, specific aspects relating to (1) assessment scope, (2) process variation in inventory balancing, (3) specification of waste and product qualities, (4) issues in prospective LCA, (5) inventory modeling consistency, and (6) interpretation and reporting specification, which is of particular relevance and importance in the LCA for CR technologies, are highlighted [7].

4.1. *Assessment scope*

With the exception of the CR process, other process steps in the chemical value chain should be considered in the assessed system. These are determined by both the applied waste fraction as well as the addressed product of the considered CR technology. Using plastic waste as an example, Fig. 10 illustrates plastic life cycle stages and varying scopes that can be applied for the CR assessment. Due to its multifunctionality as both a chemical production and waste treatment process, both *cradle-to-gate* (often referred to as "product perspective") and *end-of-life scopes* ("waste perspective") are thus relevant for the CR context.

Both perspectives include the same process steps. Depending on the investigated CR process, the functional unit will defer. In the waste perspective, the primary/normative functional unit is defined by the quantity and type of waste treated, while the product system is expanded to include the corresponding chemical product as a secondary functional unit. In contrast, under the product perspective, the functional unit is defined by the produced chemical quantity and expanded to include the corresponding waste amount.

Additive system expansion is preferable as it is compatible with Product Environmental Footprint (PEF) guidelines and avoids negative impact assessment results [49]. A combined application of the *cradle-to-gate* with the *end-of-life* scopes — as illustrated with the integrated reference system presented in Sec. 3 — can thus increase the comparability of LCA, not only between different CR technologies but also with conventional chemical production and waste treatment routes.

4.2. *Process variation in inventory balancing*

It is important to note that variations in considered processes can lead to significant differences in assessment results. To address this issue, BATs can be applied for conventional plastics production and plastic waste treatment (see Sec. 3). To ensure transparency, comprehensiveness, and comparability in the LCA of different CR pathways (see Sec. 2), differing and unique aspects associated with each CR pathway — summarized in

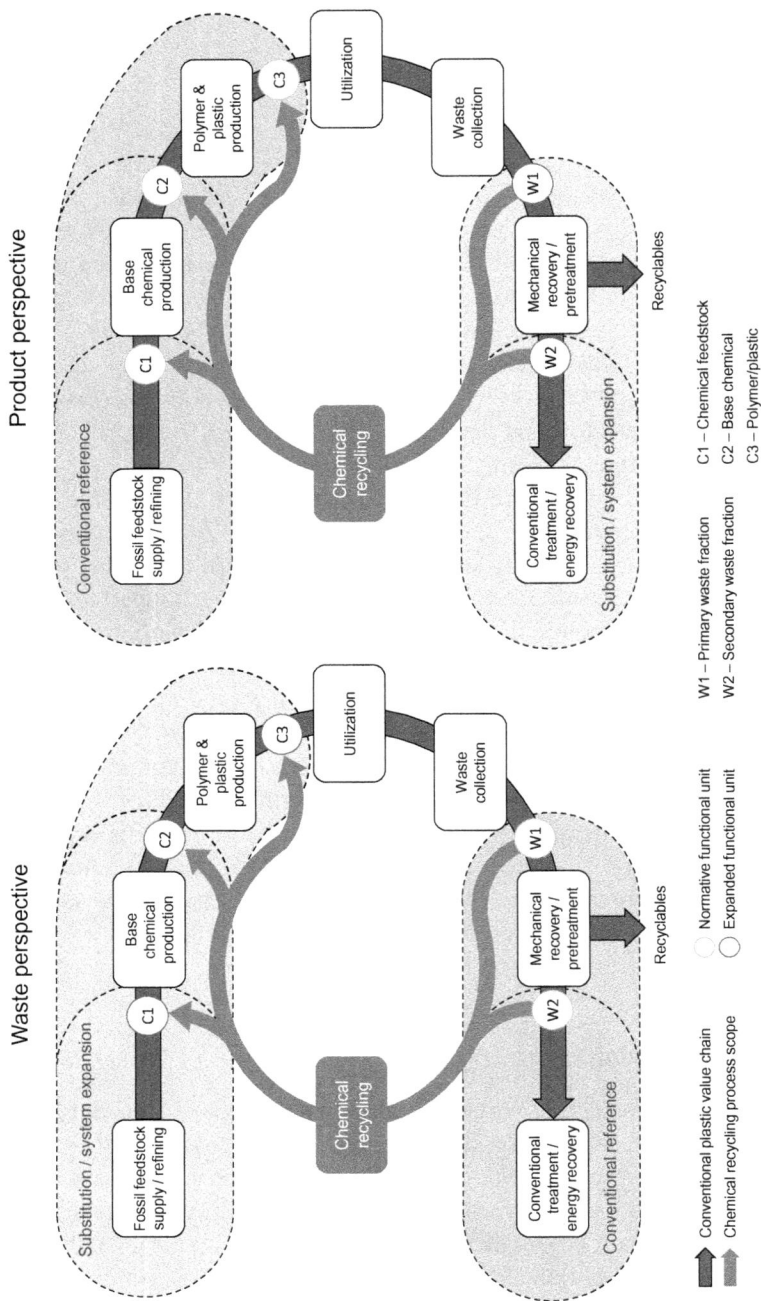

Fig. 10.   Assessment scopes — Examples of plastic waste.

Table 1.   Aspects to be considered in inventory balancing for different CR processes.

CR Pathway	Relevant Aspects for LCA
Solvent-based purification	• Demand and source of heat for the process steps • Solvent and non-solvent supply and losses • Characteristics of recovered polymer (e.g., impurities, polymer properties) • Treatment of residues and spent solvent
Depolymerization	• Solvolysis: Product yield, solvent supply and recovery, heat supply • Thermal depolymerization: Energy demand and product recovery • Catalytic depolymerization: Catalyst supply and recovery, energy supply, and product recovery • Treatment of residues and off-gas
Liquefaction	• Interdependency between feedstock composition, product oil yield and composition • Energy supply • Solid residue and off-gas treatment • Hydrogen demand and product yield in upgrading
Gasification	• Interdependency between heating value of feedstock, gasification technology, gas yield, and composition • Differentiation in energy supply (autothermal/allothermal) • Differentiation in CO_2 production (process/energy-related) • Integration into chemical production (gas treatment and processing, product synthesis)

Table 1 — should also be considered in the generation of their Life Cycle Inventory (LCI) and subsequent LCA.

4.3. *Specification of waste and product qualities*

As highlighted in Sec. 2, the waste type determines not only the applicability of different CR processes but also the need for preconditioning as well as demands on/for product qualities and upgrading. The specification of the addressed waste fraction is thus critical for the identification of applicable CR technologies and a conventional reference treatment, the definition of the functional unit, and allocation of background processes as well as subsequent quantification of its application potential. The required depth of waste specification depends on the intended CR technologies. Generally, criteria for waste specification include:

• Material and fuel composition (i.e., plastic type, non-plastic waste composition, additive contamination, heating value, elementary composition, proximate composition, etc.), and
• Waste origin (i.e., waste source, primary/secondary waste, pretreatment required, etc.).

Similar to waste specification, the specification of the addressed chemical products is essential for assessing CR technologies. Generally, it is preferable to apply process inventories that address the production of chemical intermediates at qualities that enable direct integration in the production value chains. If this is not possible or realistic, the inventory scope should be extended to include the required subsequent conversion/upgrading technologies such that the effects of diverging intermediate qualities can be quantified and included. Additionally, quantitative application limits should also be considered. At the very least, the substitution factor of conventional intermediate production should be subjected to sensitivity and uncertainty analysis if a factual determination is not feasible.

4.4. *Issues in a prospective LCA*

Conventionally, the LCA is applied to evaluate impacts of mature and established technologies with known process characteristics. This is also known as retrospective LCA. This concept is considered as a prospective LCA (also anticipatory or ex-ante LCA) [50–52], and it utilizes process data from lower development stages for the assessment. Main issues to address in prospective LCA include (1) general data availability, (2) compatibility, (3) projected scaling of technologies at lower development stages, and (4) integration of uncertainty [51, 53]:

• *Data availability*: As implementation cost increases significantly from experimental/laboratory to pilot and demonstration facilities, process data availability for validation and inventory modeling is generally the worst at high Technology Readiness Levels (TRLs) before commercialization [51]. Different methods for inventory upscaling have been proposed depending on the TRL of the considered technology [50, 54–55]. These include process simulation, manual calculations, and the application of reaction stoichiometry.

- *Compatibility*: This applies to foreground and background processes as well as the functional unit. In comparative assessments, applied process inventories should reflect similar development stages. Background processes (e.g., for energy supply) should reflect the expected implementation time frame of the developed technology [56].
- *Projected scaling of technologies*: The applied temporal scope of the assessment should be adapted to the expected development and implementation time frames. Three-time horizons for the foreground system (including the technology in focus) and the background system are meaningful, namely, the (1) current situation, (2) situation when the technology is mature for market implementation, and (3) situation when the technology is fully developed. First, as CR technologies are still developing, the development trajectory should be considered when defining the reference year. Second, the energy supply can have a significant influence on a CR assessment. As numerous countries are undergoing dynamic changes in their energy systems, the addressed energy supply should be matched with the reference year and respective expected energy mix. This should furthermore be subjected to a sensitivity analysis when determined to have a significant impact on the LCA results. In addition to the temporal scope, the applied spatial scope should also be defined (i.e., between general conceptual application without a specific application case versus a specific application focused on a defined nation/region with respective assessment conditions, e.g., for energy supply, waste composition and treatment, infrastructure conditions, etc.).
- *Uncertainty*: A consideration of uncertainty is a matter of both foreground (i.e., process characteristics of CR technologies) and background processes (e.g., relevant energy supply mix for CR implementation). Methods to incorporate uncertainty in the LCA include sensitivity analysis (single parameter variation), scenario analysis (variation of parameter sets), and Monte Carlo analysis (stochastic simultaneous parameter variation with probability distribution) [51, 57].

4.5. *Inventory modeling consistency*

Maintaining consistency in inventory modeling is essential to derive realistic and applicable results from the LCA and to prevent false conclusions [58]. Due to the generalized and interdisciplinary approaches in the LCA, detailed balancing and technological aspects are often insufficiently considered. This problem is compounded by a lack of primary data for CR technologies and/or a lack of validation for collected/reported data from CR technology providers. To maintain consistency in inventory modeling so as to derive realistic and applicable results from LCA of CR technologies and to prevent false conclusions, the following strategies are recommended:

- Check and maintain plausibility of applied process inventories during the initial inventory balancing as well as a variation of process parameters for sensitivity and uncertainty analysis.
- Check the consistency of critical parameters (i.e., elementary balance, especially carbon, as well as energy and enthalpy balances).
- Monitor consistency via measures such as Sankey diagrams for energy and carbon flows over the CR pathway.

4.6. *Interpretation and reporting specification*

General requirements for LCA interpretation and reporting are described in diverse documents and guidelines for the LCA [3–6]. To contribute to trust and credibility in the obtained LCA results for CR technologies and to increase its acceptance amongst stakeholders, it is recommended to include one reviewer with a background in chemical engineering in the appointed critical review committee, such that he/she can review the validity of the applied process balances for ensuring inventory modeling consistency.

5. Checklists for the LCA in the CR Context

To address the challenges and aspects highlighted in Sec. 4, the application of technology-specific checklists is proposed to promote transparency, comprehensiveness, and comparability in the LCA of CR technologies. Not only can LCA checklists support LCA practitioners in reviewing critical aspects in CR balancing, compliance with the checklists will also contribute to the audience's confidence in the validity of the applied balances. In view of the significant market and socio-political interest in liquefaction-based CR in the form of waste pyrolysis and gasification-based CR, i.e., waste gasification, two LCA checklists are presented in this section to support LCA evaluations of the environmental impacts of these two CR routes.

5.1. *LCA checklist for waste pyrolysis*

For liquefaction-based CR pathways, the checklist addresses specific aspects to be considered under (1) feedstock and feedstock preparation, (2) liquefaction process, (3) performance variation, (4) residue treatment, and (5) product oil processing in the LCA study.

(1) Feedstock and feedstock preparation
    (a) The waste feedstock fits the process requirements for:
        • Feeding (particle size and form and melting point).
        • Composition (including PET, PVC, and non-plastic contents).
    (b) For pre-sorting of the waste feedstock:
        • The material and enthalpy balances of pre-sorting are consistent.
        • The demands in pre-sorting process steps, energy, and utilities are considered.
        • The treatment of rejected waste fractions is considered.
    (c) Energy/facility demands for the preconditioning (i.e., drying, shredding, melting) of feedstock are considered.

(2) Liquefaction process
    (a) The supply of process utility materials necessary for continuous operation (e.g., catalyst, neutralization agent, hydrogen, feeding oil) is considered.
    (b) Material and elementary balances for the liquefaction process (including feedstock, process utilities, product oil, aquatic phase, gas, and solid residue) are consistent.
    (c) The enthalpy balance of the process (including the heating values of the reactive components, products, and the reaction enthalpy) is consistent.

(d) The energy balance of the process (including feedstock pre-heating and external energy supply) is consistent.

(e) External energy supply for endothermic processes is considered.

(f) Fuel supply and emissions from combustion flue gas for fired processes are considered.

(g) For exothermic processes: process temperature control/cooling is considered.

(3) Performance variation

(a) Documentation is provided for:
   - Current/considered Technology Readiness Level (TRL)
   - Demonstrated application scale/capacity

(b) Performance parameters (including oil yield and energy demand) are considered for:
   - The currently demonstrated process scale.
   - Optimal process performance that can be reasonably expected.

(c) The process balances (elementary, energy, enthalpy) during performance parameter variation are maintained as consistent.

(4) Residue treatment

(a) The treatment of product gas from the liquefaction process is considered.

(b) Documentation of the intended treatment of solid residue is available.

(c) Thermal treatment of solid residue (including associated emissions and possible energy recovery) is:
   - Considered.
   - Not considered. Instead, the effects of alternative treatment (e.g., application in cement kilns) on conventional process performance and application limitations are discussed. Allocation as coal substitute (i.e., only consideration of coal extraction) is avoided.

(5) Product oil processing

(a) Substitution of:
   - Crude oil-based side products (e.g., atmospheric residue or petcoke) are avoided.
   - Liquid fuel-associated products (e.g., crude oil, heavy fuel oil, diesel, feedstock for catalytic cracking) are avoided.

(b) The substitution of a large-scale commodity chemical in the quality for direct substitution — including all process steps — is addressed.

(c) For the substitution of steam cracking feedstock, adjustment of critical quality criteria for naphtha (including contents of unsaturated hydrocarbons, boiling range, and hetero atoms):
  • Is considered. By hydroprocessing (including hydrogen demand, energy demand, yields of side products, treatment of waste gases, and facility demand).
  • Is not considered. Instead, the impacts of inferior feedstock application (production impairment, application limits) are quantitatively addressed.

5.2. *LCA checklist for waste gasification*

For gasification-based CR pathways, the checklist addresses specific aspects to be considered under (1) feedstock and feedstock preparation, (2) gasification process, (3) performance variation, (4) residue treatment, and (5) product gas processing in the LCA study.

(1) Feedstock and feedstock preparation
  (a) The waste feedstock fits the process requirements for:
    • Feeding (particle size and form, melting point)
    • Composition (particle size, density, fuel properties, and elementary composition)
  (b) For pre-sorting of the waste feedstock:
    • The material and enthalpy balances of pre-sorting are consistent.
    • The demands in the pre-sorting of process steps, energy, and utilities are considered.
    • The treatment of rejected waste fractions is considered.
  (c) Energy/facility demands for the preconditioning (i.e., drying, shredding, pelletizing, torrefaction) of feedstock are considered.

(2) Gasification process
  (a) For auto-thermal gasification, the oxygen supply is considered.
  (b) For thermal gasification:
    • Energy supply is considered.
    • For fired heating, flue gas emissions are considered.

(c) The generated product gas quantity and composition are suitable to the applied waste feedstock and gasification technology.

(d) Material and elementary balances for the gasification process (including feedstock, process utilities, product gas, and solid residue) are consistent.

(e) The enthalpy balance of the process (including the heating values of the reactive components, products, and the reaction enthalpy) is consistent.

(f) The energy balance of the process (including external heat exchange) is consistent.

(g) External energy supply for endothermic processes is considered.

(3) Performance variation

(a) Documentation is provided for:
  • Current/considered TRL.
  • Demonstrated application scale/capacity.

(b) Performance parameters (including syngas yield, composition, and carbon conversion) are considered:
  • For the currently demonstrated process scale
  • For optimal process performance that can be reasonably expected

(c) The process balances (elementary, energy, and enthalpy) during performance parameter variation are maintained as consistent.

(4) Residue treatment

(a) A suitable treatment method for the solid residue, depending on the expected characteristics (especially carbon content and leachability), is considered.

(5) Product gas processing

(a) Substitution of:
  • Crude oil-based side products (e.g., atmospheric residue) are avoided.
  • Natural gas or fuel gas is avoided.
  • Liquid fuel-associated products (e.g., crude oil, heavy fuel oil, diesel, feedstock for catalytic cracking) are avoided.

(b) Suitable product gas processing steps (emissions, energy balance, waste and side product production, utility demand) are balanced for:

- Syngas composition adjustment
- Contaminant and acid gas removal
- Off-gas processing (sulfur recovery, thermal treatment)

(c) The substitution of a large-scale commodity chemical in the quality for direct substitution, including all process steps, is addressed.

(d) The assumed chemical synthesis characteristics (product yield and composition, energy balance, side products, utility demand) are realistic for the assumed syngas composition.

(e) Product refining to the required purity for the substitution of conventional commodities is considered.

(f) Off-gas thermal processing is considered.

(g) For Fischer–Tropsch-based naphtha production, improved steam cracking performance compared to conventional naphtha is considered.

6. Conclusion

This section addresses methodological issues associated with the LCA of CR technologies and provides recommendations for expanding existing LCA guidelines for their application in the CR context. To contribute to clarity in the assessment framework, it presents an integrated reference system where four CR routes, namely, solvent-based purification, depolymerization, liquefaction, and gasification, are integrated into conventional waste treatment and chemical product routes as a basis for LCA evaluations of *"Waste-to-Chemicals"*. Moreover, it identifies specific aspects relating to the assessment scope, process variation in inventory balancing, specification of waste and product qualities, issues in prospective LCA, and inventory modeling consistency, as well as interpretation and reporting specification, which are of particular relevance and importance for an LCA in the CR context. To address challenges faced in the LCA of CR technologies, two checklists are introduced (i.e., for waste pyrolysis and waste gasification) as a practical guide for practitioners. Compliance with the checklists in the LCA evaluation of CR technologies will not only support transparency, comprehensiveness, and comparability in the systemic assessment of CR's contribution to the circular and zero waste transition, it furthermore contributes to the audience's confidence in the validity of LCA results and recommendations.

Acknowledgments

This research is supported by funding from the European Climate Foundation (ECF). Any opinions, findings, conclusions and recommendations in the document are those of the authors and do not necessarily reflect the view of the ECF.

References

1. DIN EN ISO 14040: 2006, 2009: Environmental management — Life cycle assessment — Principles and framework (ISO 14040: 2006).
2. DIN EN ISO 14044: 2006, 2006: Environmental management — Life cycle assessment — Requirements and guidelines (ISO 14044: 2006).
3. European Commission, *International Reference Life Cycle Data System (ILCD) Handbook. Review Schemes for Life Cycle Assessment*, Publications Office of the European Union (EUR 24916 EN), Luxembourg (2010).
4. G. Sonnemann and B. Vigon, *Global Guidance Principles for Life Cycle Assessment Databases. A Basis for Greener Processes and Products*, UNEP, Nairobi, Kenya (2011).
5. S. Manfredi and R. Pant, *Supporting Environmentally Sound Decisions for Waste Management. A Technical guide to Life Cycle Thinking (LCT) and Life Cycle Assessment (LCA) for Waste Experts and LCA Practitioners*, Publications Office of the European Union (EUR 24916 EN), Luxembourg (2011).
6. A. Laurent, J. Clavreul, A. Bernstad, I. Bakas, M. Niero, G. Emmanuel, *et al.* Review of LCA studies of solid waste management systems — Part II: Methodological guidance for a better practice, *Waste Manag.* **34**(3), (2014).
7. F. Keller, R. Voss and R.P. Lee, Overcoming challenges of life cycle assessment (LCA) and techno-economic assessment (TEA) for chemical recycling — Recommendations for increasing transparency, comprehensiveness and comparability of LCA & TEA for chemical recycling technologies, Technical Report (2022).
8. R. Voss, R.P. Lee, L. Seidl, F. Keller and M. Fröhling, Global warming potential and economic performance of gasification-based chemical recycling and incineration pathways for residual municipal solid waste treatment in Germany, *Waste Manag.* **134**, 206–219 (2021).
9. F. Keller, R.L. Voss, R.P. Lee and B. Meyer, Life cycle assessment of global warming potential of feedstock recycling technologies: Case study of waste gasification and pyrolysis in an integrated inventory model for waste treatment and chemical production in Germany, *Resour. Conserv. Recycl.* **179**, 106106 (2022).

10. R. Voss, R.P. Lee and M. Fröhling, A consequential approach to life cycle sustainability assessment with an agent-based model to determine the potential contribution of chemical recycling to UN Sustainable Development Goals, *J. Ind. Ecol.* **27**(3), 726–745 (2022).

11. ACC, Advanced recycling, American Chemistry Council. https://www.americanchemistry.com/better-policy-regulation/plastics/advanced-recycling

12. R.P. Lee, M. Tschoepe and R. Voss, Perception of chemical recycling and its role in the transition towards a circular carbon economy: A case study in Germany, *Waste Manag.* **125**, 280–292 (2021).

13. L.P. Costa, D.M. Vaz de Miranda and J.C. Pinto, Critical evaluation of life cycle assessment analyses of plastic waste pyrolysis, *ACS Sustain. Chem. Eng.* **10**(12), 3799–3807 (2022).

14. S. Tabrizi, A.N. Rollinson, M. Hoffmann and E. Favoino. Understanding the environmental impacts of chemical recycling — Ten concerns with existing life cycle assessments (2020). https://zerowasteeurope.eu/wp-content/uploads/2020/12/zwe_jointpaper_UnderstandingEnvironmentalImpactsofCR_en.pdf

15. S. Hann and T. Connock. Chemical recycling: State of play, Eunomia Research & Consulting, United Kingdom (2020). https://chemtrust.org/wp-content/uploads/Chemical-Recycling-Eunomia.pdf

16. K. Ragaert, L. Delva and K. van Geem, Mechanical and chemical recycling of solid plastic waste, *Waste Manag.* **69**, 24–58 (2017).

17. Nova Institute, Chemical recycling — Status, trends and challenges. Technologies, sustainability. Policy and key players (2022).

18. L.G. Seidl, R.P. Lee, M. Gräbner and B. Meyer, Overview of pyrolysis and gasification technologies for chemical recycling of mixed plastic and other waste. Chemical recycling — beyond thermal use of plastic and other waste (2021).

19. S. Ügdüler, K.M. van Geem, M. Roosen, E.I.P. Delbeke and S. de Meester, Challenges and opportunities of solvent-based additive extraction methods for plastic recycling, *Waste Manag.* **104**, 148–182 (2020).

20. Y.B. Zhao, X.D. Lv and H.G. Ni, Solvent-based separation and recycling of waste plastics: A review, *Chemosphere* **209**, 707–720 (2018).

21. ECHA, Chemical recycling of polymeric materials from waste in the circular economy, European Chemicals Agency (2022). https://echa.europa.eu/documents/10162/1459379/chem_recycling_final_report_en.pdf/887c4182-8327-e197-0bc4-17a5d608de6e

22. T. Thiounn and R.C. Smith, Advances and approaches for chemical recycling of plastic waste, *J. Polym. Sci.* **58**(10), 1347–1364 (2020).

23. S. Ügdüler, K.M. van Geem, R. Denolf, M. Roosen, N. Mys, K. Ragaert, *et al.*, Towards closed-loop recycling of multilayer and coloured PET plastic waste by alkaline hydrolysis, *Green Chem.* **22**(16), 5376–5394 (2020).

24. Y. Miao, A. von Jouanne and A. Yokochi, Current technologies in depolymerization process and the road ahead, *Polymers.* **30**, 449 (2021).
25. D. Simón, A.M. Borreguero, A. Lucas and J.F. de Rodríguez, Recycling of polyurethanes from laboratory to industry, a journey towards the sustainability, *Waste Manag.* **76**, 147–171 (2018).
26. F. Liguori, C. Moreno-Marrodán and P. Barbaro, Valorisation of plastic waste via metal-catalysed depolymerisation. *Beilstein J. Org. Chem.* **17**, 589–621 (2021).
27. H. Punkkinen, A. Oasmaa, J. Laatikainen-Luntama, M. Nieminen and J. Laine-Ylijoki, Thermal conversion of plastic-containing waste: A review, VTT Research Report, D4-1-22 (2017).
28. N.I.A. Kameel, W.M.A. Wan Daud, M. Fazly, and N.W.M. Zulkifli, Influence of reaction parameters on thermal liquefaction of plastic wastes into oil: A review, *Energ. Convers. Manag. X.* **14**, 100196 (2022).
29. M.S. Qureshi, A. Oasmaa, H. Pihkola, I. Deviatkin, A. Tenhunen, J. Mannila, *et al.*, Pyrolysis of plastic waste: Opportunities and challenges, *J. Anal. Appl. Pyrolysis.* **152**, 104804 (2020).
30. U. Arena, Process and technological aspects of municipal solid waste gasification. A review, *Waste Manag.* **32**(4), 625–639 (2012).
31. BEIS, Advanced gasification technologies — Review and benchmarking. Review of current status of advanced gasification technologies, Task Report 2, Department for Business, Energy and Industrial Strategy, United Kingdom (2021). https://www.gov.uk/government/publications/advanced-gasification-technologies-review-and-benchmarking
32. R.P. Lee, L.G. Seidl, Q.-L. Huang and B. Meyer, An analysis of waste gasification and its contribution to China's transition towards carbon neutrality and zero waste cities, *J. Fuel Chem. Technol.* **49**(8), 1057–1076 (2021).
33. M. Crippa, B. de Wilde, R. Koopmans, M. De Smet, M. Linder, K. van Doorsselaer, *et al.*, A circular economy for plastics. Insights from research and innovation to inform policy and funding decisions, European Commission, Brussels (2019).
34. J. Vogel, F. Krüger and F. Matthias, Chemical recycling, Chemical Recycling Eupore (2020). https://www.chemicalrecyclingeurope.eu/
35. R. Meys, F. Frick, S. Westhues, A. Sternberg, J. Klankermayer and A. Bardow, Towards a circular economy for plastic packaging wastes — the environmental potential of chemical recycling, *Resour. Conserv. Recycl.* **162**, 105010 (2020).
36. A. Pinasseau, B. Zerger, J. Roth, M. Canova and S. Roudier, Best available techniques reference document for waste treatment, Industrial Emissions Directive 2010/75/EU, Integrated Pollution Prevention and Control (2018).
37. EUR-Lex, European directive 2008/98/EC on waste (2019). https://eur-lex.europa.eu/homepage.html

38. F. Neuwahl, G. Cusano, J.G. Benavides, S. Holbrook and S. Roudier, Best available techniques reference document for waste incineration, Industrial Emissions Directive 2010/75/EU, Integrated Pollution Prevention and Control (2019).
39. Petrochemicals Europe, Interactive flowchart (2022). https://www.petrochemistry.eu/about-petrochemistry/flowchart/
40. B. Pascal, M. Chaugny, L.D. Sancho and S. Roudier, Best available techniques (BAT) reference document for the refining of mineral oil and gas industrial emissions, Industrial Emissions Directive 2010/75/EU, Publications Office Luxembourg (2015).
41. H. Falcke, S. Holbrook, I. Clenahan, A.L. Carretero, T. Sanalan and T. Brinkmann, Best available techniques (BAT) reference document for the production of large volume organic chemicals, Industrial Emissions Directive 2010/75/EU, Publications Office (EUR, Scientific and technical research series), Luxembourg (2017).
42. I. Amghizar, J.N. Dedeyne, D.J. Brown, G.B. Marin and K.M. Van Geem, Sustainable innovations in steam cracking: CO_2 neutral olefin production, *React. Chem. Eng.* **5**(2), 239–257 (2020).
43. Z.M. Xu, Y. Zhang, C. Fang, Y. Yu and T. Ma, Analysis of China's olefin industry with a system optimization model — With different scenarios of dynamic oil and coal prices, *Energ. Policy* **135**(2), 111004 (2019).
44. C. Picuno, A. Alassali, Z.K. Chong and K. Kuchta, Flows of post-consumer plastic packaging in Germany: An MFA-aided case study. *Resour. Conserv. Recycl.* **169**, 105515 (2021).
45. M. Brouwer, T. Marieke, V. van Thoden, K. Ragaert and R. ten Klooster, Technical limits in circularity for plastic packages. *Sustainability* **12**(23), 10021 (2020).
46. M. Klotz and M. Haupt, A high-resolution dataset on the plastic material flows in Switzerland, *Data in Brief* **41**, 108001 (2022).
47. R. Geyer, J.R. Jambeck and K.L. Law, Production, use, and fate of all plastics ever made, *Sci. Adv.* **3**(7), e1700782 (2017).
48. C. Cimpan, E.L. Bjelle and A.H. Strømman, Plastic packaging flows in Europe: A hybrid input-output approach, *J. Ind. Ecol.* **25**(6), 1572–1587 (2021).
49. F. Loske, Life cycle assessment of chemical recycling: Methodology and case studies, NK2 Workshop, Sphera (2022).
50. M. Buyle, A. Audenaert, P. Billen, K. Boonen and S. Van Passel, The future of ex-ante LCA? Lessons learned and practical recommendations, *Sustainability* **11**(19), 5456 (2019).
51. N. Thonemann, A. Schulte and D. Maga, How to conduct prospective life cycle assessment for emerging technologies? A systematic review and methodological guidance, *Sustainability* **12**(3), 1192 (2020).

52. S. Cucurachi, C. van der Giesen and J. Guinée, Ex-ante LCA of emerging technologies, *Procedia CIRP* **69**, 463–468 (2018).

53. S.M. Moni, R. Mahmud, K. High and M. Carbajales-Dale, Life cycle assessment of emerging technologies: A review, *J. Ind. Ecol.* **24**(1), 52–63 (2020).

54. R.K. Rosenbaum and S.I. Olsen, Critical review, in *Life Cycle Assessment*, M.Z. Hauschild, R.K. Rosenbaum and S.I. Olsen (eds.), Springer International Publishing, pp. 335–347 (2018).

55. N. Tsoy, B. Steubing, C. van der Giesen and J. Guinée, Upscaling methods used in ex ante life cycle assessment of emerging technologies: A review, *Int. J. Life Cycle Assess.* **25**(9), 1680–1692 (2020).

56. R. Arvidsson, A.-M. Tillman, B.A. Sandén, M. Janssen, A. Nordelöf, D. Kushnir, *et al.* Environmental assessment of emerging technologies: Recommendations for prospective LCA, *J. Ind. Ecol.* **22**, 1286–1294 (2018).

57. V. Bisinella, K. Conradsen, T.H. Christensen and T.F. Astrup, A global approach for sparse representation of uncertainty in Life Cycle Assessments of waste management systems, *Int. J. Life Cycle Assess.* **21**(3), 378–394 (2016).

58. H.H. Khoo, V. Isoni and P. N. Sharratt, LCI data selection criteria for a multidisciplinary research team: LCA applied to solvents and chemicals, *Sustain. Prod. Consum.* **16**, 68–87 (2018).

Chapter 8

Evaluation of Recycling Methods: Towards Decarbonization of Chemicals and Fuels via a Circular Economy Model

**TAY Siok Wei[a]*, Warintorn THITSARTARN[a]
and Hsien Hui KHOO[b]****

[a]*Institute of Materials, Research and Engineering (IMRE),
Agency for Science, Technology and Research (A*STAR),
Singapore 138634*
[b]*Institute of Sustainability for Chemicals,
Energy and Environment (ISCE²),
Agency for Science, Technology and Research (A*STAR),
Singapore 627833*

*taysw@imre.a-star.edu.sg
**khoo_hsien_hui@isce2.a-star.edu.sg

As chemical and petrochemical industries move towards decarbonization, the re-structuring of supply chains will have to be addressed. A large aspect of transformation towards low carbon fuel or chemical products is challenged by the fundamental reliance on fossil-based resources. Recently, research developments shifted attention from linear production chain approaches and placed focus on the Circular Economy, where

161

waste materials are recycled back into the industry. Along with concerns about fossil fuel depletion, growing concerns regarding climate change and related environmental issues have encouraged considerable efforts to recycle plastic waste. Besides the recycling of plastic waste, recycling of carbon fiber reinforced polymer composite (CFRP) has also gained significant attention due to the staggering amount. Furthermore, with the increasing demand for carbon fibers, there is an increasing interest in using these CFRP wastes as feedstock for carbon fibers. This chapter combines the Life Cycle Assessment (LCA) and the Circular Economy (CE) to evaluate the circularity of chemicals/fuels from recycled plastics. Up to five case studies are investigated.

1. Introduction

Many industries are on the move to decarbonize their manufacturing and supply chain concepts [1]. Steps are taken to shift towards a more sustainable, low-carbon value chain to reduce carbon dioxide (CO_2) emissions for a range of chemicals, polymers, and fuels [2, 3]. Presently, hundreds of millions of tons of commodity chemicals, and fuels are produced annually from fossil fuel feedstock. To achieve carbon reduction in the manufacture of goods and services, the re-structuring of industrial supply chains will be needed.

A large aspect of transformation towards low-carbon fuel or chemical products are challenged by the fundamental reliance on non-renewable, fossil-based resources [4, 5]. The understanding of the carbon footprint of products for various industries can be evaluated from a system thinking perspective, combined with the Life Cycle Assessment (LCA) and Circular Economy (CE) [6, 7]. Implementing CE strategies are essential to achieve decarbonization, with significant potential to improve circular material flows for waste materials.

1.1. *A revisit to the Life Cycle Assessment approach*

Life Cycle Assessment (LCA) is a methodology commonly used to evaluate the effect on the environment caused by industrial processes and services, from raw material acquisition, processing, manufacture, and use to the final disposal of the product of interest [8]. The LCA procedure involves systematically tracking the list of material and resource flow in a production chain that leads to a focused product. Inventories of air emissions, particularly greenhouse gases (GHGs), are taken into account in the input-output flow of the defined LCA system boundary [7–9]. The LCA can be applied to quantify potential environmental impacts caused throughout the life cycle of products, including industrial, commercial, and consumer products [10, 11]. Traditionally, the LCA overall method — from "cradle-to-gate" or "end-of-life" is a linear chain of stages described in Fig. 1.

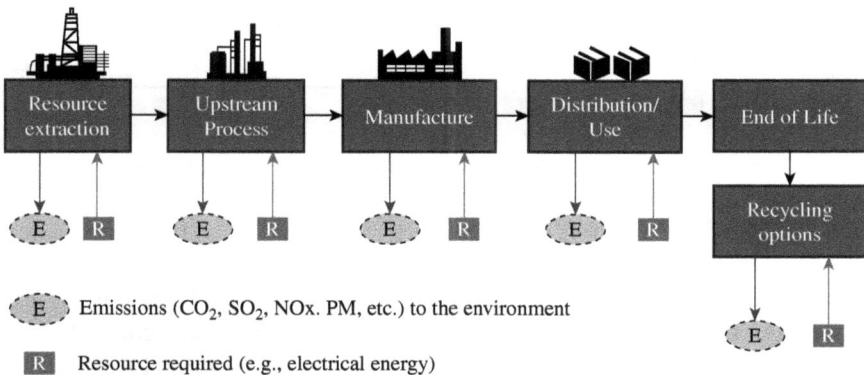

Fig. 1.   LCA system consisting of a linear chain from "cradle-to-grave" to "end-of-life" options.

The quantification of a product's life cycle carbon emissions — referred to as its carbon footprint — is a specific form of several carbon accounting methods [12]. Applications of comprehensive and appropriately designed LCA studies are imperative to provide clear evidence on the comparative Carbon Footprint of various technological options [13].

1.2. Circular economy

Recently, research developments shifted attention from linear supply chain concepts and placed focus on a Circular Economy (CE), or reverse flow of supply chains [6]. In this area, the scope for the forward flow supply chain of materials and products has been extended to include the reverse flow of (recycled) products from waste streams back to the industry [14]. The CE has been a policy initiative for supply chain looping strategies to reuse, recycle and optimize resource efficiency strategies to

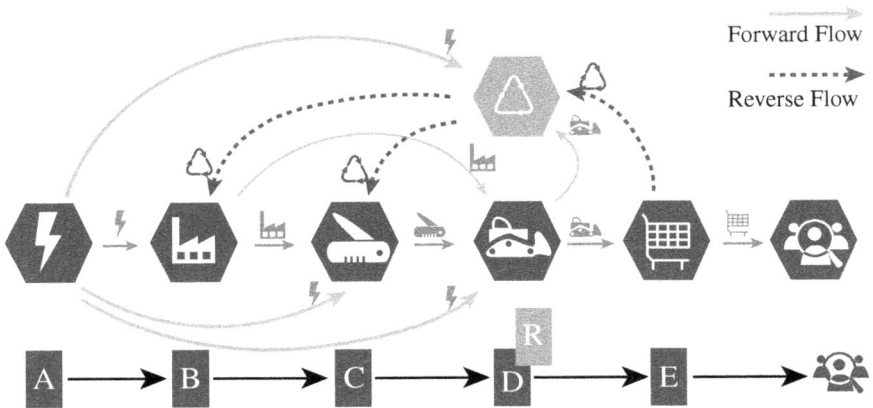

Fig. 2.   Circular economy business model. (Taken from Ref. [16] with permission from IEEE (License Number: 5392300025414).)

reduce the carbon footprint [15]. CE business models (CEBM) (Fig. 2) are recommended by Nandi and colleagues [16] for achieving competitiveness for firms and their supply chains while reducing manufacturing waste and natural resources.

As a response to the challenge of reducing fossil fuel depletion and, at the same time, reducing carbon emissions associated with polymer and hydrocarbon products, CE is gaining momentum. In one example, Geyer and colleagues [17] performed an analysis of combined plastic production data for eight different industrial product categories and modeled the lifetime of plastics before being discarded. According to the authors, 4,900 Mt (60% of all plastics produced) were discarded (in landfills or the environment), while the rest were recycled (channelled back to use) (Fig. 3).

Carbon fiber reinforced polymer (CFRP) composite has been used extensively for many industries over the last 50 years. With its light weight and high strength properties, it is used widely in the aerospace, automobile, and wind energy industries and is rapidly replacing other materials such as steel and aluminum. The European composites market size was approximated to be USD 17.88 billion in 2019. It is expected to increase at a compound growth rate of 7.5% per year from 2019 to 2025 to reach USD 27.54 billion by 2025 [18]. The global demand for carbon fiber has increased steadily and is expected to reach 180,400 MT by 2026, with the major driver being from the wind energy industry. However,

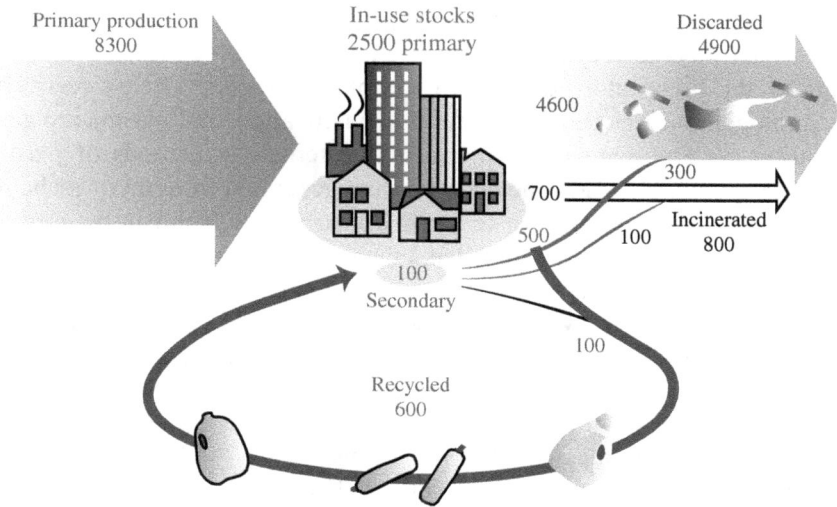

Fig. 3. Global production, use, and fate of polymer resins, synthetic fibers, and additives (1950 to 2015; in million metric tons). (Reproduced from Ref. [17] — open access under the Creative Commons CC-BY license.)

there have been environmental and economic concerns with the widespread use of CFRP. The main challenges are associated with waste production and management. The huge demand and usage led to a large amount of CFRP waste — end-of-life products and manufacturing waste. The global CFRP waste is expected to reach 20 kT per year by 2025. CFRP are difficult to recycle due to their hardness and chemical stability. The common way to dispose them is via incineration and landfill, which contrasts with the CE concept. Furthermore, carbon fibers are expensive and highly energy intensive to fabricate. From an economical and environmental point of view, it makes more sense to recycle and recover the carbon fibers [18].

2. Waste Recycling Methods

Worldwide, polymers are one of the most commonly used materials in daily life. At present, the vast majority of carbon used in polymer production is sourced from fossil fuels, with oil-derived naphtha and natural gas liquids. Growing environmental concerns associated with current linear production chains of plastic materials (produce, use and dispose) has led to various ways to recycle plastic waste. Thermal recycling takes advantage of the high calorific value of the plastic wastes. Energy recovery technologies such as catalytic pyrolysis, gasification, and plasma arc gasification are receiving more attention as alternative methods of plastic waste recycling [12, 18]. The main product generated from thermo-chemical recycling of polymer waste are hydrocarbons, which can be used as valuable fuels (e.g., diesel, gasoline, ethanol).

Pyrolysis process converts plastic waste into liquid oil, solid residue (char), and gases at high temperatures (300–900°C) via thermal decomposition. Pyrolysis has great potential to convert waste plastics into energy and other valuable products to achieve economic benefits. Studies carried out on the pyrolysis of plastic waste reported that the liquid oil generated has a high heating value (HHV) with a range of 41.7–44.2 MJ/kg, which is close to that of conventional diesel [12, 19].

Compared with mechanical methods, chemical recycling (also known as upcycling) has attracted growing attention because of its ability to convert plastic waste into high-value products, including refinery feedstock, fuel, and monomer. Upcycling of plastic wastes is foreseen as a sustainable solution to create valuable materials that can be used in petrochemical and chemical industries. Several ongoing and emerging methods have been done to upcycle plastic waste into high-value-added products [20–21] to avoid CO_2 emissions from the linear flow of plastic production and accumulated waste [22].

The use of catalysts has played an important role in polymer-recycling methods and has opened up a wide plethora of science and exploration. Tan and colleagues [23] reviewed recent advances in the development of efficient heterogeneous catalysts for the upcycling of plastic waste into liquid hydrocarbons, arene compounds, and various other carbon materials. Like the recycling method for plastics, there has been extensive research in the recycling of CFRP. The common recycling methods can be classified into mechanical, thermal and chemical recycling. Mechanical recycling involves grinding, milling, and crushing the CFRPs into smaller particles and being reused as reinforced materials in other products [24]. However, the carbon fiber will lose its economic value.

Other areas of work reported that the Ni-Co/Al_2O_3 catalyst had high catalytic activity towards the dry reforming of waste high-density polyethylene for syngas production (e.g., Ref. [25]). Among the plethora of catalyst design and applications, an inexpensive catalyst used to convert common types of polymers to low molecular mass hydrocarbons under mild conditions has been introduced [26]. The work demonstrates how a highly active earth-abundant catalyst can quickly drive hydrogenolysis of polyolefins under mild conditions. The study involved the decomposition of several polymer samples (e.g., polypropylene (PP), polyethylene (PE), etc.). One noteworthy experimental test sample demonstrated that treating 1.5 g of polyethylene with 2 atm of hydrogen at 200°C in the presence of the catalyst completely converted the polymer to liquid and gas products within 45 minutes.

In another notable example, Zichittella and colleagues [27] developed a catalyst designed to effectively break down mixed plastics into propane, which can be used as high-value fuel or for the synthesis of new polymers (see Fig. 4). The group of scientists displayed that 5 wt% cobalt supported on ZSM-5 zeolite converted solvent-free hydrogenolysis of PE and PP into propane with weight-based selectivity in a gas phase over 80 wt% after 20 hours at 523 K and 40 bar H_2.

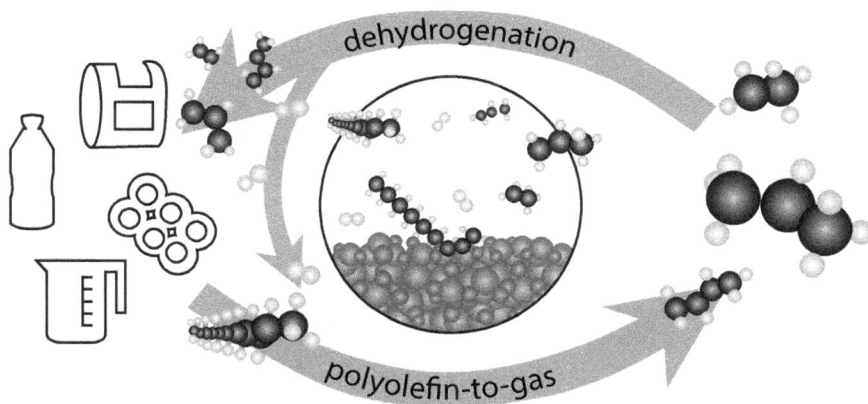

Fig. 4.   Simplified scheme of the envisioned closed-loop polyolefin cycle. (Reproduced from Ref. [27] — open access under the Creative Commons CC-BY license).

3. Recycling Methods: Combined Life Cycle Assessment and Circular Economy

While the CE is gaining global recognition for assisting to reduce climate change targets, there is, to date, no systematic method to scientific literature quantifying the reduction of GHG emission levels [28–29]. In this regard, the LCA can play an important role in evaluating the success rates of circular transition by enabling measurements of well-defined metrics such as the carbon footprint. Herein, the LCA is introduced to evaluate measurement indicators associated with the potential reduction of global warming impacts (or carbon footprint) of chemicals and fuels that are reproduced from the recycling of plastic materials.

Existing LCAs demonstrate linear chain methods (as shown in Fig. 1) for the applications of the use of fossil fuels and other resources in the production of chemicals and fuels [1–5]. For the evaluation of plastic recycling methods to reproduce valuable products, the LCA system boundary is re-designed and extended to include circular allocation of chemicals and fuels generated from waste back to the industry (i.e., reverse flow). The combined LCA and CE method is illustrated in Fig. 5.

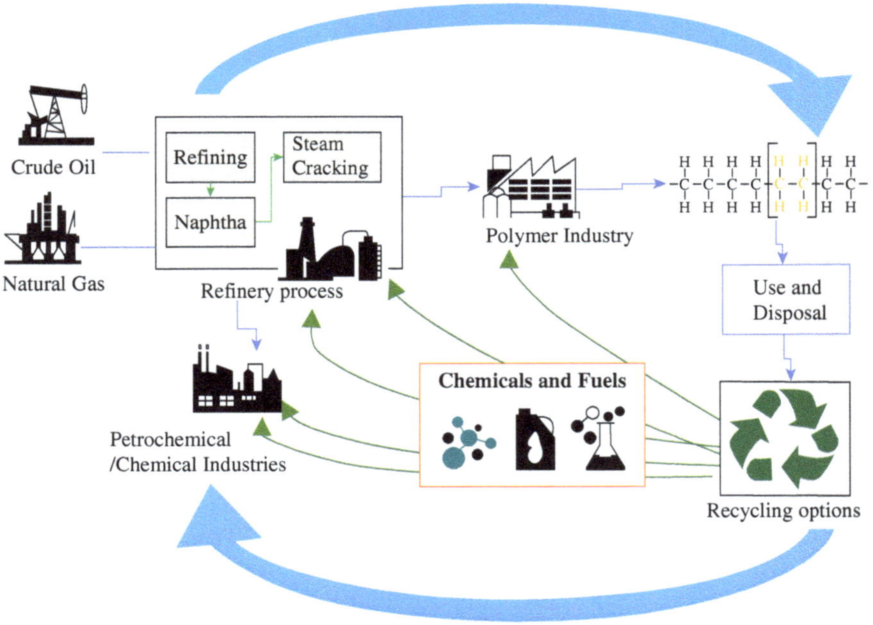

Fig. 5.   Circularity of chemicals/fuels from recycled plastics to become feedstock for conversion to useful products: the LCA approach.

3.1. *Case studies*

A total of four case studies of plastic recycling methods were compiled from the literature (cases I to IV), and one experimental method of CFRP waste recycling was done in A*STAR (Case V). The cases are all compiled in Table 1.

Table 1.   Compiled Case Studies (I–V) for 1 kg of waste materials.

Case	Recycling Details	Energy Required	Main Product Output	Comments
I	Pyrolysis [12, 30]	18.3 kJ (NG) 106 kJ (Elec. power)	Diesel (650 g) CO_2 (150 g emissions)	Inflow-outflow numbers are mass allocated to 1 kg of waste plastic input
II	Gasification [12, 30]	376 kJ (NG) 1450 kJ (Elec. power)	Ethanol (278 g) CO_2 (181 g emissions)	(No catalyst use reported)
III	Mixed plastic waste (MPW) gasification to methanol [31]	Power use = 1.98 kWh* + 3.7 kJ fuel (NG)	0.632 kg Methanol[a] 0.4 kg CO_2 Emissions	*Estimated based on the total electricity cost reported [a]Report of 63.2% conversion from 100% plastic feedstock (based on Carbon Balance)
IV	Catalytic reforming of waste polyethylene [32]	2.5 kJ + 1.5 kJ (Electrical power)[b]	H_2 (0.132 kg; precious product) CO (0.970 kg; precious product) CH_4 (0.0856 kg; air emissions)	*Est.* 1 kg plastic recycling scale [b]Energy use estimated ~95% efficiency Catalyst production and use excluded from the LCA system boundary
V	Carbon fiber recycling of amine -cured bisphenol A epoxy through oxidative recycling process using microwave heating	1499 kJ	18.0 g recycled carbon fiber and 7.7 g partially degraded epoxy	*Est.* 25 g CFRP waste

3.2. *Results and discussions*

In the LCA-CE model (Fig. 6), waste flows are eliminated, and at the same time, carbon emissions are reduced from the need to produce chemicals or fuels from fossil fuel resources. Negative Carbon Footprint results can be obtained due to the avoidance of manufacturing stages that emit carbon emissions.

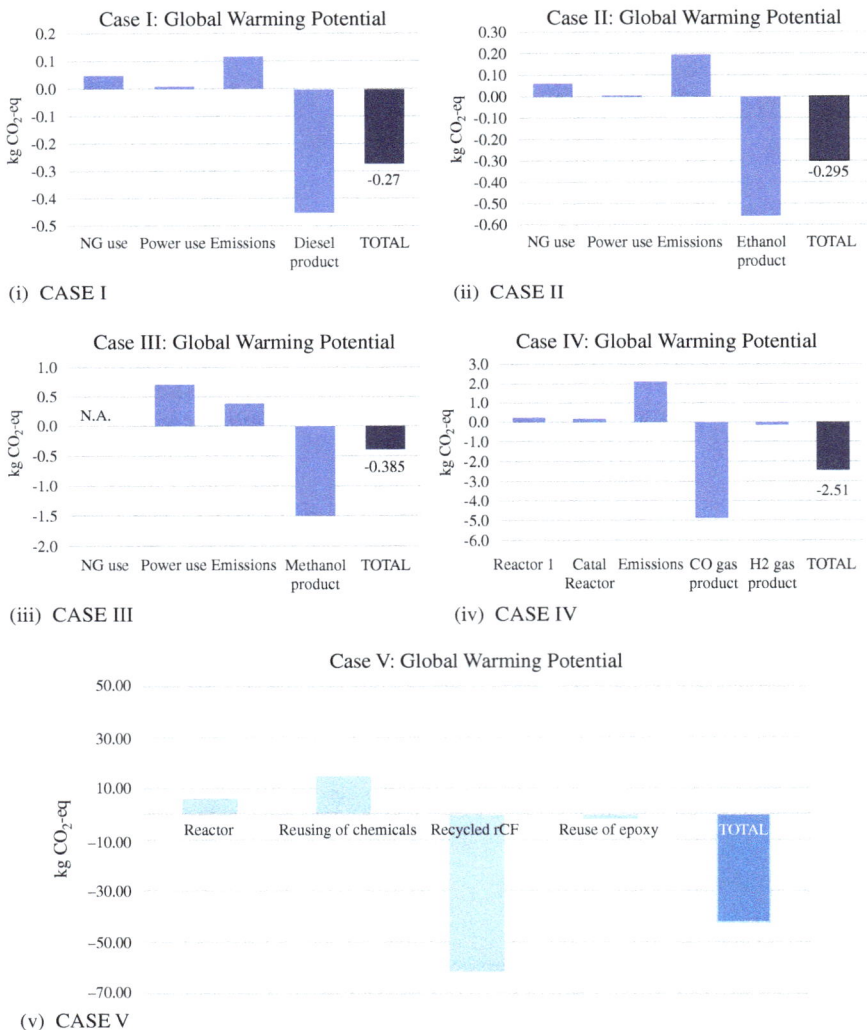

(i) CASE I

(ii) CASE II

(iii) CASE III

(iv) CASE IV

(v) CASE V

Fig. 6.   Carbon footprint results for Cases I to V.

4. Discussions

Chemical recycling has been heavily researched to recover carbon fiber, and many ways have been proposed over the years. Solvolysis is one of them. It normally involves chemical treatment using a solvent to degrade the resin to recover both the clean fibers and depolymerized matrix in the form of monomers or petrochemical feedstock. Solvolysis can be classified according to the conditions used during the treatment. When the solvents involved are relatively environmentally friendly, the conditions required are usually more stringent. Green solvents like water and ethanol have been reported to be effective in removing polymers from carbon fiber when in a supercritical state. For most of the systems reported, the recovered carbon fiber retained 90% of its tensile strength when compared to the pristine ones [31].

Chemical recycling under mild conditions has also been researched. One popular way is to employ the oxidation method, in which an oxidant, such as nitric acid, oxygen, hydrogen peroxide, and peracetic acid, is used. This method is effective for amine-cured epoxy. Usually, an acid is used to swell the matrix, and the oxidant is used to degrade the epoxy resin. The recycling can also be conducted in a one-step reaction in which, peracetic acid, which is generated when acetic acid is mixed with hydrogen peroxide, is used to oxidize the CFRP. The degradation percentage of resin is reported to be as high as 97% at 65°C for 5 hours, and clean long carbon fibers of comparable strength with their pristine counterparts are recovered [33]. Besides acetic acid, other organic acids, such as tartaric acid, were also reported to be effective in resin removal, with an efficiency reaching more than 90% [34]. In our experimental set-up, we recycled CFRP using a peracetic acid system. Microwave radiation was used as the heating source, and the carbon fiber recovered was 91% pristine.

5. Concluding Remarks

Recently, research developments shifted attention from linear production chain approaches and placed their focus on the CE, where waste materials are recycled back into industry. Along with concerns of fossil fuel depletion, growing concerns regarding climate change and related environmental issues have encouraged considerable efforts to recycle plastic waste. As a response to the challenge of reducing fossil fuel depletion and, at the same time, reducing carbon emissions associated with polymer and hydrocarbon products, the CE is gaining momentum. While the CE is gaining global recognition in assisting to reduce climate change targets, there is, to date, no systematic method to scientific literature quantifying the reduction of GHG emission levels [28–29].

Besides the recycling of plastic waste, the recycling of carbon fiber reinforced polymer composite (CFRP) has also attracted significant attention due to the staggering amount. Furthermore, with the increasing demand for carbon fiber, the degradation percentage of resin is reported to be as high as 97% at 65°C for 5 hours, and clean long carbon fibers of comparable strength with their pristine counterparts are recovered [33]. Besides acetic acid, other organic acids, such as tartaric acid, were also reported to be effective in resin removal, with an efficiency of reaching more than 90% [34]. In our system, we recycled CFRP using a peracetic acid system. Microwave radiation was used as the heating source, and the carbon fiber recovered was able to achieve 91% that of pristine.

References

1. D. Saygin and D. Gielen, Zero-emission pathway for the global chemical and petrochemical sector, *Energies* **14**, 3772 (2021).
2. Á. Galán-Martín, V. Tulus, I. Díaz, C. Pozo, J. Pérez-Ramírez and G. Guillén-Gosálbez, Sustainability footprints of a renewable carbon transition for the petrochemical sector within planetary boundaries, *One Earth* **4**(4), 565–583 (2021).
3. K. Kümmerer, J.H. Clark and V.G Zuin, Rethinking chemistry for a circular economy, *Science* **367**, 369–370 (2020).
4. M. Stork, J. de Beer, N. Lintmeijer and B. den Ouden, Chemistry for climate: Acting on the need for speed, roadmap for the Dutch Chemical Industry towards 2050, Ecofys and Berenschot, Utrecht, The Netherlands (2018).
5. M. Broeren, D. Saygin and M. Patel, Forecasting global developments in the basic chemical industry for environmental policy analysis, *Energ. Policy* **64**, 273–287 (2014).
6. S. Agrawal, R.K. Singh and Q. Murtaza, A literature review and perspectives in reverse logistics, *Resour. Conserv. Recycl.* **97**, 76–92 (2015).
7. H.H. Khoo, R.M. Eufrasio-Espinosa, L.S.C Koh, P.N Sharratt and V. Isoni, Sustainability assessment of biorefinery production chains: A combined LCA-supply chain approach, *J. Clean. Prod.* **235**, 1116–1137 (2019).
8. G. Bishop, D. Styles and P.N.L. Lens, Environmental performance comparison of bioplastics and petrochemical plastics: A review of life cycle

assessment (LCA) methodological decisions, *Resour. Conserv. Recycl.* **168**, 105451 (2021).

9. E.J.P. Martin, D.S.B.L. Oliveira, L.S.B.L. Oliveira and B.S. Bezerra, Life cycle comparative assessment of pet bottle waste management options: A case study for the city of Bauru, Brazil, *Waste Manag.* **119**, 226–234 (2021).

10. L. Shen, E. Worrell and M.K. Patel, Open-loop recycling: A LCA case study of PET bottle-to-fibre recycling, *Resour. Conserv. Recycl.* **55**(1), 34–52 (2010).

11. A.D. La Rosa, S. Greco, C. Tosto and G. Cicala, LCA and LCC of a chemical recycling process of waste CF-thermoset composites for the production of novel CF-thermoplastic composites. Open loop and closed loop scenarios, *J. Clean. Prod.* **304**, 127158 (2021).

12. H.H. Khoo, LCA of plastic waste recovery into recycled materials, energy and fuels in Singapore, *Resour. Conserv. Recycl.* **145**, 67–77 (2019).

13. D. Guo, H. Hou, J. Long, X. Guo and H. Xu, Underestimated environmental benefits of tailings resource utilization: Evidence from a life cycle perspective, *Environ. Impact Assess. Rev.* **96**, 106832 (2022).

14. S.A. Alumur, S. Nickel, F. Saldanha-da-Gama and V. Verter, Multi-period reverse logistics network design, *Eur. J. Oper. Res.* **220**(1), 67–78 (2012).

15. M. Christis, A. Athanassiadis and A. Vercalsteren, Implementation at a city level of circular economy strategies and climate change mitigation — the case of Brussels, *J. Clean. Prod.* **218**, 511–520 (2019).

16. S. Nandi, A.A. Hervani and M.M. Helms, Circular economy business models — Supply chain perspectives, *IEEE Eng. Manag. Rev.* **48**(2) 193–201 (2020).

17. R. Geyer, J.R. Jambeck and K.L. Law, Production, use, and fate of all plastics ever made, *Sci. Adv.* **3**, e170782 (2017).

18. Europe Composites Market Size; Industry Growth Report, 2019–2025. https://www.grandviewresearch.com/industry-analysis/europe-composites-market (accessed: 27 January 2021)

19. R. Miandad, M.A. Barakat, M. Rehan, A.S. Aburiazaiza, I.M.I. Ismail and A.S. Nizami, Plastic waste to liquid oil through catalytic pyrolysis using natural and synthetic zeolite catalysts, *Waste Manag.* **69**, 66–78 (2017).

20. P.S. Roy, G. Garnier, F. Allais and K. Saito, Strategic approach towards plastic waste valorization: Challenges and promising chemical upcycling possibilities, *ChemSusChem.* **14**(9), 4007–4027 (2021).

21. P.T. Helmer and F. Conti, Improving the circular economy via hydrothermal processing of high-density waste plastics, *Waste Manag.* **68**, 24–31 (2017).

22. M.R. Johansen, T.B. Christensen, T.M. Ramos and K. Syberg, A review of the plastic value chain from a circular economy perspective, *J. Environ. Manage.* **302**, 113975 (2022).

23. T. Tan, W. Wang, K. Zhang, Z. Zhan, W. Deng, Q. Zhang, *et al.*, Upcycling plastic wastes into value-added products by heterogeneous catalysis, *ChemSusChem.* **15**(14), e202200522 (2022).
24. Y. Peng, Y. Wang, L. Ke, L. Dai, Q. Wu, K. Cobb, *et al.*, A review on catalytic pyrolysis of plastic wastes to high-value products, *Energ. Convers. Manag.* **254**, 115243 (2022).
24. J. Howarth, S.S.R. Mareddy and P.T. Mativenga, Energy intensity and environmental analysis of mechanical recycling of carbon fibre composite, *J. Clean. Prod.* **81**, 46–50 (2014).
25. J.M. Saad, M.A. Nahil, C. Wu and P.T. Williams, Influence of nickel-based catalysts on syngas production from carbon dioxide reforming of waste high density polyethylene, *Fuel Process. Technol.* **138**, 156–163 (2015).
26. A.H. Mason, A. Motta, A. Das, Q. Ma, M. J. Bedzyk, Y. Kratish, *et al.*, Rapid atom-efficient polyolefin plastics hydrogenolysis mediated by a well-defined single-site electrophilic/cationic organo-zirconium catalyst, *Nat. Commun.* **13**, 7187 (2022).
27. G. Zichittella, A.M. Ebrahim, J. Zhu, A.E. Brenner, G. Drake, G.T. Beckham, *et al.*, Hydrogenolysis of polyethylene and polypropylene into propane over cobalt-based catalysts, *JACS Au.* **2**, 2259–2268 (2022).
28. S.J. Pickering, H. Yip, J.R. Kennerley, R. Kelly and C.D. Rudd. *The Recycling of Carbon Fibre Composites Using a Fluidized Bed Process, FRC 2000: Composites for the Millennium*, Elsevier, Amsterdam, The Netherlands, pp. 565–572 (2000).
29. L. Giorgini, T. Benelli, L. Mazzocchetti, C. Leonardi, G. Zattini, G. Minak, *et al.* Pyrolysis as a way to close a CFRC life cycle: Carbon fibers recovery and their use as feedstock for a new composite production, *AIP Conf. Proc.* **1599**, 354–357 (2014).
30. RTI International, Environmental and economic analysis of emerging plastics conversion technologies, RTI Project No. 0212876.000 (2012). http://energy.cleartheair.org.hk/wp-content/uploads/2012/05/Environmental-and-Economic-Analysis-of-Emerging-Plastics-Conversion-Technologies.pdf
31. Z.-S. Tian, Y.-Q. Wang and X.-L. Hou, Review of chemical recycling and reuse of carbon fiber reinforced epoxy resin composites, *New Carbon Mater.* **37**(6), 1021–1045 (2022).
32. D.D. Yao, H.P. Yang, H.P. Chen and P.T. Williams, Investigation of nickel-impregnated zeolite catalysts for hydrogen/syngas production from the catalytic reforming of waste polyethylene, *Appl. Catal. B.* **227**, 477–487 (2018).
33. M. Das, R. Chacko and S. Varughese, An efficient method of recycling of CFRP waste using peracetic acid, *ACS Sustain. Chem. Eng.* **6**(2), 1564–1571 (2018).
34. O. Zabihi, M. Ahmadi, C. Liu, R. Mahmoodi, Q. Li and M. Naebe, Development of a low cost and green microwave assisted approach towards the circular carbon fibre composites, *Compos. B. Eng.* **184**, 107750 (2020).

Chapter 9

Feasibility Review of Hydrogen Production Options: Life Cycle Assessment and Economic Assessment

Pancy ANG* and Hsien Hui KHOO**

*Institute of Sustainability for Chemicals,
Energy and Environment (ISCE²),
Agency for Science, Technology and Research (A*STAR),
Singapore 627833*

**pancy_ang@isce2.a-star.edu.sg
**khoo_hsien_hui@isce2.a-star.edu.sg*

As part of Singapore's international climate commitment to achieve net-zero carbon emissions by 2050, the production and use of hydrogen is one of the nation's major decarbonization pathways. Conventional technologies such as Steam Methane Reforming (SMR) and Coal Gasification (CG) depend on fossil fuels for hydrogen generation. The usage of fossil fuels has led to environmental issues. To combat environmental impacts due to rising hydrogen demands, a review is done to identify and evaluate the sustainability of hydrogen production options based on environmental and cost indicators. The sustainable usage of SMR and CG can be potentially integrated with Carbon Capture and Storage (CCS) or substituting fossil fuels with renewable energy. Electrolysis and Biomass Gasification (BG) show potential as alternatives to SMR and CG. Although the application of renewable energy has lower environmental impacts, the hydrogen production cost is higher than SMR and CG. Other promising hydrogen production options include white and aquamarine hydrogen. In the foreseeable future, further research and development are required for a judicious evaluation based on the environment and cost of advanced methods for hydrogen production.

1. Introduction

Conventional energy generation systems rely heavily on fossil fuels use, which leads to the release of atmospheric pollutants such as nitrogen oxides, sulfur oxides, particulate matter, and carbon dioxide. Climate change is the resulting environmental impact of carbon dioxide. Additionally, the use of fossil fuels accounts for 82% of the world's energy [1], and many efforts are spent investigating alternative energy resources to combat environmental problems associated with fossil fuels. Low-carbon energy pathways have since gained traction among scientific communities and industries. Hydrogen is a clean, renewable, secondary energy source and is receiving significant attention as a versatile energy carrier. It has potential in energy generation, chemicals, transportation, and iron and steel production applications [2–5]. As part of Singapore's international climate commitment to achieve net-zero carbon emissions by 2050, the production and use of hydrogen is one of the nation's major decarbonization pathways [6]. The global hydrogen demand is projected to increase from 70 million tonnes in 2019 to 120 million tonnes by 2024 [5, 7]. Hydrogen is also expected to become an important energy source in Japan, China, and the United Kingdom (Chaps. 3, 5 and 6). With its growing requirement, one should consider the whole life cycle of hydrogen production options to evaluate its true sustainability in terms of environment and cost.

In this chapter, we review the conventional and emerging hydrogen production options. These include the thermochemical, electrolytic, direct solar water splitting, and biological processes. The feasibility of these hydrogen production options via the Life Cycle Assessment (LCA) and Economic Assessment, i.e., Techno-Economic Assessment (TEA) and Life Cycle Costing (LCC), are also reviewed.

2. Hydrogen Production Options

Hydrogen production options are classified into different color codes based on the production route's cleanness, without considering the entire life cycle [5]. Grey, blue and green hydrogen are the main categories of hydrogen production options to be discussed in this book chapter. Grey hydrogen is produced from Steam Methane Reforming (SMR) or Coal Gasification (CG). Blue hydrogen is formed by integrating Carbon Capture and Storage (CCS) with SMR or CG. Green hydrogen production routes include Electrolysis and Biomass Gasification, accompanied by power sourced from renewable energy sources [8]. This chapter provides an overview of various technologies for hydrogen production from renewable and non-renewable resources.

2.1. *Steam Methane Reforming*

Steam Methane Reforming (SMR) is a developed technology pathway for worldwide hydrogen production (Fig. 1). SMR has a process efficiency of 60–85% [9]. In 2021, 47% of the global hydrogen production is from natural gas [10]. Since hydrogen from this mainstream approach is produced from fossil fuels, it resulted in high carbon dioxide emissions [11]. SMR can be coupled with CCS for lower carbon dioxide emissions.

In the initial processing stage of SMR, sulfur is removed from the natural gas [12]. Next, methane reacts with steam under 3–25 bar pressure in the presence of a catalyst to produce syngas, as shown in Eq. (1) [13].

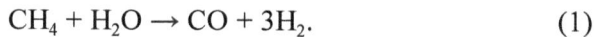

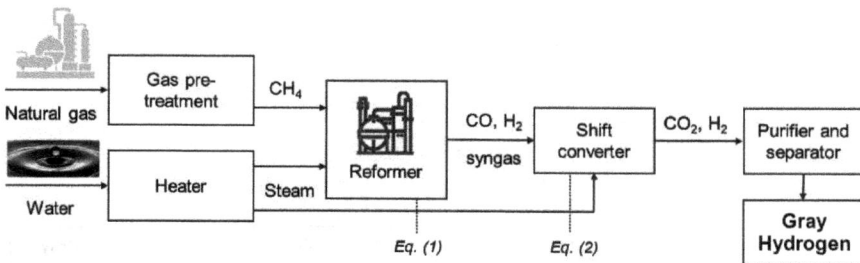

$$CH_4 + H_2O \rightarrow CO + 3H_2. \tag{1}$$

Fig. 1. The schematic production process of Steam Methane Reforming. (Reproduced from Ref. [14]; open access article — Creative Commons CC BY license.)

Carbon monoxide and steam are then reacted in the presence of a catalyst to produce carbon dioxide and hydrogen, as depicted in Eq. (2). This is known as the "water–gas shift reaction" [13].

$$CO + H_2O \rightarrow CO_2 + H_2. \tag{2}$$

In the final step, carbon dioxide is removed to obtain purified hydrogen [13].

2.2. *Coal Gasification*

Coal Gasification (CG) is another widespread hydrogen production technology, having a process efficiency of 74–85% (Fig. 2) [9]. In 2021, coal contributed 27% of the global hydrogen production [10]. Similarly to SMR, this technology also results in high carbon dioxide emissions due to the usage of fossil fuels. CG consists of two phases. In the first phase, oxygen obtained from air separation is supplied into the gasifier to oxidize a fraction of coal into carbon dioxide (Eq. (3)) [14].

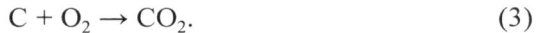

$$C + O_2 \rightarrow CO_2. \tag{3}$$

In the second phase, steam is injected to react with a fraction of coal to form carbon dioxide and hydrogen (Eq. (4)). After the heat is depleted to a certain threshold, oxygen is introduced into the gasifier again. These two steps alternate until coal is completely converted [14].

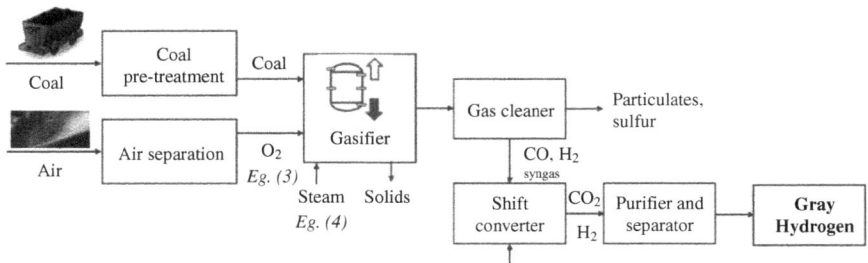

$$C + H_2O \rightarrow CO_2 + H_2. \tag{4}$$

Fig. 2.   The schematic production process of Coal Gasification. (Reproduced from Ref. [14]; open access article — Creative Commons CC BY license.)

2.3. *Electrolysis*

Electrolysis separates water into hydrogen and oxygen using electricity. Electrolysis is a promising hydrogen production option that renders lower carbon dioxide emissions when renewable energy resources are employed. These renewable energy resources include wind, solar, nuclear, and biomass [15]. Other advantages of electrolysis include high hydrogen purity >99.95% and the absence of unwanted byproducts such as sulfates, carbon oxides, and nitrogen oxides that cause environmental pollution [6]. Although electrolysis shows capability as a renewable route, only 4% of global hydrogen demand comes from it [10, 16]. This is due to challenges related to high capital costs, energy efficiency, and operational issues [13].

Electrolysis can be grouped into three significant types — Polymer Electrolyte Membrane (PEM), Alkaline and Solid Oxide electrolysis [13]. PEM electrolysis is the most preferred option owing to its higher hydrogen production rate, better energy efficiency, compact and small design, easy integration with renewable energy, and low maintenance [12].

Referring to Fig. 3, the PEM electrolysis process is as follows: At the anode, water reacts and forms oxygen and positively charged hydrogen

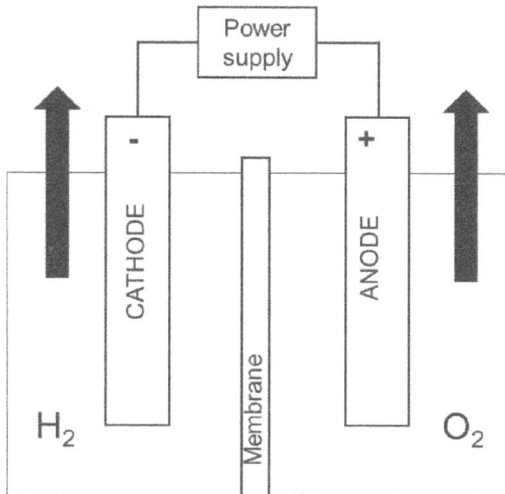

Fig. 3. The schematic production process of Polymer Electrolyte Membrane. (Reproduced from Ref. [14]; open access article — Creative Commons CC BY license.)

ions (protons) (Eq. (5)). At the cathode, hydrogen ions combine with the electrons to form hydrogen gas (Eq. (6)).

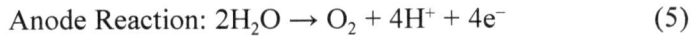

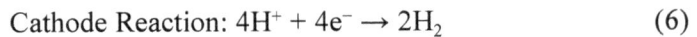

$$\text{Anode Reaction: } 2H_2O \rightarrow O_2 + 4H^+ + 4e^- \tag{5}$$

$$\text{Cathode Reaction: } 4H^+ + 4e^- \rightarrow 2H_2 \tag{6}$$

2.4. *Biomass Gasification*

Biomass Gasification (BG) is a potential hydrogen production technology since biomass is an abundant and renewable energy resource. Furthermore, net carbon dioxide emission is lower than SMR and CG since plants consume carbon dioxide while they make biomass. However, only around 1% of the global hydrogen demand is produced by biomass [16]. This may be associated with limitations such as lower efficiency, higher capital costs, and biomass feedstock costs [13, 16, 17]. Other reasons include aggressive molten slag, tar formation at low temperatures, and potential biomass supply issues [18]. Sources of biomass include sustainable wood waste, agricultural residues, organic municipal solid waste, and animal waste [13, 19].

The process of BG is very similar to CG; however, biomass does not gasify easily compared to coal. As a result, other hydrocarbon compounds are produced. These hydrocarbon compounds must undergo reforming with a catalyst to yield clean hydrogen, carbon monoxide, and carbon dioxide. The water–gas shift reaction will then convert carbon monoxide to carbon dioxide. Purified hydrogen is then obtained with the adsorbers or special membranes [13].

3. Life Cycle Assessment

To achieve near-zero carbon emissions, a systematic environmental tool is needed for evaluating the list of hydrogen production options. An LCA helps identify the environmental impacts and ascertain environmental feasibility by considering the entire life cycle of the process chains. The suitability of using emerging hydrogen production options is assessed.

Antonini and colleagues [20] perform a "cradle-to-gate" LCA of natural gas reforming and biomethane-based hydrogen production with and without the integration of CCS. Global Warming Potential (GWP) is lowered by 45–85% with the addition of CCS in different natural gas reforming-based hydrogen production configurations. However, other environmental impacts worsen due to higher electricity consumption, carbon dioxide transport, and storage requirements. Alternatively, in all investigated cases of biomethane-based hydrogen production, the addition of CCS leads to net-negative emissions. With soil carbon sequestration, bio-based hydrogen achieves net-negative life cycle GHGs emissions without CCS. The authors also deemed electrolysis an equally valid option to reforming-based hydrogen with CCS if electricity is generated from abundant intermittent renewables.

Amaya-Santos and colleagues [21] evaluated the LCA of biohydrogen, blue hydrogen and green hydrogen production technologies. In current and future scenarios, biohydrogen production shows potential with its negative climate change impacts when CCS and biogenic carbon content are considered (Fig. 4). The team established that environmental impacts are greatly dependent on electricity sources. Biohydrogen is a more sustainable option if electricity is sourced from renewables (e.g., solar or wind). The benefits of achieving decarbonization via solar hydrogen have been illustrated in Chap. 5.

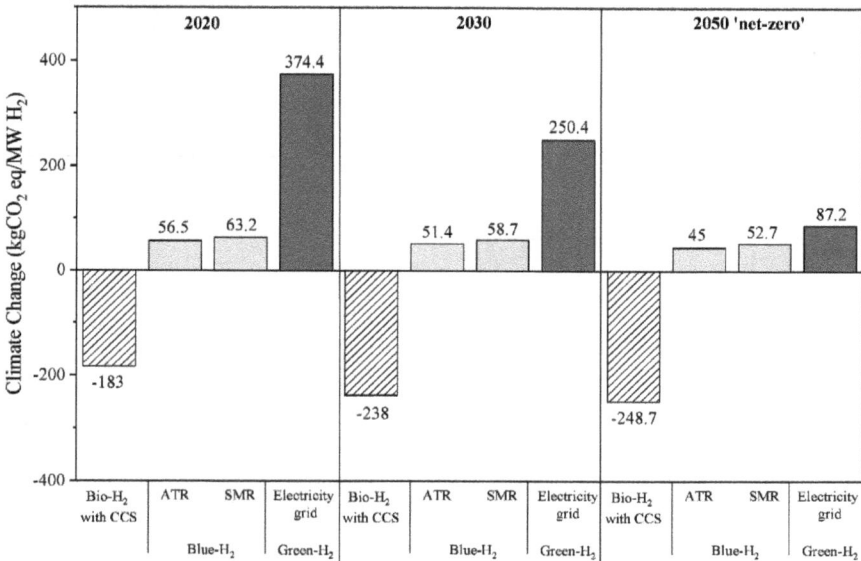

Fig. 4.   Climate Change impacts of current (2020) and future (2030 and 2050) electricity grid mix scenarios. (Reproduced from Ref. [21] with permission from Elsevier.)

A comparative LCA of coal-to-hydrogen (CTH) with biomass-to-hydrogen (BTH) processes were performed by Li and colleagues [22]. Results revealed that energy consumption and GHG emissions are respectively 75.4% (Fig. 5) and 89.6% (Fig. 6) lower in the BTH process in comparison with the CTH process. To further lower GHG emissions and improve energy efficiency, the authors suggest gasification temperatures ranging from 1,400–1,500°C and pipeline transport. Energy consumption of pipeline transport increases at a slower rate as compared to rail and road (Fig. 7). Hence, the pipeline is the desired transport mode. The sustainability benefits of hydrogen from biomass and waste were also described in Chap. 6.

Bareiß and colleagues [23] performed an LCA comparison study of proton exchange membrane water electrolysis and SMR. A carbon emission reduction of up to 75% is obtained from the electrolytic system when

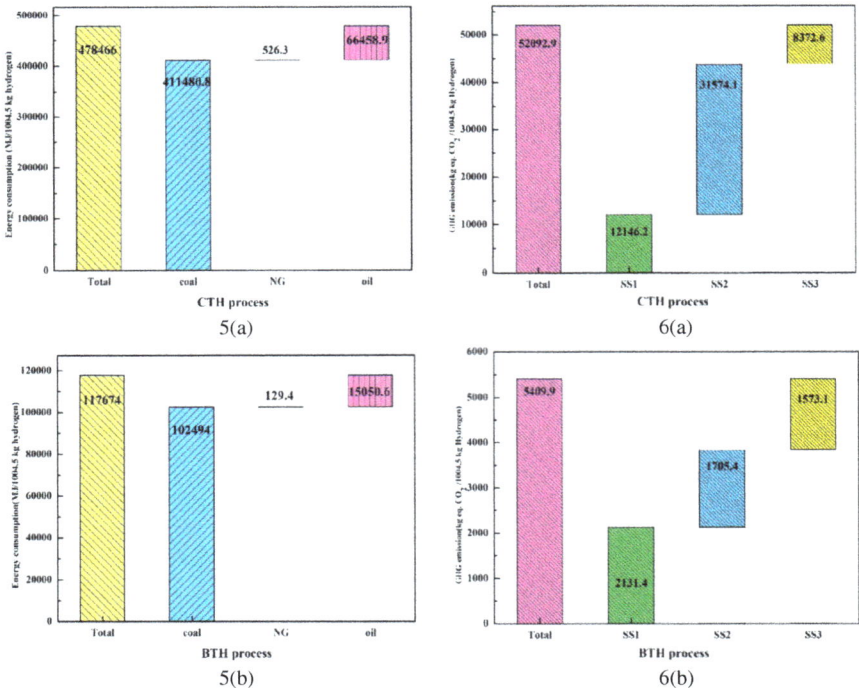

Figs. 5 and 6.   Energy consumption and GHG emissions of (a) CTH process, and (b) BTH process. (Reproduced from Ref. [22] with permission from Elsevier.)

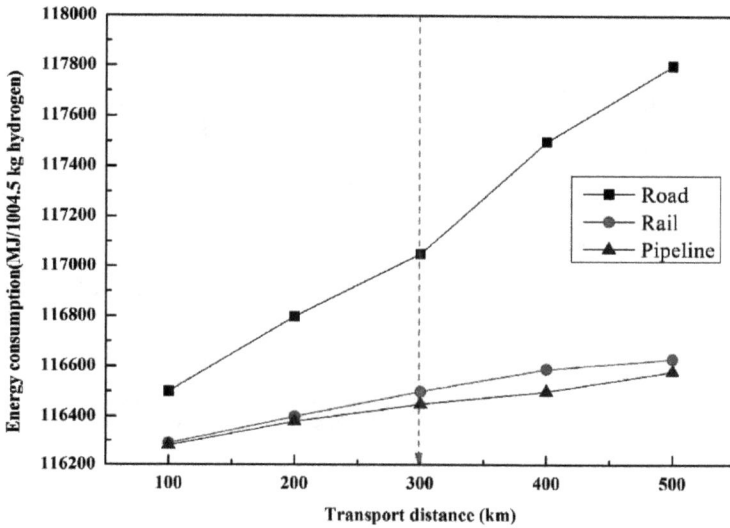

(a)

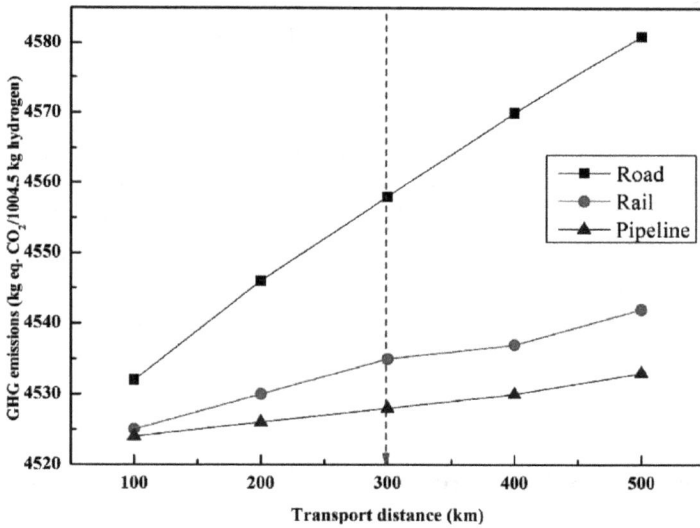

(b)

Fig. 7. Effects of hydrogen transport mode and hydrogen transport distance on energy consumption and GHG emissions. (Reproduced from Ref. [22] with permission from Elsevier.)

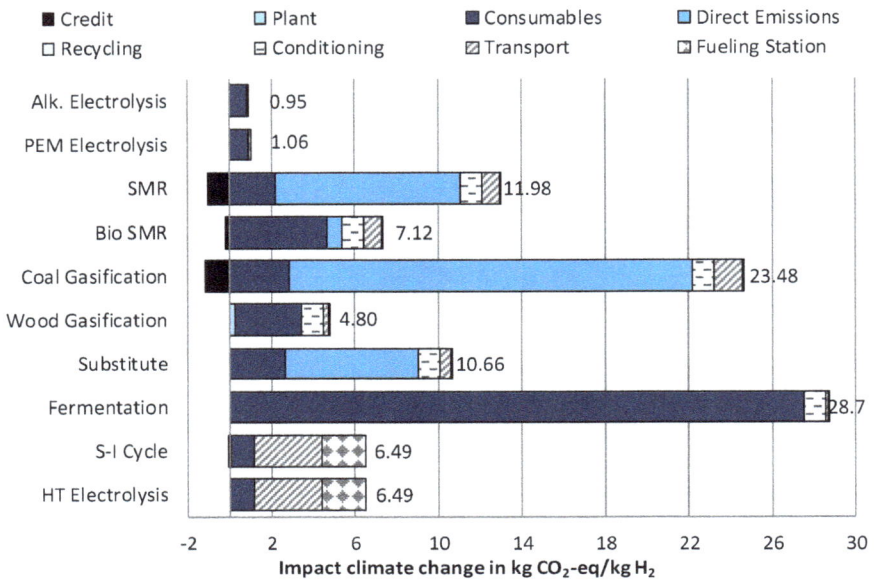

Fig. 8. Impact on Climate Change. (Reproduced from Ref. [17]; open access article —
Creative Commons CC BY license.)

electricity is generated from renewable energy sources. Wulf and colleagues [17] reported that PEM and alkaline electrolysis has the lowest impacts on Climate Change (Fig. 8). Apart from Climate Change, other environmental impact categories were investigated, i.e., Eutrophication, Particulate Matter, Photochemical Ozone Creation, and Acidification (Figs. 9–12). Further studies have shown that switching from fossil fuel-based electricity mix to renewable sources can reduce the global warming impact [23, 24].

4. Economic Assessment

Besides the environmental concerns of hydrogen production, other sustainability indicators involve cost or economic aspects. Techno-Economic Assessment (TEA) and Life Cycle Costing (LCC) are different methods to analyze the economic performance of hydrogen production options. Herein, we assess conventional and emerging hydrogen production options based on economic costs.

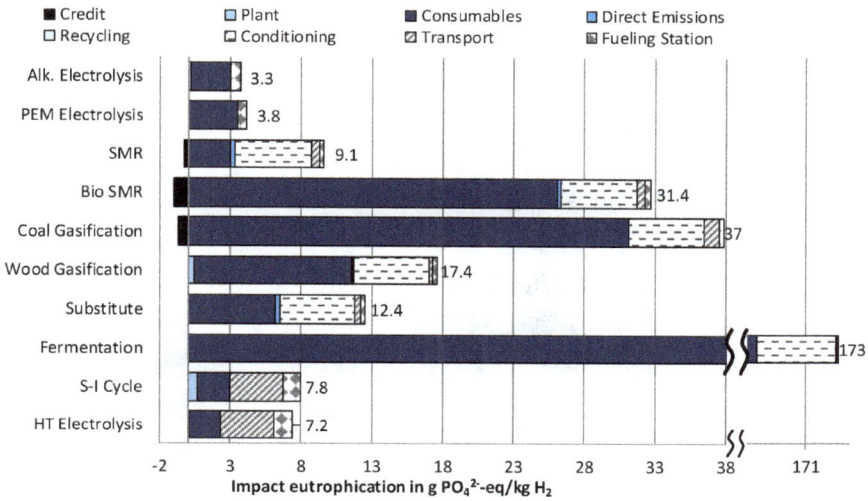

Fig. 9.   Impact on Eutrophication. (Reproduced from Ref. [17]; open access article —
Creative Commons CC BY licence.)

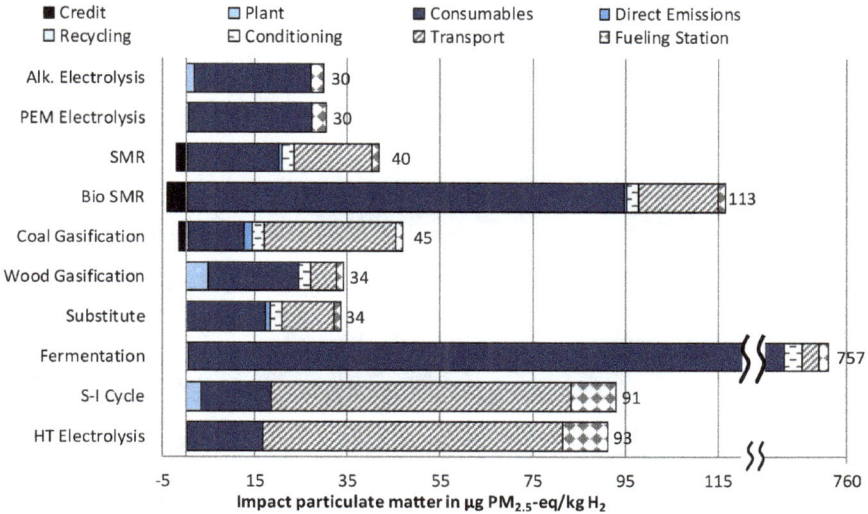

Fig. 10.   Impact on Particulate Matter. (Reproduced from Ref. [17]; open access
article — Creative Commons CC BY license.)

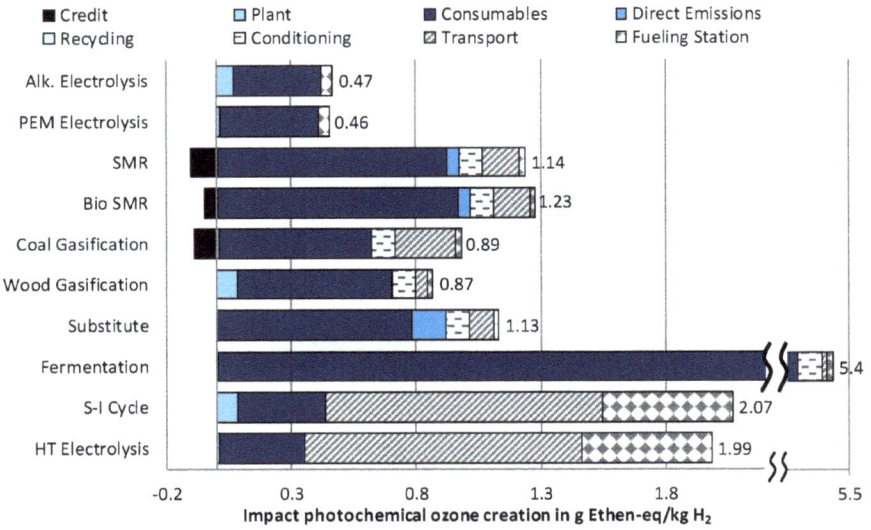

Fig. 11.   Impact on Photochemical Ozone Creation. (Reproduced from Ref. [17]; open access article — Creative Commons CC BY license.)

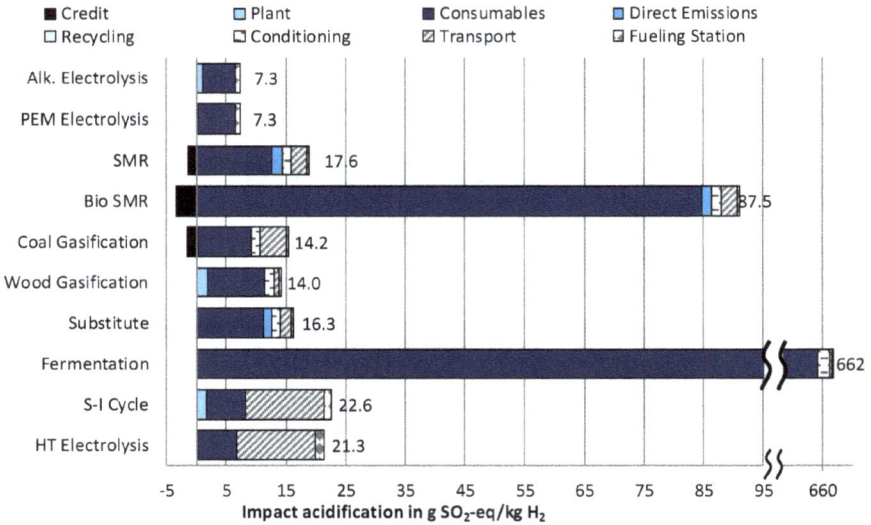

Fig. 12.   Impact on Acidification. (Reproduced from Ref. [17]; open access article — Creative Commons CC BY license.)

Currently, conventional technologies (SMR and CG) have lower costs of hydrogen production without carbon dioxide capture [17, 25–28]. Khan and colleagues [29] developed a spatial techno-economic framework to evaluate SMR with Carbon Capture, Utilization and Storage (CCUS). Natural gas contributed the highest to hydrogen production costs, followed by CCS capital costs. Conversion of carbon dioxide emissions into formic acid via carbon dioxide electro-reduction decreases costs associated with carbon storage and hydrogen production costs by 4–9%. A reduction in natural gas and emission mitigation costs can lower blue hydrogen costs.

Yan and colleagues [30] investigated the techno-economic performance of Sorption-enhanced Steam Methane Reforming (SE-SMR) processes. The economic performance is comparable with SMR integrated with CCS. As seen in Fig. 13, fuel costs are the hot spot in influencing economic performance, followed by net efficiency, total direct equipment capital costs, and carbon dioxide storage costs.

Nguyen and colleagues [31] show that electrolytic hydrogen production with underground storage costs is comparable to SMR with CCS. Electrolytic systems with high-capacity factors, integrated with wholesale electricity markets have high feasibility potential, as depicted in Fig. 14.

LCC analysis is conducted on PEM electrolysis by Zhao and colleagues [32]. The results show that hydrogen costs for energy applications are higher than fossil fuels. The scale and geography may affect the costs of the system since research is conducted on the Orkney Islands. Other reasons include high electricity consumption and lesser renewable energy sources for the power grid (Fig. 15).

Pozo and colleagues [18] found that the co-gasification of coal and biomass is potentially a viable alternative to natural gas or electrolysis. Approximately 30% cost reduction can also be generated by recovering waste heat from hot water production.

The study by Valente and colleagues [33] incorporated an overall evaluation of environmental, cost and social indicators. The levelized cost of hydrogen (LCOH) from biomass gasification is approximately 1.7 times higher than SMR (Fig. 16). Improvements to system efficiency, feedstock, and labor reduction can improve the economic performance of biomass gasification.

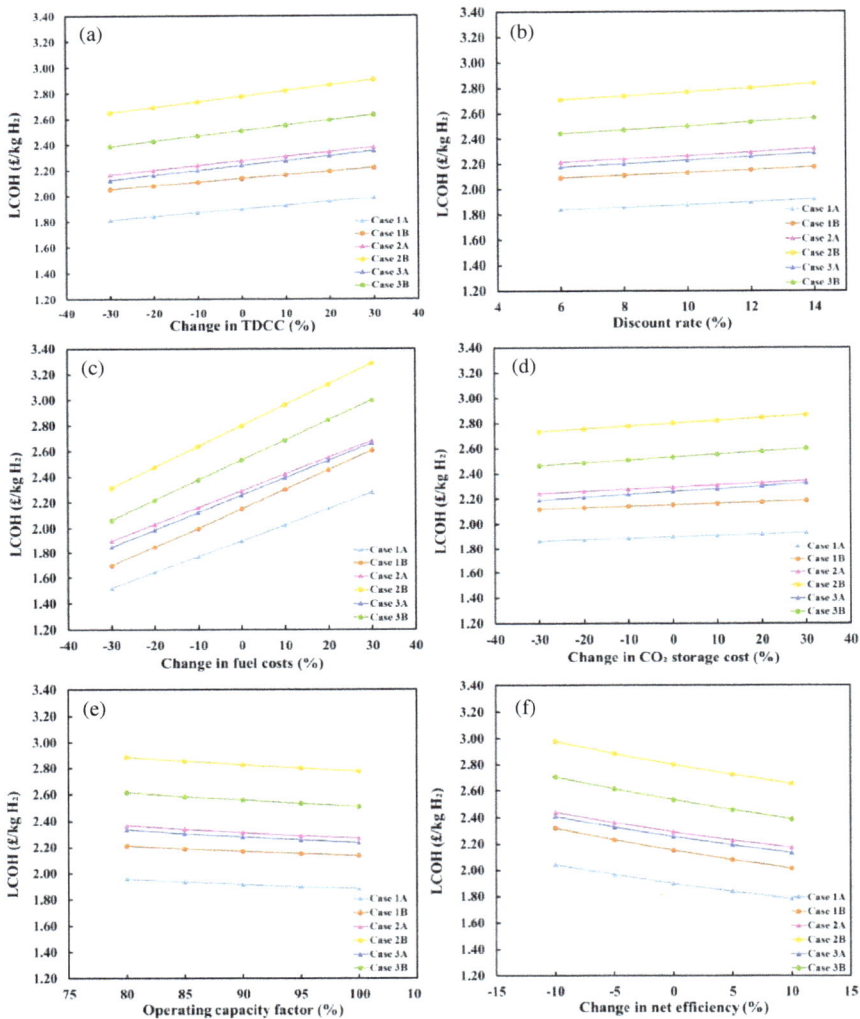

Fig. 13.   Sensitivity analyses: (a) Total direct capital cost, (b) Discount rate, (c) Fuel costs, (d) CO_2 storage costs, (e) Operating capacity factor, and (f) Net efficiency vs. levelized cost of hydrogen. (Reproduced from Ref. [30] with permission from Elsevier.)

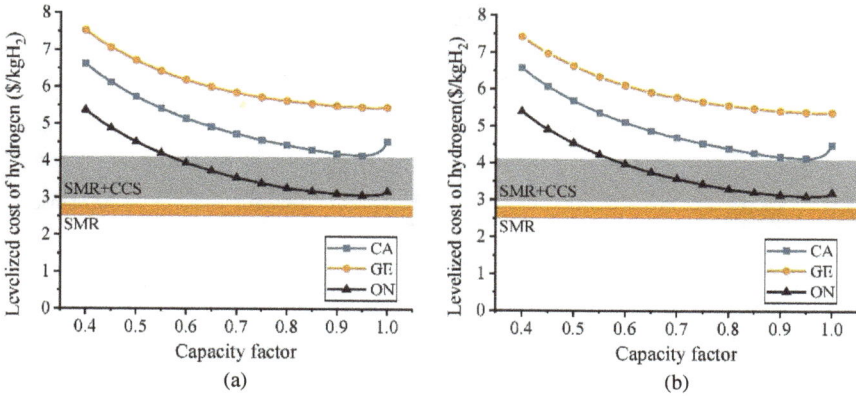

Fig. 14.   Levelized cost of hydrogen in three wholesale electricity markets as a function of capacity factor: (a) Alkaline electrolysis underground storage, and (b) Proton exchange membrane electrolysis underground storage (CA = California; GE = Germany; ON = Ontario). (Reproduced from Ref. [31] with permission from Elsevier.)

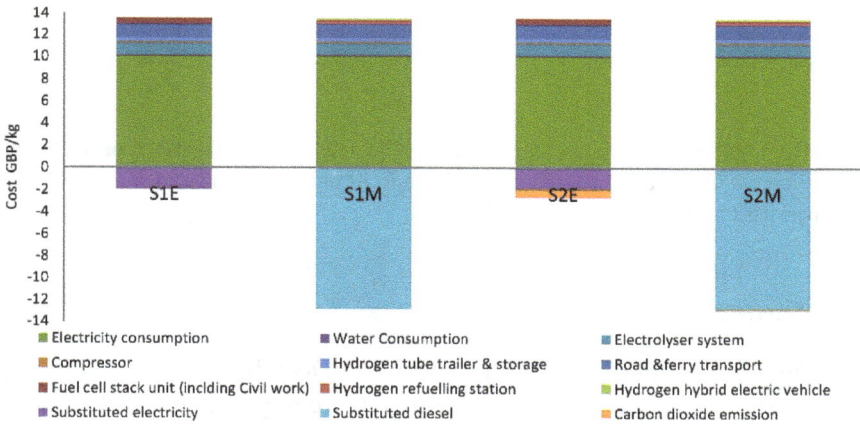

Fig. 15.   Life Cycle Cost analysis of hydrogen in four scenarios. (Reproduced from Ref. [32] with permission from Elsevier.)

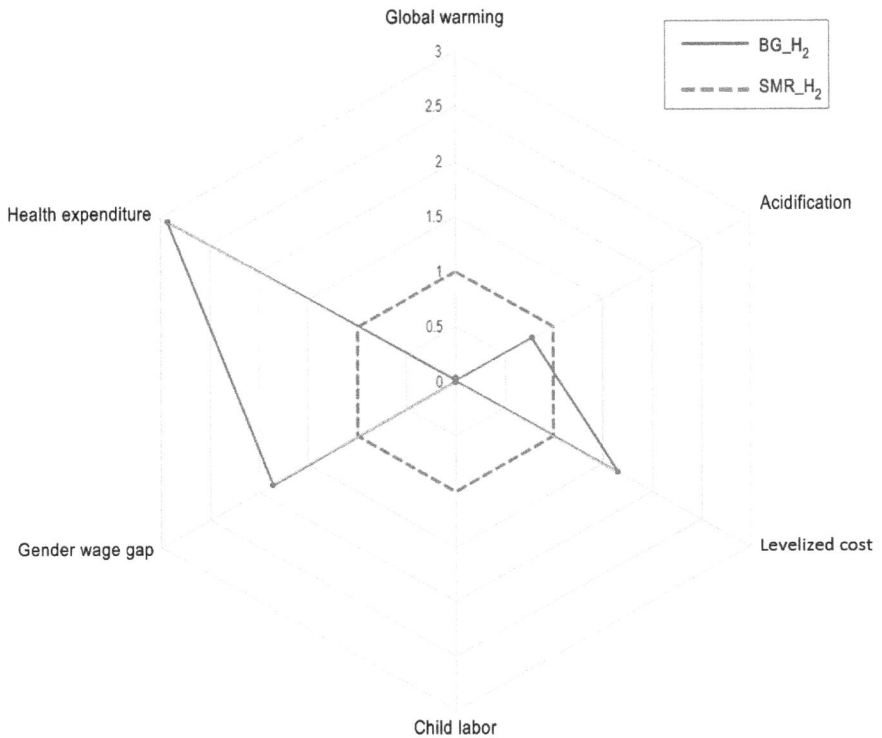

Fig. 16.   Comparison of environmental, economic and social performance of hydrogen from Biomass Gasification and Steam-Methane Reforming. (Reproduced from Ref. [33] with permission from Elsevier.)

5. Conclusion

Hydrogen is rapidly gaining momentum. Decarbonization efforts in advanced methods for hydrogen production are ongoing in countries like Singapore, Japan, China, and the United Kingdom. Challenges relating to technical and non-technical barriers relating to its production and sustainability are still present. The current mainstream technologies (SMR and CG) are to obtain hydrogen from natural gas and coal. However, hydrogen produced from fossil fuels resulted in high carbon dioxide emissions. SMR and CG can be coupled with CCS for lower carbon dioxide emissions. With the use of renewable energy sources, other environmental impacts can potentially be reduced.

Electrolysis and BG are potential emerging hydrogen production options. Although they have lower environmental impacts with renewable energy usage, the hydrogen production cost is higher than SMR and CG. Factors influencing the production cost include investment costs, location, system efficiency, electricity consumption, labor, and operational requirements.

Other than the above-mentioned pathways, there are other promising pathways, such as white and aquamarine hydrogen. White hydrogen is made by solar catalytic thermochemical water splitting with concentrated solar energy, and aquamarine hydrogen is made by methane solar thermal pyrolysis in the presence of a carbon catalyst [34].

This review shows the importance of investigating the entire life cycle of hydrogen production options for a well-rounded sustainability assessment in terms of environment and cost. All in all, more research and development of emerging hydrogen production options are required at a larger scale to establish their potential for better economic and environmental costs.

References

1. BP, 2021, At a glance — BP statistical review of world energy 2022, British Petroleum (2022). https://www.bp.com/content/dam/bp/business-sites/en/global/corporate/pdfs/energy-economics/statistical-review/bp-stats-review-2022-at-a-glance.pdf
2. H. Nazir, N. Muthuswamy, C. Louis, S. Jose, J. Prakash, M.E.M. Buan, *et al.*, Is the H_2 economy realizable in the foreseeable future? Part III: H_2 usage technologies, applications, and challenges and opportunities, *Int. J. Hydrog. Energy* **45**, 28217–28239 (2020).
3. World Economic Forum, Everything you need to know about hydrogen in the clean energy transition (2023). https://www.weforum.org/agenda/2023/01/hydrogen-clean-energy-transition-2023/
4. C. Acar and I. Dincer, Review and evaluation of hydrogen production options for better environment, *J. Clean. Prod.* **218**, 835–849 (2019).
5. A.I. Osman, N. Mehta, A.M. Elgarahy, M. Hefny, A. Al-Hinai, A.H. Al-Muhtaseb, *et al.* Hydrogen production, storage, utilisation and environmental impacts: A review, *Environ. Chem. Lett.* **20**, 153–188 (2022).
6. MTI, Singapore launches national hydrogen strategy to accelerate transition to net zero emissions and strengthen energy security, Ministry of Trade and Industry, Singapore (2022). https://www.ema.gov.sg/media_release.aspx?news_sid=202210246JoalvsFYBOc

7. S. Atilhan, S. Park, M.M. El-Halwagi, M. Atilhan, M. Moore and R.B. Nielsen, Green hydrogen as an alternative fuel for the shipping industry, *Curr. Opin. Chem. Eng.* **31**, 100668 (2021).
8. F. Safari and I. Dincer, A review and comparative evaluation of thermo-chemical water splitting cycles for hydrogen production, *Energy Convers. Manag.* **205**, 112182 (2020).
9. F. Dawood, M. Anda and G.M. Shafiullah, Hydrogen production for energy: An overview, *Int. J. Hydrog. Energy* **44**, 3847–3869 (2019).
10. IRENA, Hydrogen, The International Renewable Energy Agency (2022). https://www.irena.org/Energy-Transition/Technology/Hydrogen
11. A.I. Osman, T.J. Deka, D.C. Baruah and D.W. Rooney, Critical challenges in biohydrogen production processes from the organic feedstocks, *Biomass Convers. Biorefin.* (2020).
12. H. Ishaq, I. Dincer and C. Crawford, A review on hydrogen production and utilization: Challenges and opportunities, *Int. J. Hydrog. Energy* **47**, 26238–26264 (2022).
13. DOE, Hydrogen production processes, U.S. Department of Energy (2023). https://www.energy.gov/eere/fuelcells/hydrogen-production-processes
14. A. Ajanovic, M. Sayer and R. Haas, The economics and the environmental benignity of different colors of hydrogen, *Int. J. Hydrog. Energy* **47**, 24136–24154 (2022).
15. M. Ji and J. Wang, Review and comparison of various hydrogen production methods based on costs and life cycle impact assessment indicators, *Int. J. Hydrog. Energy* **46**, 38612–38635 (2021).
16. D. Das, N. Khanna and T.N. Veziroğlu, Recent developments in biological hydrogen production processes, *Chem. Ind. Chem. Eng. Q.* **14**(2), 57–67 (2008).
17. C. Wulf and M. Kaltschmitt, Hydrogen supply chains for mobility — Environmental and economic assessment, *Sustainability*, **10**(6), 1699 (2018).
18. C.A. Pozo, S. Cloete and A.J. Álvaro, Carbon-negative hydrogen: Exploring the techno-economic potential of biomass co-gasification with CO_2 capture, *Energy Convers. Manag.* **247**, 114712 (2021).
19. DOE, Biomass Energy Basics, National Renewable Energy Laboratory, U.S. Department of Energy (2023). https://www.nrel.gov/research/re-biomass.html
20. C. Antonini, K. Treyer, A. Streb, M. van der Spek, C. Bauer and M. Mazzotti, Hydrogen production from natural gas and biomethane with carbon capture and storage — A techno-economic analysis, *Sustain. Energy Fuels* **4**, 2967 (2020).
21. G. Amaya-Santos, S. Chari, A. Sebastiani, F. Grimaldi, P. Lettieri and M. Materazzi, Biohydrogen: A life cycle assessment and comparison with alternative low-carbon production routes in UK, *J. Clean. Prod.* **319**, 128886 (2021).

22. G. Li, P. Cui, Y. Wang, Z. Liu, Z. Zhu and S. Yang, Life cycle energy consumption and GHG emissions of biomass-to-hydrogen process in comparison with coal-to-hydrogen process, *Energy* **191**, 116588 (2020).
23. K. Bareiß, C. de la Rua, M. Möckl and T. Hamacher, Life cycle assessment of hydrogen from proton exchange membrane water electrolysis in future energy systems, *Appl. Energy* **237**, 862–872 (2019).
24. D. Chisalita, L. Petrescu and C. Cormos, Environmental evaluation of European ammonia production considering various hydrogen supply chains, *Renew. Sust. Energ. Rev.* **130**, 109964 (2020).
25. B. Parkinson, M. Tabatabaei, D.C. Upham, B. Ballinger, C. Greig, S. Smart, *et al.* Hydrogen production using methane: Techno-economics of decarbonizing fuels and chemicals, *Int. J. Hydrog. Energy* **43**, 2540–2555 (2018).
26. Z. Navas-Anguita, D. García-Gusano, J. Dufour and D. Iribarren, Revisiting the role of steam methane reforming with CO_2 capture and storage for long-term hydrogen production, *Sci. Total Environ.* **771**, (2021).
27. S. Sadeghi, S. Ghandehariun and M.A. Rosen, Comparative economic and life cycle assessment of solar-based hydrogen production for oil and gas industries, *Energy* **208**, (2020).
28. Z. Navas-Anguita, D. García-Gusano, J. Dufour and D. Iribarren, Revisiting the role of steam methane reforming with CO_2 capture and storage for long-term hydrogen production, *Sci. Total Environ.* **771**, 145432 (2021).
29. M.H.A. Khan, R. Daiyan, P. Neal, N. Haque, I. MacGill and R. Amal, A framework for assessing economics of blue hydrogen production from steam methane reforming using carbon capture storage & utilisation, *Int. J. Hydrog. Energy* **46**, 22685–22706 (2021).
30. Y. Yan, V. Manovic, E.J. Anthony and P.T. Clough, Techno-economic analysis of low-carbon hydrogen production by sorption enhanced steam methane reforming (SE-SMR) processes, *Energy Convers. Manag.* **226**, 113530 (2020).
31. T. Nguyen, Z. Abdin, T. Holm and W. Mérida, Grid-connected hydrogen production via large-scale water electrolysis, *Energy Convers. Manag.* **200**, 112108 (2019).
32. G. Zhao, E.R. Nielsen, E. Troncoso, K. Hyde, J.S. Romeo and M. Diderich, Life cycle cost analysis: A case study of hydrogen energy application on the Orkney Islands, *Int. J. Hydrog. Energy* **44**, 9517–9528 (2019).
33. A. Valente, D. Iribarren and J. Dufour, Life cycle sustainability assessment of hydrogen from biomass gasification: A comparison with conventional hydrogen, *Int. J. Hydrog. Energy* **44**, 21193–21203 (2019).
34. A. Boretti, There are hydrogen production pathways with better than green hydrogen economic and environmental costs, *Int. J. Hydrog. Energy* **46**, 23988–23995 (2021).

Chapter 10

Life Cycle Assessment Digitalization to Assess the Performance of Linear Infrastructure Projects

Koji NEGISHI

Oris Connect,
54 Avenue Hoche, 75008 Paris, France

koji.negishi@oris-connect.com

Linear infrastructures are material-intensive systems with ever-increasing material demands. Evidence shows 30 to 40% of construction materials are used in linear infrastructure globally every year, while the impact of construction materials manufacturing is estimated at 9% of energy-related CO_2 emissions. We estimate that the carbon footprint from linear infrastructure is 85% linked to the materials chosen, meaning that the optimization of construction materials used in such projects presents a significant climate change mitigation lever. Moreover, the particularly long service life of several decades requires maintenance; with this comes additional material and energy consumption to ensure the infrastructure is fit for purpose, offering user safety and structural integrity. Material sourcing is of utmost importance for eliminating unnecessities, a reduction in natural resource use, and low-carbon infrastructure. Connecting digitally infrastructure projects to real-time material databases at a very early project stage enables one to measure, predict, and support informed decision-making, based on environmental and economic performances in the infrastructure sector. This chapter introduces such digitized Life Cycle Assessment (LCA) capabilities to predict and improve the infrastructure's performance and greatly contribute to informed decision-making at the early stage of projects.

1. Introduction

Globally, 30 to 40% of construction materials are used in land-based logistical infrastructures (e.g., roads, railways). Materials and infrastructure design choices impact up to 60% of the cost of a project and about 85% of its overall greenhouse gas (GHG) emissions [1]. Most infrastructures are designed according to national standards and historical design methods, where materials availability and adequacy are considered only later during the project construction phase due to the difficulty in obtaining localized material properties and performance data. Additionally, the lack of a systematic approach in performing such an evaluation exists. Such an approach from the infrastructure project partner organizations (owners, investors, designers, contractors) lacks integration of sustainability requirements such as climate change mitigation, natural resources scarcity, or social disparities. Based on 12 different pavement solutions from 8 projects, we estimate the average results of life cycle CO_2 emissions, cost, and material consumptions to be 275 ton CO_2-eq/km/lane, 320 k USD/km/lane and 5.4 k ton/km/lane, respectively [2]. However, Fig. 1 shows clearly that there is significant variability in results due to the diversity of types of projects. This fact demonstrates the need to look at projects individually to ensure that the best decisions are made for each individual project.

Connecting real-time materials databases at a very early project life cycle stage would enable one to measure and predict CO_2 emissions and cost and support informed decision-making. Analyzing projects through a digital approach could allow more sustainable infrastructure with optimized solutions for each local sourcing environment. In such a platform, users would be from the infrastructure project team comprising: infrastructure owners, material suppliers, contractors, engineering firms, financial institutions, and/or infrastructure financiers.

This ambition to create a digital platform to support and segment engineering capacity is at the origin of some of the most advanced solutions, such as the ORIS Material Intelligence platform. In ORIS, firstly,

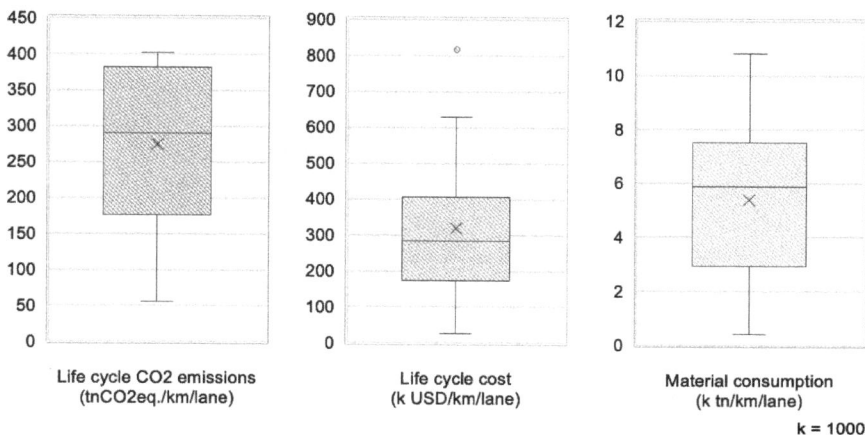

Fig. 1. The results of CO_2 emissions (left), cost (center), and material consumption (right) through the whole service life of pavement construction projects, as calculated by ORIS [1].

the project alignment is loaded onto the platform that geolocalizes project and material sourcing opportunities. Secondly, by embedding local design catalogs and specifications, the user has the capacity to select several structures from all the available designs in order to compare and choose their preferred design according to their priorities: carbon footprint mitigation, increased durability, local materials, and cost reduction. Beyond design choice and type of construction, i.e., flexural or rigid pavement or high-speed railway, the user can select several maintenance scenarios according to the infrastructure types to be compared and assess the impact on the use phase's CO_2 emissions and resource consumptions. The key to success is the capability to combine planning, execution, maintenance and recycling on one platform. Such a global approach assesses the impact at every stage of the infrastructure's life cycle. Furthermore, this data-driven approach and abundant geolocalized material databases enables one to extend to any linear infrastructure projects around the world. These are important in regard to public expenses for any country's budget.

A recent review work revealed a significant increase in research on Life Cycle Assessments (LCAs) applied to infrastructure projects in the last 20 years [3–6]. In fact, this shows the enhancement of a sustainable strategy on the infrastructure, with emphasized focus on reducing overall CO_2 emissions. Meanwhile, also highlighted was the fact that the majority of published studies contain an inconsistency in the functional unit and an incomplete system boundary. Subsequently, the main challenge in data collection and a significant execution time of an LCA study led to the unavailability of results and decision-making early on in the project's life, i.e., at the design stage. To the best of our knowledge, none of the existing studies observed provided a systematic approach that enabled full and easier infrastructure LCAs. In order to improve sustainability in infrastructure, the swiftness and data availability using data science techniques are of great importance and allow a project assessment to be performed at a very early stage.

To meet these challenges, an exhaustive and data-driven approach is required to promptly and seamlessly perform an extended level analysis of design performance during its life cycle stages, conforming to LCA standards [7–10]. Data-driven LCA studies, accompanied by CO_2 emission information, can make decisions to reduce global warming impacts in land infrastructure design phases.

From raw material extraction and acquisition to material production, transport and construction operations, with customized options in terms of materials, actual properties, and design solutions, a digitized LCA solution expands its analysis to the use and maintenance phases, up to the end-of-life stage, i.e., "Cradle-to-Grave" as per LCA standards, compared to previously developed infrastructure LCA approaches that, in most cases, only consider production and construction stages, i.e., "Cradle-to-Gate".

In road applications, state-of-the-art models are used to incorporate the impact of road use due to specific pavement properties (albedo, carbonation and lighting) and the impact due to Pavement-Vehicle Interactions (PVI). The modules associated with PVI take into account not only pavement mechanical characteristics and pavement aging but also the evolution and growth of annual average daily traffic (AADT) and local climate conditions over the road service's lifetime in order to calculate increased vehicle fuel consumptions.

To comply with the above referenced LCA standards, CO_2 emissions are calculated by applying the Life Cycle Impact Assessment (LCIA) method, which converts the resource use and emissions contained in the Life Cycle Inventory (LCI) to the Global Warming Potential (GWP) indicator measured in kilograms of carbon dioxide equivalent (kg CO_2-eq). Moreover, given the early design focus, data quality assessment should be carried out using inventory datasets to incorporate uncertainty in results.

Using the ORIS digital platform, a case study is presented to demonstrate the potential in uncovering the environmental impact of the different decisions made during the design of a project as well as the indirect effects that the decisions made in one phase (e.g., construction or maintenance) can have on the subsequent use phase. Compared to most conventional road LCA approaches only covering a part of the life cycle, ORIS's solution is scientifically based on data and proves that holistic assessment over the full life cycle is a key to improve the overall sustainability of the road infrastructure. Besides the CO_2 emissions criteria, material circularity (recycled materials used in the project and potential benefits from demolition waste) is monitored to help users make even more sustainable and economical driven decision-making.

2. Using a Digitized Framework to Model the Life Cycle Carbon Footprint of Infrastructure

Today there is a huge trade-off between high construction material demand and climate concerns. Brundtland [11] introduced the idea of sustainable development with a focus on a proper management of material supply to mitigate resource uses and CO_2 emissions. A digital platform like ORIS gathering construction materials information and knowledge makes it accessible for the entire project's organizations. Such a tool paves the way to sustainable infrastructure, providing circular, low-carbon and resource-optimized solutions for safer and more resilient infrastructure. These evaluations are performed on the basis of the LCA applied to infrastructure projects. ORIS is a digital platform of which its Application Programming Interface (API) established was deployed in the web-based tool. Supported by artificial intelligence techniques and abundant material data, ORIS assesses the impact of linear infrastructure designs in a multi-dimensional view. The platform aggregates and computes data around materials properties, geolocations, and expected traffic and weather conditions for effective and sustainable construction. Based on this data-driven analysis, users have the ability to make informed decisions on optimizing materials consumption, overall carbon emissions, and cost on the basis of the LCA methodology.

Figure 2 shows the developed digital solutions for LCA calculations and their links with background databases. In total, five core functions are separately developed, each corresponding to a life cycle phase investigating

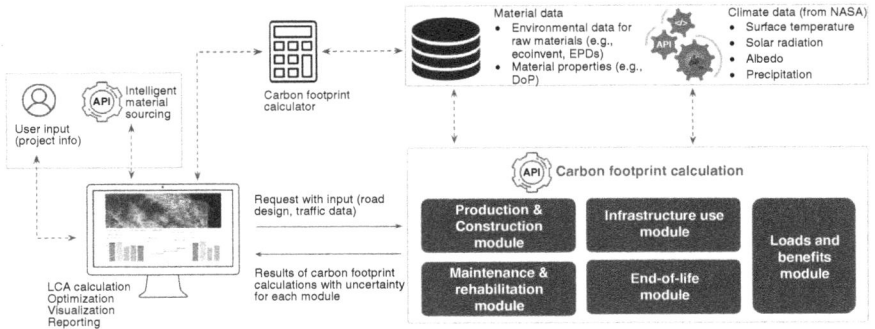

Fig. 2.   Digitized framework of the infrastructure LCA in the ORIS platform.

each calculation request for material and CO_2 emission factors stored in background Structured Query Language (SQL) databases. The calculation is completed by climate data via the POWER web service from the National Aeronautics and Space Administration (NASA). The result of the GWP indicator is assessed by applying characterization factors from the CML (Centrum voor Milieukunde Leiden) method [12] as a reference method to be in accordance with LCA standards.

The whole digitized process of the LCA conducted in the ORIS platform is depicted in Fig. 3. This advanced digital technology allows innovative data management systems and data science techniques to rapidly

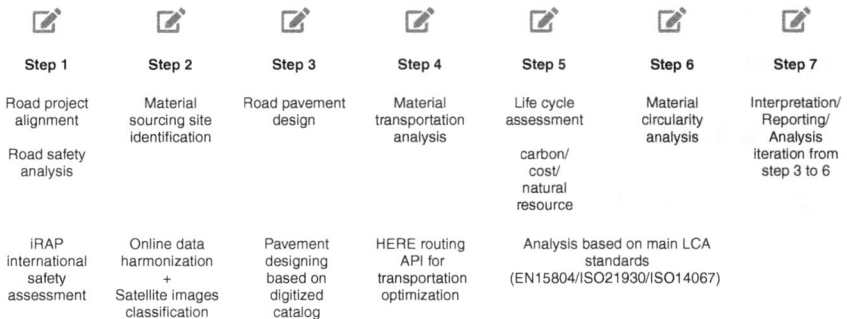

Fig. 3.   Life cycle stages of a road pavement in accordance with the ORIS platform methodology.

assess the key project indicators: project alignment, identification of local material sourcing sites, infrastructure designing from digitized catalogs and specifications or user customized standards, material transportation analysis, life cycle carbon and cost analysis, natural resource consumption analysis, and material circularity analysis. Its capability of an easy extension of databases serves to embed user data in the platform and thus allows multiple simulations and scenarios, including a sensitivity analysis of construction projects, i.e., the effects of applying a different material service lives over the analysis period. The platform that has been developed and aligned with this methodology is therefore capable of combining the planning, execution, maintenance, and recycling phases on one platform, hence greatly contributing to efficiency by reducing the effort and assessment execution time and early stage decision-making, which would otherwise require lengthy manual calculations by trained and competent engineers and the subsequent checking of these calculations following any minor alterations to the design or material choices.

The extensive external database and the automated calculations allow users to test, compare, and visualize multiple solutions from the design (Fig. 2, step 3) to the analysis (steps 4 to 6) and come back to the design phase (step 3). The ease of this scenario analysis reveals the huge potential of carbon reduction, cost reduction, and material optimizations.

2.1. *Project alignment*

Collecting project and material data is a crucial element of a data-driven approach, as it establishes the basis for any further analysis and decision-making. Any general project's general data (e.g., location, infrastructure type, local standards) are the first elements to be input into ORIS in order to identify local material properties and sources that meet local infrastructure designs and specifications.

In road construction projects, reducing fatalities on road sections is of the utmost importance for any road owner. The ORIS platform integrates safety evaluation using the international Road Assessment Program (iRAP) star rating system [13] regarding the safety target proposed by the United Nations. iRAP star ratings represent the relative risk of death or serious injury for individual road user populations comprising motorists, motorcyclists, cyclists, and pedestrians. The iRAP star rating system evaluates more than 52 road attributes. These parameters influence the safety for the above user populations. The iRAP star rating evaluation considers the following parameters:

- Likelihood: referring to the risk of a crash being initiated,
- Severity: referring to the severity of crash when it happens,
- Operating speed: referring to the degree to which risk changes with speed,
- External flow influence factors: accounting for the degree to which a person's risk of being involved in a crash is a function of another person's use of the road, and
- Median ability to traverse factors: accounting for the potential that an errant vehicle will cross a median.

The safety assessment is undertaken after the pavement and footway designs are selected, as the surface material will impact road and user group performance.

2.2. *Building a database of local material sourcing sites*

The next phase of an infrastructure construction project is the identification of local material sourcing capabilities. Each material sourcing site is recognized as a material production site, such as asphalt, bitumen, concrete, cement, steel, and any other raw materials needed for a project. ORIS integrates an advanced level of database where geolocalized material sourcing sites are stored. The information of these sites is verified and completed by using data science techniques. Once material sourcing sites are identified, the attributes of these materials required for the successful completion of the database are collected from various data sources, including verified data provided by clients and data available in the public domain, such as a Declaration of Performance (DoP) to accompany each product to specify its mechanical or physical characteristics. Material prices in ORIS are sourced from contractors and are introduced in a cost table dedicated to a project, while associated environmental datasets are from sources aligned with LCA standards and specific to the geographical context (e.g., ecoinvent, Base Carbone in France, industrial Environmental Product Declarations (EPD)).

2.3. *Linear infrastructure design choice*

In ORIS, linear infrastructure can be designed either manually by entering the description of the structure (e.g., thickness of each pavement layer) or automatically by selecting from the digitized local catalogs. Manual input requires the platform to be fed with information specific to each element of the pavement structure and materials used (e.g., dimension — pavement thickness and width, material type and density, internal transport costs, etc.). When it comes to the use of local design catalogs, the user needs to feed the platform with the general information required by the local regulations (e.g., country, pavement structure type (i.e., flexible or rigid), traffic type and volume, soil classification, local climate conditions, and materials transport modes). Based on these parameters, ORIS automatically selects compliant designs prepared and stored in the database (Fig. 4). It is important to note that the method of structural design, criteria of layer sizing, and available materials differ from one country to another. From the solutions selected, the basic information on the dimensions (width and thickness, angles) and material density allows the calculation of the reference material to be retrieved from the design catalog.

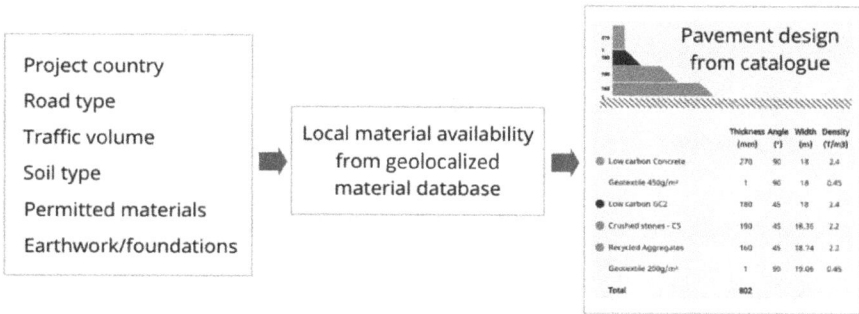

Fig. 4.   Example of the pavement design process from project alignment to design selection.

The capacity to digitize road infrastructure design makes it possible to perform an LCA study very early on in a project; this is because it solves the unavailability of LCI data for the modeling. An early assessment will bring higher potential of cost and carbon footprint reduction than an assessment in a later phase of the project, enabling project efficiencies to be identified to the road infrastructure owners/financers early on. To better anticipate how materials can perform under local weather conditions and over their lifetime, multiple material properties and factors are included in the model, in particular, flexural strength of concrete, fatigue behavior of concrete, elasticity, subgrade reaction, and albedo. These parameters, combined with maintenance programs, are used in use phase models to compute these effects on vehicle or train performance.

2.4. *Optimization of material transportation*

In the life cycle for a construction project, material transportation is needed in separate phases. Once a material sourcing sites database is built and a project location is generated, the digital platform automatically simulates the transport distance and associated carbon footprint and cost. The capacity to geographically connect between projects and material sourcing sites helps to identify the right route from multiple possible solutions by considering road constraints (e.g., restricted truck dimensions and maximum gross weights, environmental avoidance zones) and traffic conditions.

2.5. *Life cycle carbon, cost and material consumption assessment*

The assessment of carbon footprints follows LCA standards covering a full scope of the infrastructure life cycle. From the designs selected, maintenance scenarios, and end-of-life (EoL) scenarios, an LCI analysis is conducted to calculate material and energy consumptions through a full life cycle. In road applications, the use phase modules are divided into two categories.

The first is related to the properties of road surface material, and it calculates the albedo effect on the global balance of atmospheric radiative forcing, the carbonation effect, and electricity consumption for road lighting.

The second is related to the PVI effects by measuring the vehicle's fuel consumption due to the fatigue and surface rolling resistance of the pavement over the pavement use phase. As the condition of the pavement is a key factor to calculate the use phase modules, pavement degradation and the rehabilitation undertaken through periodic maintenance programs are taken into account for use phase calculations. In railway applications, one can assume the modal shift of transport passengers from vehicles to trains. This would create a significant potential of carbon reductions due to fuel consumptions, although it is assumed that the construction of a high-speed rail network would have substantial CO_2 emissions.

The methodology also includes a systematic estimation of benefit in terms of carbon emissions based on default material recovery scenarios. The assessment of the use phase increasingly contributes to the overall results, helping to better identify the potential for a reduction in carbon footprint and avoid ramifications in the decision-making. Figure 5 gives an overview of life cycle stages in the infrastructure LCA.

Life cycle stages	Product stage			Construction stage		Use stage							End of life				Benefits and loads beyond the system boundary
Modules	A1	A2	A3	A4	A5	B1	B2	B3	B4	B5	B6	B7	C1	C2	C3	C4	D
	Raw material supply	Transport	Manufacturing	Transport	Construction	Use	Maintenance	Repair	Replacement	Refurbishment	Operational energy use	Operational water use	Demolition	Transport	Waste processing	Disposal	Reuse / Recovery/ Recycling potential

Fig. 5.   Life cycle stages and modules of a built asset applied to the infrastructure LCA as per LCA standards.

Life cycle cost analysis takes into account construction and maintenance operations and subsequent material consumptions among the different stages of road projects and transforms them into costs. The material production stage considers the cost related to raw material supply, transport, and manufacturing. The transportation cost includes the finished product transportation (from the final material sourcing sites) to the project main access point, while the internal transport cost is also included to account for the product transportation from the main access point at the site to other work locations as needed. Then, the construction stage considers the energy consumption from the plant and equipment used converted into cost. The maintenance costs are also included in the analysis and are considered as a key indicator for the overall whole LCA, as maintenance interventions affect pavement life cycle costs and the environmental and overall performance of the asset.

Finally, natural resource consumptions are calculated according to the project section length, layer widths, thickness, and material densities.

By monitoring virgin and recycled materials incorporated into each section of the designs, it is possible to evaluate the effect of recycled materials regarding carbon and cost performances.

2.6. *Material circularity analysis*

After the EoL of an asset (i.e., road pavement) comes the measurement of material circularity potential from demolished waste. This gives additional information beyond the life cycle, and indicates the environmental benefit obtained from the extension of a material's service life by either reusing or recycling them. The scenario of demolished materials valorization is specific to each project, and it requires identifying potential transport to recycling facilities and further processing, depending on project requirements and design scenarios.

2.7. *Results interpretation*

The final phase of an LCA is a visual interpretation of results, a design comparison, and evaluation. ORIS's digital tool helps decision-making, thanks to the different options of design comparison and scoring systems from the economic, environmental, and material point of view through a whole road life cycle. Besides, the digital platform and its extensive database offer easy implementation of alternative pavement scenarios, so projects can return back to any previous steps.

3. Case Study: Reconstruction of a 4-km Flexible Pavement in the United Kingdom

The case study is based on a road improvement project to rehabilitate an existing dual carriageway in the United Kingdom (UK). Due to heavy traffic loading, the asphalt surface of the carriageway had come to the end of its serviceable life, with large potholes that resulted in an uneven surface, which posed a safety hazard for drivers. Safety, quick construction, and minimal disruption to road users traveling on the local road network was paramount, though another significant challenge was to reduce embodied carbon through reduced emissions, which is now mandated in the UK's Net Zero plan. For the assessment in this case study, the ORIS digital platform's LCA module was used.

The design life in accordance with the UK's design standards [14] is 40 years, and a maintenance plan was also developed for all pavement design options, factoring in periodic replacement of the upper (bound) asphalt pavement layers. For all maintenance scenarios in the 14th year after construction, it is proposed that minor maintenance be undertaken, comprising the removal of the top 40 mm of the pavement surface by milling, and replacing this layer with the same thickness of the Thin Surface Course System (TSCS). In the 26th year, major maintenance is planned, with the removal and replacement of the top 100 mm, comprising 40 mm of TSCS and 60 mm of the Asphalt Concrete binder course. From the material database embedded in the ORIS platform, the project found the material sourcing sites of 10 asphalt plants and 13 quarries as well as 2 recycling facilities, to which cold recycled foamix was assigned in this project. The ORIS data science technique is used to locate the best materials from locally available sources, hence, avoiding haulage of materials by heavy goods vehicles traveling long distances. The transport distance of these materials is automatically calculated by ORIS factoring in any restrictions for heavy goods vehicles along the route, for example, maximum gross weight, vehicle height, and width. The scheme climate data was obtained from the POWER Project's daily 2.2.20 version and climatology v2.2.18 version on 14 March 2022.

The case study considers an AADT of 7,000 vehicles, with 90% being vehicles below 7.5 tons or 6.6 m in length, and the remaining 10% being Heavy Goods Vehicles (HGVs), buses or coaches. A growth rate of 5% per year over 40 years was applied to the AADT.

The construction and maintenance activities for the pavement rehabilitation, estimated fuel consumptions of paving equipment and plants were used for laying each asphalt layer. It was assumed that the material properties such as the International Roughness Index (IRI) and albedo values of the pavement surface were restored to their initial values when replacement activities were performed in the 14th and 26th year. These effects on the pavement structure and resultant changes are combined with the AADT, which are all evolving over time, are then reflected into the use phase modules.

For the EoL module and for the sake of simplicity, it is assumed that 100% of the replacement (milled) asphalt is transported to a waste recycling center with a default transport distance of 50 km and is subsequently blended with fresh asphalt and reused in new pavement construction. The processing of reclaimed asphalt processing (sorting and grading) is modeled using an average recycling process profile for this case study. Table 1 summarizes the project alignment information.

Table 1.   Project alignment information.

Type of Data	Value
Country	United Kingdom
Road type	Principal Highway
Project type	Resurfacing
Road length	3,875 m
Road section width	16 m
Average daily traffic	7,000
Percentage of commercial vehicles, busses and coaches	10%
Traffic growth rate	5% / year
Truck capacity	27 tons
Transport fuel type	Diesel
Truck utilization rate	100%
Truck return utilization rate	0%

Based on the general road project alignment information, the digital platform was able to generate multiple pavement design solutions based on different asphalt materials compliant with the UK's pavement standards and specifications. Table 2 summarizes two pavement designs, which have been selected respectively based on the Design Manual for

Table 2.   Selected pavement designs.

	Reference Design	Thickness (mm)	Alternative Design	Thickness (mm)
Surface course	Thin Surface course System	40	Thin Surface Course System	40
Binder course	Hot mix asphalt concrete	60	Warm mix asphalt concrete	60
Base course 1	Hot mix asphalt concrete	110	Cold recycled foamix	220
Base course 2	Hot mix asphalt concrete	110		
Total thickness		320		320

Roads and Bridges (DMRB) [14] and Transport Research Laboratory (TRL) Report 611 [15]. The first design, a fully flexible pavement, was chosen as the reference design. Compared to that, the second design adopts a different asphalt technology, which is a Warm Mix Asphalt (WMA); WMAs employ chemical technology that allows production temperatures to be reduced by around 40°C; therefore, WMAs require less amount of thermal energy for their production at the mixing site. A further low-carbon initiative was considered for the base layer — this asphalt incorporated recycled asphalt, resulting in a complete base course having the same thickness as the reference design. The production site using these technologies was identified within the project search radius of 50 km.

The cumulative global warming impacts expressed in tons of CO_2 equivalent and the results of the life cycle cost for the two pavement options are shown in Fig. 6. The flexible foamix design appears as the best solution because of its optimum balance between life cycle carbon performance and cost performance. Compared to the reference design, the alternative design reduces the total GWP impact by 24% and the total construction cost by 14%, which is equivalent to £790,000 in savings. GWP impacts are dominated by the material production (phases A1–A3), the maintenance operation (phases B2–B4), and the PVI effect (phase B6) for both designs. The use of foamix and WMA technology in the alternative design resulted in 56% reduction of carbon during the production (phases A1–A3) compared to the reference design

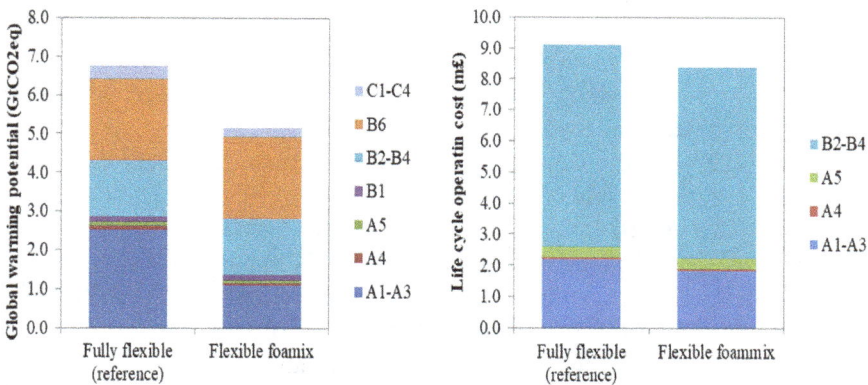

Fig. 6.   GWP impact assessment (left) and life cycle cost assessment (right) for the two pavement designs.

using hot mix asphalt. Employing low-carbon incentives makes the use phase and maintenance phase even more viable in terms of carbon reduction. Also, the choice of the pavements service life has a significant effect on the overall contribution of carbon emissions produced during the use and maintenance phases. A longer service life requires fewer maintenance interventions over the analysis period of the pavement design, leading to a clearer contrast in long-term effect factors such as traffic growth and IRI.

4. Chapter Summary

Early engagement with low-carbon design strategies leads to the development of a low-carbon pavement solution in linear infrastructure. The key to the success of a significant reduction in GWP impacts lies in the combination of a materials-efficient pavement structure, local and optimum material sourcing, low-carbon materials technology, optimum maintenance operations, and timely interventions. In addition to the reduction in GWP impact, cost and material savings, while ensuring safer and smoother pavement solutions extensively, meet the sustainable strategy in the infrastructure sector. This paper first presented the overall ORIS's digital platform linking real material markets with road construction projects. The required carbon and cost data and a catalog of multiple pavement solutions, depending on structural and local requirements, are all embedded into the background databases, enabling an automatized data-driven approach. This disposition of data allows the easing of the pavement design process at a very early stage. Second, a case study involving two pavement solutions (reference and alternative ones) was demonstrated to show how the ORIS assessment helps early-stage decision-making for sustainable roads.

As shown with the case study, considering materials from the early stage of road designs makes a difference in terms of carbon and cost performance. The use of low-carbon materials, for example, cold recycled foamix, was locally identified thanks to ORIS's material sourcing technique, which led to a tangible solution of GWP reductions by 24% compared to the standard pavement solution. According to our internal research, the choice of designs in road construction is a major factor on costs (60% of the total construction costs) and carbon footprint (85% of carbon emissions). Each road is unique and is dependent on its traffic volume and type, climatic conditions, and geographic location. For increased sustainability, as shown in this example in the UK, each road should be designed taking into account the following elements:

(1) Local sourcing: to optimize material supplies and avoid long transport/haulage distances,
(2) Circular Economy: using recycled materials to preserve natural resources being part of the local equation from the early stages of the design,
(3) Carbon footprint: given the climate urgency, carbon emissions reduction with adapted design is a must for all sectors, and
(4) Durability: the longevity of a road is key to avoid extra maintenance work and to ensure long, safe and reliable service to its linked communities.

Bringing materials engineering upfront in road construction and maintenance is a unique capability for the market and will become necessary in the future. A digital platform such as ORIS will make a major difference in improving the carbon performance of linear infrastructure.

References

1. K. Negishi, R. De Montaignac and N. Miravalls, Materials role in pavement design and its impacts in LCA of road construction and use phase, *IOP Conf. Ser.: Earth Environ. Sci.* **1122**, 012037 (2022).
2. ORIS, Statistics on CO_2, cost and natural resource consumptions through a whole life cycle of 12 different pavement solutions realized with ORIS material intelligence platform (2022).
3. H. Azarijafari, A. Yahia and M. Ben Amor, Life cycle assessment of pavements: Reviewing research challenges and opportunities, *J. Clean. Prod.* **112**, 2187–2197 (2016).
4. P. Babashamsi, N.I. Md Yusoff, H. Ceylan and N.G. Md Nor, Life cycle assessment for pavement sustainable development: Critical review, *Appl. Mech. Mater.* **802**, 333–338 (2015).
5. A. Balaguer, G.I. Carvajal, J. Albertí and P. Fullana-i-Palmer, Life cycle assessment of road construction alternative materials: A literature review, *Resour. Conserv. Recycl.* **132**, 37–48 (2018).
6. N.J. Santero, E. Masanet and A. Horvath, Life-cycle assessment of pavements. Part I: Critical review, *Resour. Conserv. Recycl.* **55**, 801–809 (2011).

7. CEN, ISO 21930:2017 — Sustainability in buildings and civil engineering works — Core rules for environmental product declarations of construction products and services (2017).

8. CEN, EN 15804:2012+A2:2019 — Sustainability of construction works — Environmental product declarations — Core rules for the product category of construction product (2019).

9. ISO, ISO 14040:2006 — Environmental management — Life cycle assessment — Principles and framework (2006).

10. ISO, ISO 14040:2006 — Environmental management — Life cycle assessment — Requirements and guidelines (2006).

11. G.H. Brundtland and M. Khalid, *Our Common Future: Report of the 1987*, World Commission on Environment and Development (1987).

12. R. Heijungs, J.B. Guinée, G. Huppes, R.M. Lankreijer, H.A. Udo de Haes, W. Sleeswijk, *et al.*, Environmental life cycle assessment of products: Guide and backgrounds (1992).

13. L. Rogers, iRAP star rating and investment plan implementation support guide, International Road Assessment Program (2017).

14. Highways Agency, Design Manual for Roads and Bridges (HD26/06) (2006).

15. D. Merrill, M. Nunn and I. Carswell, A guide to the use and specification of cold recycled materials for the maintenance of road pavements (TRL 611) (2004).

Chapter 11

Biodiesel Production from Palm Oil Feedstock: Techno-Economic Assessment and Life Cycle Assessment Applications

Iskandar HALIM[a] and Hsien Hui KHOO[b]

*Institute of Sustainability for Chemicals,
Energy and Environment (ISCE²),
Agency for Science, Technology and Research (A*STAR),
Singapore 627833*

[a]*Iskandar_Halim@isce2.a-star.edu.sg*
[b]*khoo_hsien_hui@isce2.a-star.edu.sg*

Massive consumption of fossil fuels has led to rapid depletion of petroleum-based fuel supplies, and at the same time, increased global warming effects. In a move towards the transition of low-carbon economy strategies, one recommendation is to substitute the use of fossil fuels with biofuels in order to meet growing energy demands. Presently, production and utilization of biofuels have become a key part of our decarbonization efforts. One such example is biodiesel, which is fast gaining prominence as an alternative transportation fuel to replace the petroleum-based diesel. This chapter explores the cost and carbon emissions of palm-oil-based biodiesel production using the Techno-Economic Assessment (TEA) and Life Cycle Assessment (LCA). The Aspen Plus process simulation tool was used to evaluate the sustainability of an alkali-based transesterification of virgin palm oil to biodiesel. Additional carbon emissions incurred through land clearance needed for palm oil cultivation, as well as various social implications, were also highlighted.

1. Introduction

The scientific consensus on climate change is that human activities over the past 100 years have tremendously increased greenhouse gas (GHG) concentrations in the atmosphere, leading to a rise in the Earth's average surface temperature. It is projected that between 2030 and 2052, the average temperature will reach 1.5°C above the pre-industrial levels [1]. In order to curb global temperature rise, the United Nations (UN) has repeatedly urged governments around the world, especially those of large emitting countries, to sharply reduce their GHG emission levels. Today, the Glasgow Climate Pact has become the most comprehensive agreement that has been signed by 197 countries with a view of tackling climate change challenges [2].

One of the largest share of anthropogenic GHG emissions comes from the burning of fossil fuels in the transport sector. This accounts for about a fifth of global GHG emissions, with road transport contributing up to 75% of the share [3]. One way to mitigate climate change effects is to reduce our reliance on fossil fuels by switching to renewable sources such as biofuels. In this regard, the emergence of research trends focusing on bio-based fuels to replace fossil-based gasoline and diesel in the transport sector is currently on the rise.

Sustainably produced biofuels or biodiesels can play a critical role in our efforts to decarbonize the transportation sector [4, 5]. Studies associated with different methods of producing biodiesel have been reported in the literature. For example, Brahma and colleagues [6] reviewed the synthesis of biodiesel from various oil sources and applied machine-learning algorithms to optimize the process parameters for producing biodiesel efficiently (see Fig. 1).

Apart from waste cooking oil sources, biodiesel can be produced from a wide variety of virgin oil feedstock. While biodiesel derived from waste oils has lower production costs and lower environmental impacts than that of virgin oils, it poses major challenges in terms of feedstock quality, reliability and security [7–9]. This chapter will focus on the production of biodiesel from virgin palm oil. Compared to other vegetable oil sources

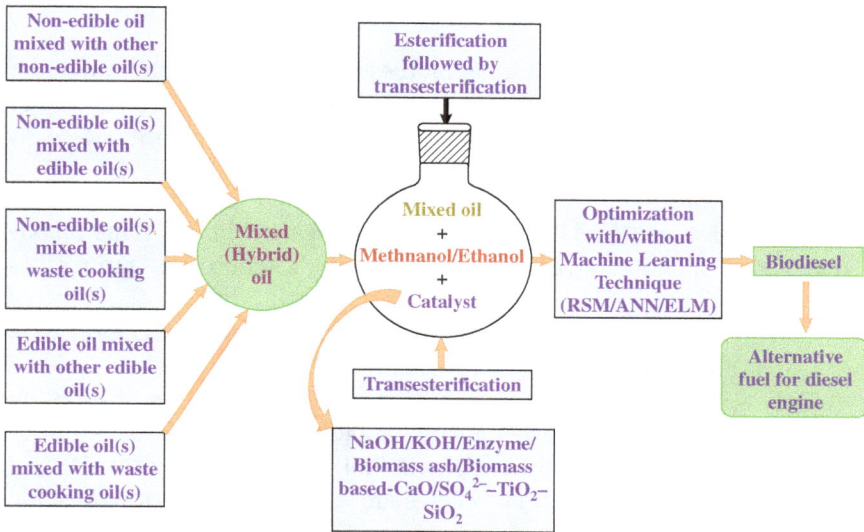

Fig. 1.   Sustainable approach towards industrial biofuel production. (Reproduced from Ref. [6] under a Creative Commons license.)

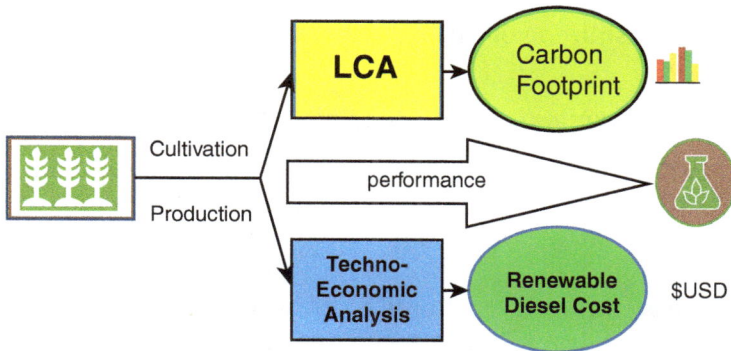

Fig. 2.   The Techno-Economic Analysis and Life Cycle Assessment approaches to evaluate the sustainability of palm-oil-based diesel production. (Adapted from Ref. [9].)

such as rapeseed and soybean, palm oil-based biodiesel is known to have the highest oil yield per land size [10]. We will particularly discuss the applications of the Techno-Economic Analysis (TEA) and Life Cycle Assessment (LCA) to evaluate the sustainability of this type of biodiesel production (see Fig. 2).

2. Biodiesel Production from Palm Oil

One common method of producing biodiesel from palm oil is through the alkali-based transesterification process. Figure 3 shows the flowsheet of the biodiesel production, which involves the following operation steps:

- Transesterification: Stream of fresh methanol and sodium hydroxide is mixed with a recycled stream for reaction with palm oil. This alkali-catalyzed reaction converts the oil into a biodiesel product and a glycerol byproduct.
- Methanol recovery: From the reactor unit, the product is sent to a vacuum distillation process to recover the excess methanol.
- Water washing: This step involves washing the bottom distillation stream with water in an extraction column to separate the biodiesel product from the mixture of glycerol, methanol, and sodium hydroxide.
- Alkali removal: This step involves neutralizing the sodium hydroxide component by mixing it with phosphoric acid to produce sodium phosphate salt and water.
- Purification: In the final step, the biodiesel mixture from the washing process is distilled further under a vacuum condition to obtain a high purity product; meanwhile, the glycerol mixture from the filtration unit is vacuum distilled to obtained high-purity glycerol.

2.1. *Techno-Economic Analysis of biodiesel production*

Aspen Plus process simulation software (version 12) was used to perform the TEA of this biodiesel process. The six main unit operations used in the process are as follows:

- Continuous stirred tank reactor (CSTR), which operates at 4 bar, 60°C and 1 hour residence time. A total of 96 reactions involving palm oil components and a sodium hydroxide catalyst are modeled using the power law kinetic expression. Table 1 highlights ten of these important reactions that take place inside the reactor.
- Methanol recovery column is a vacuum distillation column with seven stages and operated at 0.2 bar with 29°C condenser and 60°C reboiler temperature. About 94% of the methanol can be recovered and sent back to the reactor.

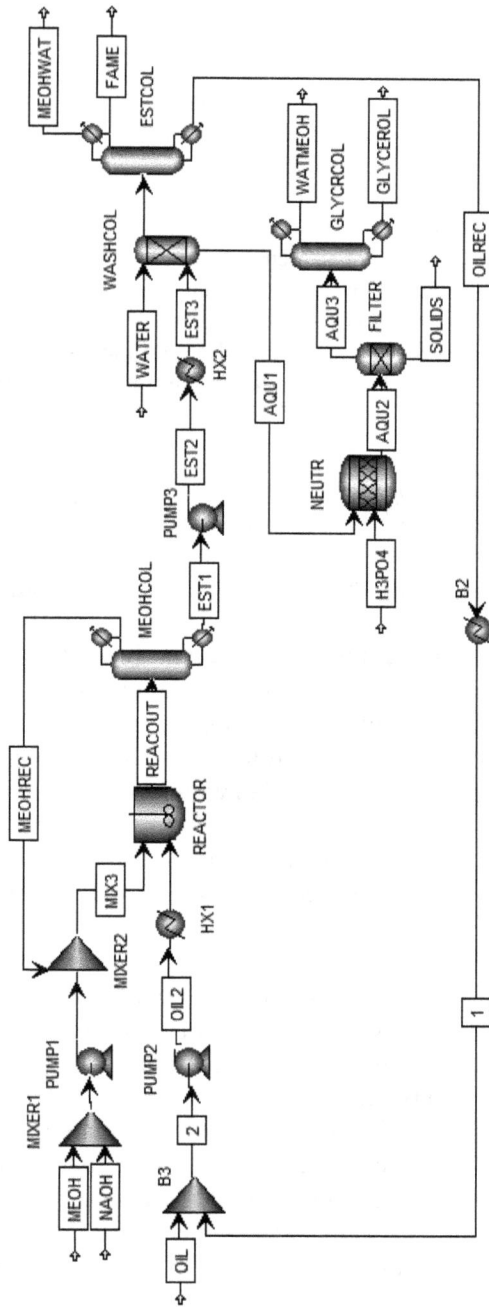

Fig. 3.   Flowsheet of the alkali-catalyzed biodiesel production process.

Table 1.   Reaction kinetics involving important palm oil components.

Reaction Stoichiometry
Triolein + Methanol → Methyl oleate + 1,2-diolein
1,2-diolein + Methanol → Methyl oleate + Monoolein
Trimyristin + Methanol → Methyl myristate + 1,3-dimyristin
1,3-dimyristin + Methanol → Methyl myristate + 1-monomyristin
1-monopalmitin + Methanol → Glycerol + Methyl palmitate
Triacylglycerol-PPS + Methanol → Methyl stearate + 1,3-dipalmitin
Triacylglycerol-PPS + Methanol → Methyl palmitate + SN-1-palmito-3-stearin
Triacylglycerol-PLIO + Methanol → Methyl linoleate + SN-1-palmito-3-olein
SN-1-oleo-3-stearin + Methanol → Methyl oleate + 1-monostearin
Triacylglycerol-MLIP + Methanol → Methyl palmitate + SN-1-myristo-3-linolein

- Biodiesel washing column is an extraction column that operates at atmospheric pressure and ambient temperature, using water as solvent. The bottom part of the column consists mainly of glycerol, methanol, and acid catalyst, while the top column contains biodiesel, methanol, and water.
- Biodiesel purification column is a vacuum distillation unit operated at 0.15 bar pressure with 148°C condenser temperature.
- Neutralizing column that uses phosphoric acid solution to neutralize the sodium hydroxide solution from the washing column. The salt that forms during this operation is settled while the liquid mixture is sent for purification.
- Glycerol purification column is a vacuum distillation column operated at 0.4 bar with 263°C reboiler temperature to obtain high-purity glycerol (100%).

Some of the reactions involved in the production process are listed in Table 1. Table 2 shows the input-output material and energy flows. The biodiesel production rate is 1051.82 kg per hour, which means almost all of the palm oil gets converted to biodiesel.

Table 3 shows the estimated equipment installation cost for this biodiesel plant, which has been obtained using the Aspen Process Economic Analyzer. To calculate the required capital (investment) cost of this plant, the cost parameters proposed by Kiani and colleagues [11] were used. The plant investment cost was calculated to be US$4,639,003, as shown in Table 4.

Table 2.   Input-output material and energy flows.

Stream	Status	Flow Rate	Unit
MeOH	Feed	127.12	kg/h
NaOH	Feed	50	kg/h
Palm oil	Feed	1050	kg/h
Water	Feed	50	kg/h
H_3PO_4	Feed	40.84	kg/h
Waste MeOH	Waste	133.43	kg/h
Waste water	Waste	4.85	kg/h
Solids	Waste	13.68	kg/h
Electricity	Utility	71.61	kW
Cooling water	Utility	44.55	t/h
Refrigerant	Utility	2.66	t/h
Steam	Utility	0.24	t/h
Glycerol	Byproduct	114.17	kg/h
Biodiesel	Product	1051.82	kg/h

Table 3.   Equipment installation cost for biodiesel plant.

Equipment List	Equipment Type	Installed Equipment Cost (U$)
Reactor	CSTR	251,400
Pump1	Centrifugal pump	29,700
Pump2	Centrifugal pump	30,900
Pump3	Centrifugal pump	30,700
HX1	Heat exchanger	51,400
HX2	Heat exchanger	51,400
B2	Heat exchanger	75,500
Neutr	Mixing vessel	178,800
Meohcol	Distillation column	493,700
Glycrcol	Distillation column	554,000
Estcol	Distillation column	733,400
Filter	Filtration column	112,500
Washcol	Extraction column	208,000
Total equipment installation costs		**2,801,400**

Table 4.   Investment cost estimation.

Item	Description	Amount (U$)
Total direct cost (TDC)	Equipment installation cost from Table 3	2,801,400
Total indirect cost (TIC)	19.4% of TDC	543,472
Engineering and construction	10% of (TDC+TIC)	334,487
Contractor fees	3% of (TDC+TIC)	100,346
Contingencies	25% of (TDC+TIC)	836,218
Total plant cost (TPC)	Sum of above	4,615,923
Spare parts cost	0.5% of TPC	23,080
Total investment cost (TIC)	**TPC + spare parts**	**4,639,003**

2.2. *Life Cycle Assessment of biodiesel production*

The Life Cycle Assessment (LCA) has been used to assess the environmental implications of biodiesel production [12–14] and will be applied to evaluate the net CO_2 emissions (carbon footprint) of the palm-oil-based biodiesel process case study. The boundaries of LCA stages start with plantation and palm oil production before reaching the transesterification process (as displayed in Fig. 3).

Figure 4 shows the "cradle-to-gate" LCA model of the biodiesel production, which has been developed using Gabi life cycle process engineering software. The amount of GHG emissions associated with the flow of material and energy resources is shown in Fig. 5.

The LCA of this biodiesel process shows a carbon footprint of 599.8 kg CO_2-eq for the production of 1051.82 kg of biodiesel. The carbon footprint value is normalized according to 1 kg biodiesel, and compared with other types of biodiesel production obtained from several reports [5, 14–18]. Figure 6 compares these values on the basis of an estimated calorific value of 42.5 MJ/kg biodiesel. It shows a carbon footprint of 0.57 kg CO_2-eq per kg of biodiesel, making it comparable to waste cooking oil (WCO)-based diesel [17]. However, it should be highlighted that the LCA boundary of this value does not take into account any emissions related to land use and other emissions from the transportation stage.

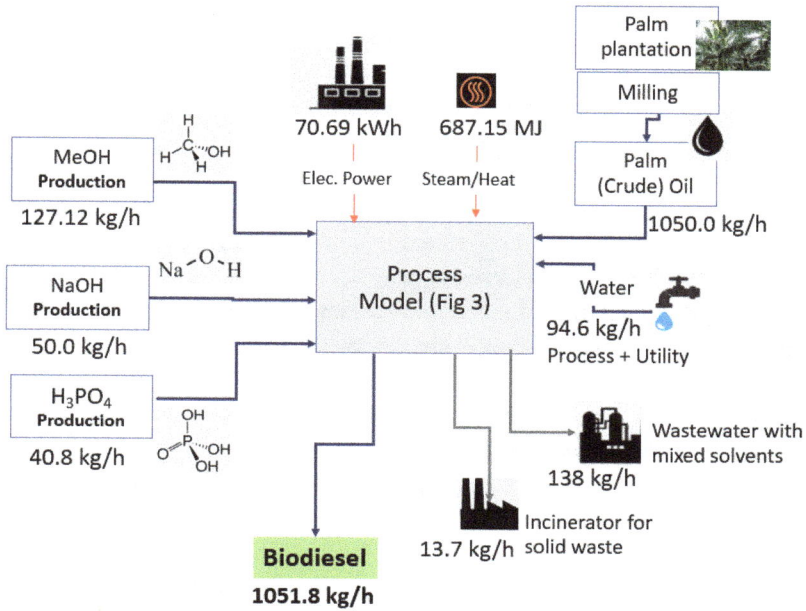

Fig. 4. "Cradle-to-gate" LCA model — from feedstock to biodiesel.

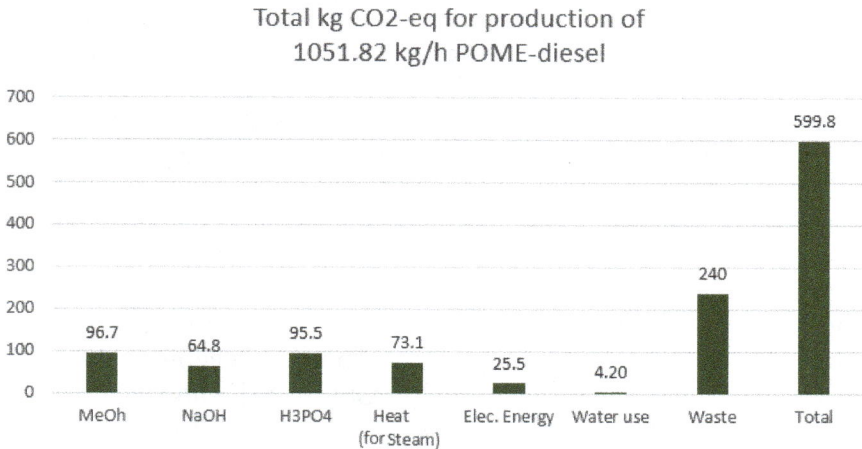

Fig. 5. Carbon footprint of biodiesel production.

Normalized results: comparison of kg CO_2 per 1 kg biodiesel

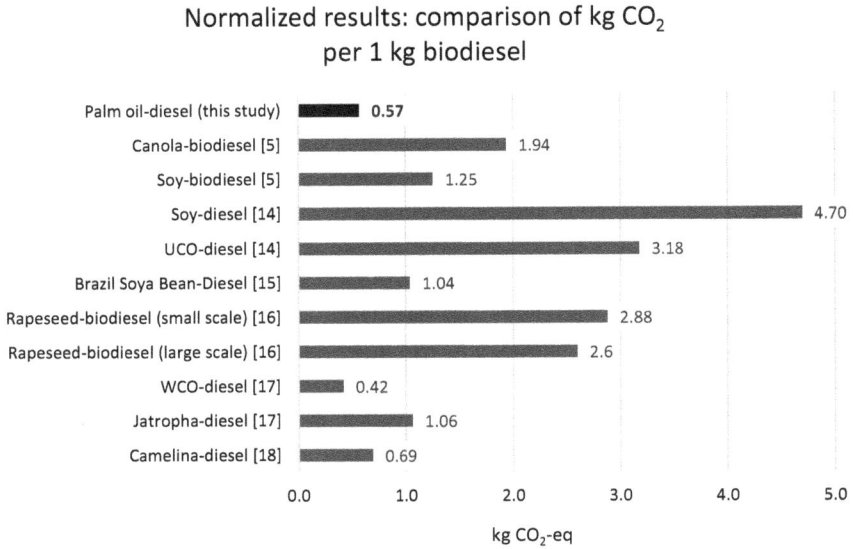

Fig. 6.   Normalized carbon footprint results: kg CO_2-eq per kg biodiesel.

3. Further Discussions

The tremendous growth of fossil-based energy consumption has led to a rapid depletion of fossil fuels, increased global warming effects, and other types of pollution. Decarbonization efforts to meet growing energy demands through the utilization of biomass-based fuels have led to expanding research to produce biodiesel in a sustainable manner to replace petroleum-based diesel. For this, numerous studies have been performed to evaluate the feasibility of biodiesel production by considering both the environmental and economic aspects [8–9].

Palm oil has played a significant role in the global supply of biomass sources, surpassing soybean, rapeseed, and sunflower oil feedstock [19]. However, besides providing economic benefits to the local population, palm oil cultivation has been associated with various environmental problems, including deforestation, freshwater depletion, soil erosion, and loss of biodiversity [20]. Therefore, it is critical for any LCA study to also consider GHG emissions and other environmental aspects related to a palm oil plantation and milling alongside the biodiesel production.

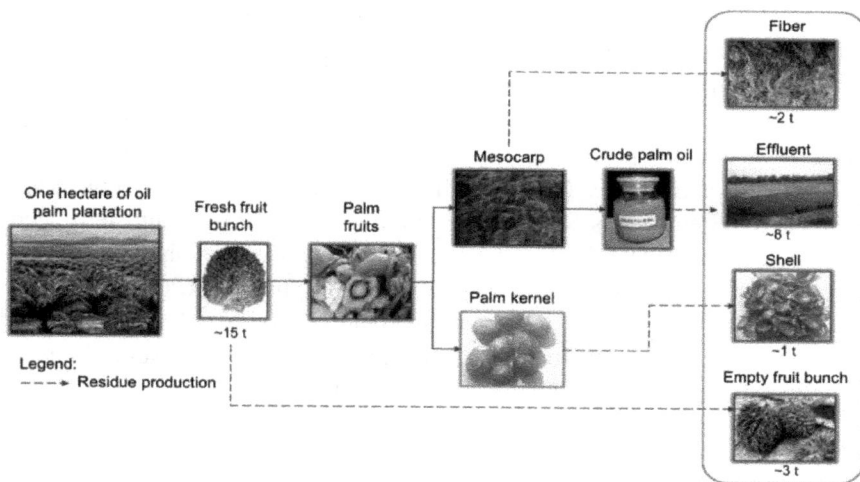

Fig. 7.   Simplified illustration of palm biomass residue generated from one hectare of palm oil plantation. Note: Conversion rate per one-hectare of plantation was obtained from the fieldwork data of one palm oil mill in Sumatra in 2016. (Reproduced from Ref. [21] via free access article.)

Other sustainability concerns facing palm oil producers involve issues pertinent to biomass harvesting, collection, pretreatment, storage, transport, conversion to products, and generation of waste byproducts. Harahap and colleagues [21] evaluated different ways of utilizing palm oil wastes by converting them into useful resources (see Fig. 7). Their work can be applied to enhance the sustainability of palm oil production and supply chain via optimal biomass utilization, GHGs reduction, and selection of the most economical conversion technologies.

3.1. *Land use change issues*

Biofuels are currently being promoted as a low-carbon alternative to fossil fuels to help reduce GHG emissions and other related climate change impacts in the transport sector. However, there are several concerns associated with their wider deployment, which could lead to unintended environmental consequences. One such concern is related to the practice of land clearing through deforestation, which is still happening till now. Besides the additional release of GHGs caused by land clearance, deforestation remains a hotly contended issue associated with palm oil cultivation, which is challenging to tackle [22].

One of the well-known research areas for addressing land use change (LUC) issues involves the use of LCA models to assess the carbon storage capacity of forest areas and to estimate the amount of GHGs released from LUC [23]. For example, Xu and colleagues [24] took into account LUC-induced GHG emissions for the production of bio-oil from palm fatty acid distillate (PFAD) and refined palm oil (RPO). Their results showed that for high Induced Land Use Change (ILUC) emissions, neither RPO- nor PFAD-derived bio-oil would result in significant GHGs emission reductions relative to conventional petroleum-based diesel. Their results, which were obtained through well-to-wheels (WTW) modeling, are highlighted in Fig. 8.

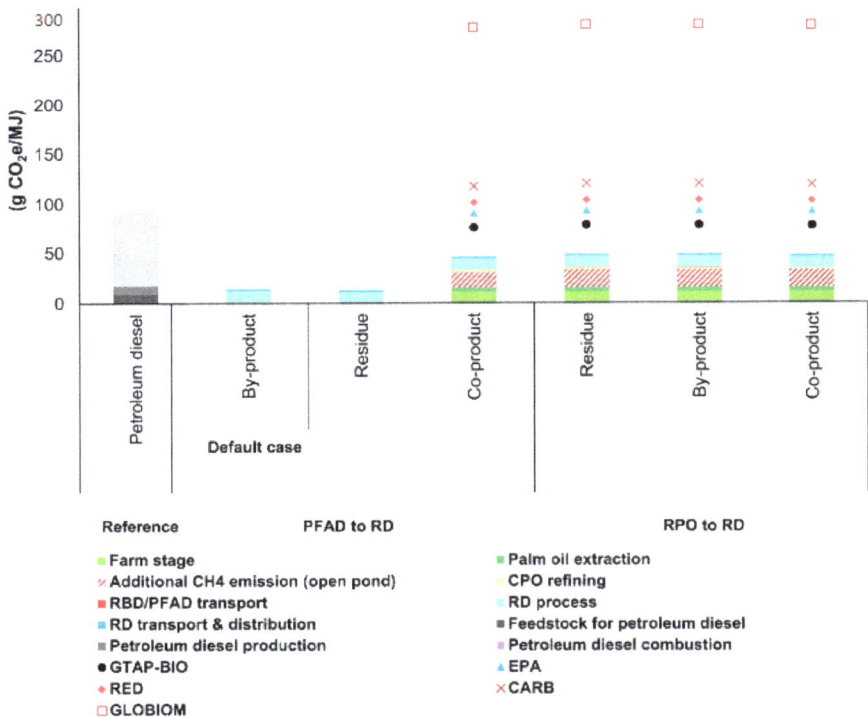

Fig. 8.   Comparison of well-to-wheels (WTW) of GHG emissions of petroleum diesel, bio-diesel produced from palm fatty acid distillate (PFAD), and refined palm oil (RPO). Dots represent land use change (ILUC) emissions, and stacked bars represent WTW emissions without ILUC emissions. (Reproduced from Ref. [24] via free access article.)

To address the concerns of sustainable biodiesel production, several ways can be taken. One proven strategy is through certification. For this, non-profit global organizations such as RSPO (Roundtable on Sustainable Palm Oil) have developed a set of criteria (environmental and social) for palm oil producing companies to comply with in order to be certified as sustainable palm oil producers [25]. Meanwhile, implementing sustainable biodiesel production throughout the entire value chain based on environmentally acceptable practices needs to be pursued as well, in order to achieve low-carbon GHG emissions for biodiesel production.

4. Concluding Remarks

Rapid depletion of fossil fuels as energy sources and rising GHG emissions from their burning have led to the era of biodiesel utilization as an alternative fuel for the transport sector. Among the available biomass sources, palm oil contains a high amount of oleic acids, which makes it suitable to be used as raw material feedstock. Moreover, compared to other agricultural crops, palm oil exhibits the most versatile and efficiency in terms of oil yield per land size. In order to evaluate the sustainability of palm oil for biodiesel production, this chapter describes its TEA and LCA. Using the production of 1051.82 kg biodiesel per hour as the basis, the total carbon footprint is calculated to be 599.8 kg CO_2-eq kg, which translates to 0.57 kg CO_2-eq per kg biodiesel. The total equipment installation cost is US\$2,801,400, and the plant investment cost, excluding the land purchase cost, is estimated to be US\$4,639,003. GHG emissions of palm oil plantation associated with LUC was also discussed.

References

1. Intergovernmental Panel on Climate Change, Special report: Global warming of 1.5°C (2018). https://www.ipcc.ch/sr15/chapter/spm
2. World Economic Forum, Glasgow climate pact: Where do all the words and numbers we heard at COP26 leaves us? (2021). https://www.weforum.org/agenda/2021/11/glasgow-climate-pact-cop26
3. J.-P. Rodrique. Global greenhouse gas emissions by the transportation sector (2014). https://transportgeography.org/contents/chapter4/transportation-and-environment/greenhouse-gas-emissions-transportation
4. S. Morais, T.M. Mata, A.A. Martins, G.A. Pinto and C.A.V. Costa, Simulation and life cycle assessment of process design alternatives for

biodiesel production from waste vegetable oils, *J. Clean. Prod.* **18**(13), 1251–1259 (2010).

5. R. Chen, Z. Qin, J. Han, M. Wang, F. Taheripour, W. Tyner, *et al.*, Life cycle energy and greenhouse gas emission effects of biodiesel in the United States with induced land use change impacts, *Bioresour. Technol.* **251**, 249–258 (2018).

6. S. Brahma, B. Nath, B. Basumatary, B. Das, P. Saikia, K. Patir, *et al.*, Biodiesel production from mixed oils: A sustainable approach towards industrial biofuel production, *Chem. Eng. J. Adv.* **10**, 100284 (2022).

7. J. Poudel, S. Karki, N. Sanjel, M. Shah and S.C. Oh, Comparison of biodiesel obtained from virgin cooking oil and waste cooking oil using supercritical and catalytic transesterification, *Energies* **10**(4), 546 (2017).

8. D. Greer, Recycling local waste oil and grease into biodiesel (2010). https://www.biocycle.net/recycling-local-waste-oil-and-grease-into-biodiesel

9. R.E. Davis, D.B. Fishman, E.D. Frank, M.C. Johnson, S.B. Jones, C.M. Kinchin, *et al.*, Integrated evaluation of cost, emissions, and resource potential for algal biofuels at the national scale, *Environ. Sci. Technol.* **48**(10), 6035–6042 (2014).

10. Gro Intelligence, Palm oil: Growth in Southeast Asia comes with a high price tag (2016). https://gro-intelligence.com/insights/palm-oil-production-and-demand

11. A. Kiani, K, Jiang and P. Feron, Techno-economic assessment for CO_2 capture from air using a conventional liquid-based absorption process, *Front. Energy Res.* **8**(92) (2020).

12. A.N. Evanthia and C.J. Koroneos, Comparative LCA of the use of biodiesel, diesel and gasoline for transportation, *J. Clean. Prod.* **20**, 14–19 (2012).

13. H. Xu, L. W. Ou, Y. Li, T.R. Hawkins and M. Wang, Life cycle greenhouse gas emissions of biodiesel and renewable diesel production in the United States, *Environ. Sci. Technol.* **56**, 7512–7521 (2022).

14. X. Ou, X. Zhang, S. Chang and Q. Guo, Energy consumption and GHG emissions of six biofuel pathways by LCA in (the) People's Republic of China, *App. Energy.* **86**, S197–S208 (2009).

15. C.E.P Cerri, X. You, M.R. Cherubin, C.S. Moreira, G.S. Raucci, B.A. Castigioni, *et al.*, Assessing the greenhouse gas emissions of Brazilian soybean biodiesel production, *PLoS ONE* **12**(5), e0176948 (2017)

16. R. Gupta, R. McRoberts, Z. Yu, C. Smith, W. Sloan and S. You, Life cycle assessment of biodiesel production from rapeseed oil: Influence of process parameters and scale, *Bioresour. Technol.* **360**, 127532 (2022).

17. M.K. Pasha, L. Dai, D. Liu, M. Guo and W. Du, An overview to process design, simulation and sustainability evaluation of biodiesel production, *Biotechnol. Biofuels* **14**, 129 (2021).

18. X. Li and E. Mupondwa, Life cycle assessment of camelina oil derived biodiesel and jet fuel in the Canadian Prairies, *Sci. Total Environ.* **15**(481), 7–26 (2014).
19. M.N.A.M. Yusoff, N.W.M. Zulkifli, N.L. Sukiman, O.H. Chyuan, M.H. Hassan, M.H. Hasnul, *et al.*, Sustainability of palm biodiesel in transportation: A review on biofuel standard, policy and international collaboration between Malaysia and Colombia, *Bioenergy Res.* **14**, 43–60 (2021).
20. V. Vijay, S.L. Pimm, C.N. Jenkins and S.J. Smith, The impacts of oil palm on recent deforestation and biodiversity loss, *PLoS ONE* **11**(7), e01596688 (2016).
21. F. Harahap, S. Leduc, S. Mesfun, D. Khatiwada, F. Kraxner and S. Silveira, Opportunities to optimize the palm oil supply chain in Sumatra, Indonesia, *Energies.* **12**(3), 420 (2019).
22. C. Malins, R. Plevin and R. Edwards, How robust are reductions in modeled estimates from GTAP-BIO of the indirect land use change induced by conventional biofuels? *J. Clean. Prod.* **258**, 120716 (2020).
23. S.B. Hansen, S.I. Olsen and Z. Ujang, Carbon balance impacts of land use changes related to the life cycle of Malaysian palm oil-derived biodiesel. *Int. J. Life Cycle Assess.* **19**, 558–566 (2014).
24. H. Xu, U. Lee and M. Wang, Life-cycle energy use and greenhouse gas emissions of palm fatty acid distillate derived renewable diesel, *Renew. Sust. Energ. Rev.* **134**, 110144 (2020).
25. Roundtable on Sustainable Palm Oil. (2022). https://rspo.org/about

Chapter 12

Progress on Carbon Balancing in Life Cycle Assessment with a Focus on the Ecosystem Service Contribution

Benedetto RUGANI[a],* and Marco ALLOCCO[b],**

[a]*RDI Unit on Environmental Sustainability Assessment and Circularity (SUSTAIN), Environmental Research & Innovation (ERIN) Department, Luxembourg Institute of Science and Technology (LIST), Maison de l'Innovation, 5 avenue des Hauts-Fourneaux, L-4362 Esch-sur-Alzette, Luxembourg*
[b]*SEAcoop STP, Corso Palestro 9, I-10122 Turin, Italy*

*benedetto.rugani@gmail.com
**allocco@seacoop.com

It is undoubtful that anthropogenic greenhouse gas (GHG) emissions largely contribute to current impacts on climate change. However, biogenic carbon emissions may also play a meaningful role to the global balance of GHGs, which should not be neglected. This is an important factor for the development of carbon management plans to decrease the carbon footprint generated by production life cycles at various scales: product design, operations, and territorial dynamics. Current evidence, standards, and guidelines on GHGs inventories recommend to duly incorporate information of carbon removal and storage. To allow

organizations achieve net-zero GHGs balance, those accounting protocols also lead towards market-based voluntary schemes for carbon
credits acquisition. However, some uncertainty and a lack of transparency still occur in the way carbon credits are quantified and traced. The
hypothesis proposed and discussed in this chapter is about designing a
carbon management scheme based on established "mitigation hierarchy"
frameworks successful in the field of ecosystem services and biodiversity offsetting. The rationale is to couple carbon footprint results via
Life Cycle Assessment evaluations with on-field measurements of the
increase in carbon sequestration from restoring natural capital. The delta
of gain in such an ecosystem service can be used to compensate residual
emissions and reach carbon neutrality.

1. Introduction

A quick and progressive phase-out from carbon-intensive industries will
be necessary to achieve a climate-neutral economy by 2050, as announced
by many countries around the world, including the European Union [1].
For such an ambitious commitment to be successful, however, public and
private organizations need to rely on robust, transparent, and shared
instruments for carbon management that may cover both carbon footprint
mitigation strategies, and evidence-based measures to compensate for
unavoidable greenhouse gas (GHG) releases. Achieving net-zero GHG
emissions, or carbon neutrality, is already possible nowadays through the
adoption of carbon neutrality certificates that usually follow this procedure: (i) quantification of the GHG emissions generated by the activity
processes, (ii) implementation of technical/management strategies to
decrease the GHGs along those processes, and (iii) offseting the residual
emissions (those that cannot be avoid) by acquiring carbon credits from
voluntary carbon markets.

The term "offsetting" specifically refers to a mechanism compensating for the carbon footprint of a product through the prevention of the
following: release of, reduction in, or removal of, an equivalent amount of
GHG emissions in a process outside the boundary of the product system,
such as external investment in renewable energy technologies, energy
efficiency measures, or afforestation/reforestation activities [2]. Examples
of standards-based voluntary schemes built according to these principles

exist either at the national scale or economic sector in different countries, such as Montenegro [3] and Costarica [4]. The latter case is of particular interest since it is where a domestic Carbon Neutrality label has been recently developed and successfully applied in different economic sectors, such as in the coffee production sector [5]. Moreover, the literature shows that the ISO 14064 standard for GHG emission inventories can be used to certify carbon neutrality at a territorial scale, as successfully demonstrated with the Siena Carbon Neutral label [6]. Voluntary offset markets for carbon credits acquisition are mostly developed and established in the forestry sector. However, recent discussions are also promoting soil organic carbon sequestration as a basis for carbon credit certificates [7]. It is uncertain, however, whether private soil carbon certificates sold as voluntary emission offsets are a suitable instrument [8]. Drawbacks of adopting carbon credits or emission reduction strategies are pointed out in the literature, such as for the case of deforestation in Brazil made to compensate for GHGs in the steel industry with the use of charcoal [9].

Rooting on the ISO 14040 [10] and ISO 14044 [11] norms, which only broadly encompass the theme of GHGs, the more recent ISO 14067 [12] specifically focuses on product carbon footprint, accounting as one of the impact categories that can be included in Life Cycle Assessment (LCA) studies. ISO 14067 addresses only a single impact category, i.e., climate change, while carbon offsetting is considered outside the scope of norm. Similar approaches are established in other well-known LCA standards and guidelines, such as the Product Environmental Footprint. Overall, however, there is a global consensus about the accuracy, representativeness and pertinence of using the LCA and life-cycle-based approaches for quantifying the carbon footprint of products, organizations, and even territorial systems. In contrast, the way carbon sequestration shall be accounted for and incorporated in the GHG emissions balance to neutralize the carbon footprint is still debated.

This chapter aims to shed some light on how carbon management plans can be appropriately designed to incorporate the notion of ecosystem services (ES) through carbon sequestration and how this can complement the LCA model for the achievement of net-zero GHGs. As illustrated in the chapter, an impact mitigation hierarchy method rigorously based on a stepwise combination between the LCA and ES evaluation may offer a replicable and transparent methodological reference for organizations to avoid the pitfalls of controversial carbon credits purchasing activities.

2. Carbon Management in Life Cycle Assessment: State-of-the-Art

The Carbon Footprint (CF) is one of the most known and targeted indicators of environmental impact in LCA, accounting for the total GHG emissions generated directly and indirectly by a life cycle activity. Such an account can also consider the carbon removals from the atmosphere, which usually occur as a result of photosynthetic processes. Several methodological standards and guidelines exist to support companies and public authorities in conducting CF analyses at product, organizational and territorial levels (Table 1).

Table 1.   Most relevant international standards and methodological guidelines to conduct carbon footprint analyses for different scopes and business scales (product, organization, territory[∂]).

Scope[∂]	Full Reference	Further Information[#] *(text retrieved from the main web page of each item)*
Ps	**PAS 2050:2011** (Specification for the assessment of the life cycle greenhouse gas emissions of goods and services) — The British Standards Institution (BSI)	This Publicly Available Specification (PAS) specifies requirements for the assessment of the life cycle GHG emissions of goods and services (collectively referred to as "products") based on key life cycle assessment techniques and principles. This PAS is applicable to organizations assessing the GHG emissions of products across their life cycle, and to organizations assessing the "cradle-to-gate" GHG emissions of products.
Ps	**PAS 2050-2:2012** (Assessment of life cycle greenhouse gas emissions — Supplementary requirements for the application of PAS 2050:2011 to seafood and other aquatic food products) — The British Standards Institution (BSI)	PAS 2050-2 is appropriate for use by organizations operating in the seafood and other aquatic food product sectors that are intending to undertake a programme of GHG emissions reduction of their product life cycle, or those needing to provide information on the GHG emissions from their products or processes to downstream business partners or other stakeholders (e.g., environmental regulators).

Table 1.   (*Continued*)

Scope[∂]	Full Reference	Further Information[#] *(text retrieved from the main web page of each item)*
Ps, OCs	**PAS 2060:2014** (Specification for the demonstration of carbon neutrality) — The British Standards Institution (BSI)	PAS 2060 helps organizations demonstrate the carbon neutrality of a specific product, entity, or activity. It underpins reliable, credible claims that the subject of such a claim can indeed be considered carbon neutral. It specifies the requirements to be met by any entity seeking to demonstrate carbon neutrality through the quantification, reduction and offsetting of GHG emissions from a uniquely identified subject.
Ps, OCs	**PAS 2080** (Requirements of carbon management in infrastructure) — The British Standards Institution (BSI)	PAS 2080 discusses carbon management in infrastructure. Carbon management is about taking steps to measure and manage GHG emissions within an organization and extend the reduction of emissions across its supply chain. PAS 2080 thus specifies requirements for the management of whole life carbon in infrastructure — defined as the transport, energy, water, waste and communications sectors — both in the provision of new infrastructure assets and programmes of work and the refurbishment of existing infrastructure.
Ps	**PAS 2395:2014** (Specification for the assessment of greenhouse gas emissions from the whole life cycle of textile products) — The British Standards Institution (BSI)	PAS 2395 gives supplementary requirements for the assessment of GHG emissions from the whole life cycle of any products manufactured substantially from textiles and can be beneficially applied wherever textile products are manufactured and used. When used with the other specified methodologies, it can also deliver credible GHG emissions assessments that are optimized for textiles and textile-based products.

(*Continued*)

Table 1.   (*Continued*)

Scope[θ]	Full Reference	Further Information[#] (*text retrieved from the main web page of each item*)
Ps	**Product Life Cycle Accounting and Reporting Standard** — World Resources Institute (WRI) & World Business Council for Sustainable Development (WBCSD); the Greenhouse Gas Protocol Initiative	The Product Life Cycle Accounting and Reporting Standard can be used to understand the full life cycle emissions of a product and focus efforts on the greatest GHGs reduction opportunities. The Product Standard can be used to understand the full life cycle emissions of a product and focus efforts on the greatest GHGs reduction opportunities.
Ps	**Product Environmental Footprint (PEF)** — European Commission (EC), Directorate-General for Environment	One of the aims of the PEF is to establish a common methodological approach to enable Member States and the private sector to assess, display and benchmark the environmental performance of products, services and companies based on a comprehensive assessment of environmental impacts over the life cycle ("environmental footprint"). The PEF may thus encompass GHGs accounting and the characterization of the contribution of GHGs to the impact on climate change.
OCs	**Organization Environmental Footprint (OEF)** — European Commission (EC), Directorate-General for Environment	The OEF is a multi-criteria measure of the environmental performance of a goods/services-providing organization from a life cycle perspective. That includes companies, public administrative entities, and other bodies. Based on a LCA, the OEF allows modeling and quantifying the physical environmental impacts of the flows of material/energy and resulting emissions and waste streams associated with organizational activities from a supply-chain perspective. The OEF may thus encompass GHGs accounting and the characterization of the contribution of GHGs to the impact on climate change.

Table 1. (*Continued*)

Scope[θ]	Full Reference	Further Information[#] (*text retrieved from the main web page of each item*)
Ps	**ISO 22526-1:2020** (Plastics — Carbon and environmental footprint of biobased plastics — Part 1: General principles) — The International Organization for Standardization (ISO)	ISO 22526-1 specifies the general principles and the system boundaries for the carbon and environmental footprint of biobased plastic products. It is an introduction and a guidance document to the other parts of the ISO 22526 series.
Ps	**ISO 22526-2:2020** (Plastics — Carbon and environmental footprint of biobased plastics — Part 2: Material carbon footprint, amount (mass) of CO_2 removed from the air and incorporated into polymer molecule) — The International Organization for Standardization (ISO)	ISO 22526-2 defines the material carbon footprint as the amount (mass) of CO_2 removed from the air and incorporated into plastic and specifies a determination method to quantify it. This document is applicable to plastic products, plastic materials, and polymer resins that are partly or wholly based on biobased constituents.
Ps	**ISO 22526-3:2020** (Plastics — Carbon and environmental footprint of biobased plastics — Part 3: Process carbon footprint, requirements and guidelines for quantification) — The International Organization for Standardization (ISO)	ISO 22526-3 specifies requirements and guidelines for the quantification and reporting of the process carbon footprint of biobased plastics (see ISO 22526-1), being a partial carbon footprint of a bioplastic product, based on ISO 14067 and consistent with International Standards on LCAs (ISO 14040 and ISO 14044).
Ps	**ISO 14044:2006** (Environmental management — Life cycle assessment — Requirements and guidelines)	ISO 14044:2006 specifies requirements and provides guidelines for the LCA. Accordingly, the norm also encompasses the evaluation of GHGs, allowing for carbon accounting and the characterization of the contribution of GHGs to the impact on climate change.

(*Continued*)

Table 1.   (*Continued*)

Scope[a]	Full Reference	Further Information[#] *(text retrieved from the main web page of each item)*
OCs	**ISO 14064-1:2018** (Greenhouse gases — Part 1: Specification with guidance at the organisation level for quantification and reporting of greenhouse gas emissions and removals) — The International Organization for Standardization (ISO)	ISO 14064-1 includes requirements for determining GHG emission and removal boundaries, quantifying an organization's GHG emissions and removals, and identifying specific company actions or activities aimed at improving GHG management. It also includes requirements and guidance on inventory quality management, reporting, internal auditing, and the organization's responsibilities in verification activities.
PTCs	**ISO 14064-2:2019** (Greenhouse gases — Part 2: Specification with guidance at the project level for quantification, monitoring and reporting of greenhouse gas emission reductions or removal enhancements) — The International Organization for Standardization (ISO)	ISO 14064-2 details principles and requirements for determining baselines and the monitoring, quantifying, and reporting of project emissions. It focuses on GHG projects or project-based activities specifically designed to reduce GHG emissions and/or enhance GHG removals. It provides the basis for GHG projects to be verified and validated.
Ps, OCs, PTCs	**ISO 14064-3:2019** (Greenhouse gases — Part 3: Specification with guidance for the verification and validation of greenhouse gas statements) — The International Organization for Standardization (ISO)	ISO 14064-3 details requirements for verifying GHG statements related to GHG inventories, GHG projects, and carbon footprints of products. It describes the process for verification or validation, including verification or validation planning, assessment procedures, and the evaluation of organizational, project and product GHG statements.
Ps	**ISO 14067:2018** (Greenhouse gases — Carbon footprint of products — Requirements and guidelines for quantification) — The International Organization for Standardization (ISO)	ISO 14067 defines the principles, requirements, and guidelines for the quantification of the carbon footprint of products. The aim of ISO 14067 is to quantify GHG emissions associated with the life cycle stages of a product, beginning with resource extraction and raw material sourcing and extending through the production, use, and end-of-life phases of the product.

Table 1. (*Continued*)

Scope[a]	Full Reference	Further Information[#] (*text retrieved from the main web page of each item*)
OCs	**ISO/TS 14072:2014** (Environmental management — Life cycle assessment — Requirements and guidelines for organizational life cycle assessment)	ISO/TS 14072:2014 provides additional requirements and guidelines for an effective application of ISO 14040 and ISO 14044 to organizations. The document is not intended for the interpretation of ISO 14001 and specifically covers the goals of ISO 14040 and ISO 14044.
PTCs	**ISO 14080:2018** (Greenhouse gas management and related activities — Framework and principles for methodologies on climate actions) — The International Organization for Standardization (ISO)	ISO 14080 is guidance to countries and other interested parties on a consistent, comparable and transparent approach to selecting, proposing, using, revising, and maintaining methodologies on climate action. The standard is applicable to climate actions for addressing climate change, including adaptation to its impacts and GHG mitigation in support of sustainability. Such actions can be used by or for projects, organizations, jurisdictions, economic sectors, technologies and products, policies, programmes, and non-government activities.
OCs	**A Corporate Accounting and Reporting Standard, Revised Edition** — World Resources Institute (WRI) & World Business Council for Sustainable Development (WBCSD); the Greenhouse gas Protocol Initiative	The GHG Protocol Corporate Accounting and Reporting Standard provides requirements and guidance for companies and other organizations preparing a GHG emissions inventory. This standard is written primarily from the perspective of a business developing a GHG inventory. However, it applies equally to other types of organizations with operations that give rise to GHG emissions, e.g., NGOs, government agencies, and universities.

(*Continued*)

Table 1. (*Continued*)

Scope[a]	Full Reference	Further Information[#] *(text retrieved from the main web page of each item)*
PTCs	**The GHG Protocol for Project Accounting** — World Resources Institute (WRI) & World Business Council for Sustainable Development (WBCSD); the Greenhouse gas Protocol Initiative	The GHG Protocol for Project Accounting is the most comprehensive, policy-neutral accounting tool for quantifying the GHG benefits of climate change mitigation projects. The Project Protocol provides specific principles, concepts, and methods for quantifying and reporting GHG reductions — i.e., the decreases in GHG emissions, or increases in removals and/or storage — from climate change mitigation projects (GHG projects).
OCs, PTCs	**IWA (International Workshop Agreement) 42:2022** Net zero guidelines — Accelerating the transition to net zero	The IWA 42:2022 provides guiding principles and recommendations to enable a common, global approach to achieving net-zero GHG emissions through alignment of voluntary initiatives and the adoption of standards, policies, and national and international regulation. When used in combination with applicable science-based pathways, the document provides guidance for organizations seeking to set robust climate strategies and thus how to effectively contribute to global efforts to limit warming to 1.5°C by achieving net zero no later than 2050.
OCs	**GRI 305: Emissions 2016** (GRI Sustainability Reporting Standards)	GRI 305: Emissions 2016 is part of the GRI Standards and specifically contains disclosures for organizations to report information about their emissions-related impacts and how they manage these impacts.

[a]Scope of the carbon footprint analysis: Product scale (**Ps**); Organization/Corporate scale (**OCs**); Project/Territorial/Country scale (**PTCs**).
[#]Web-sources (accessed on Jan 2022): BSI standards; GHG Protocol standards; ISO standards; EC guidelines; GRI standards.

The most established and applied ones at international scale are the reporting standards from the GHG Protocol, some specific series from the International Organization for Standardization (ISO), and the standards developed by the British Standards Institute (BSI). Despite differing slightly in terms of inventory requirements, scope of the analysis, and coverage of emission and removal flows, they are all built on life cycle thinking principles, and they recommend incorporating the latest global warming potentials calculated by the Intergovernmental Panel on Climate Change (IPCC).

A short state-of-the-art about the application of those standards as well as other methods and solutions for carbon footprinting (such as classical LCA tools and product/organization environmental footprint methodologies) is illustrated in the next two sections.

2.1. *Carbon management in products and organizational LCAs*

GHGs accounting has been historically promoted by the IPCC as a tool to trace and monitor the release of anthropogenic carbon into the atmosphere [13, 14]. In parallel, the LCA has served, among other aims, as a method for detailed GHGs accounting of production systems, allowing for transparent, reproducible and comparative assertions of product and organizational activities.

Over the last 15 years, research on the LCA has substantially evolved with regard to the assessment of the dynamic behavior of GHGs according to the latest IPCC and science evidence. The method and support tools (e.g., inventory databases, impact characterization models, etc.) have also become more sophisticated and can nowadays incorporate or move towards including information about biogenic contributions to the overall carbon balance of the life cycle system. In so doing, the LCA is turning out to be a robust and reliable methodology to perform either simplified or detailed production and consumption CF assessments, both at product and organizational scales. As anticipated in Table 1, several standards and methodological frameworks have been developed to support companies towards the rigorous application of CF assessment requirements and principles, such as ISO 14067, PAS 2050, or the GHG Protocol. Most of the available guidelines focus on defining methodological procedures to account for carbon emissions and removals along the activities of a product life cycle.

While all of them provide detailed guidance on how to build the life cycle GHGs inventory without neglecting to include carbon uptake in the balance, they may slightly differ in several aspects of the methodology, such as in the way system boundaries are defined, emissions calculated, or the data quality checked. In-depth investigations and comparisons of the most used methodologies and standards can be found in previous studies, e.g., Ref. [15]. Recent developments in the LCA have pushed the standardization of the method to assess not only the impact of products (i.e., goods and services) but also the whole activities of an organization taken as a functional unit (e.g., through the development of ISO 14072 and the OEF). This was likely a consequence of the growing interest of government and private organizations towards implementing tools to manage their environmental responsibilities. A striking example is the well-known ISO 14001 environmental management system, whose worldwide adoption has seen an average annual increase of certificates by ~14% from 1999, reaching almost 500,000 certificates in 2021 [16]; see Fig. 1.

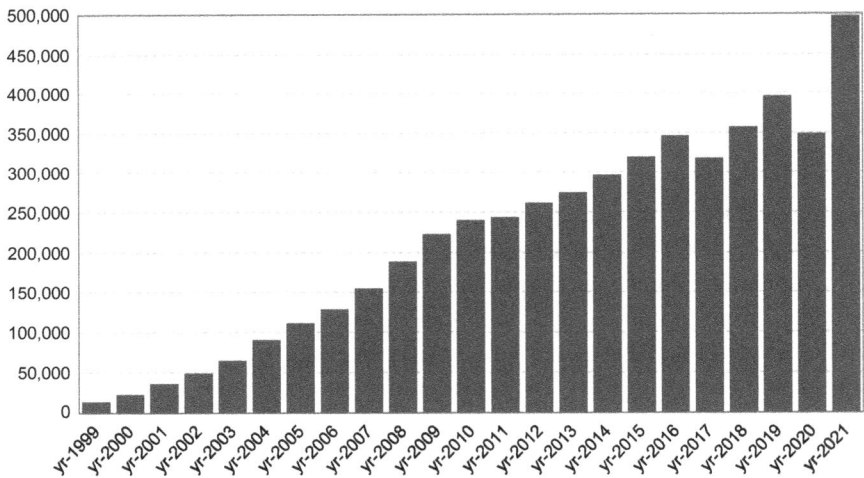

Fig. 1.   Evolution of ISO 14001:2015 certifications worldwide. Source of data: Ref. [16].

Likewise, the specific ISO norm on carbon footprint for organizations, traditionally ISO 14064-1, has also been very successful in recent years [17], together with other similar standards and guidelines (e.g., the GHG protocol or the GRI 305). In all these cases, however, the LCA is not explicitly contemplated as core methodology to account for GHGs. Other approaches, notably those relying on the use of then IPCC emission factors, so-called "global warming potentials", or GWPs [13], are usually recommended instead of the LCA, even if their accuracy and representativeness are questioned by some [18, 19].

The difference between the use of pre-calculated IPCC emission factors and the application of LCA principles and tools to determine GHGs is merely associated with the accounting perspective and consequent attribution of environmental responsibility for the release of GHGs: a "producer" perspective usually applies in the former, while the use of an LCA approach typically implies the adoption of a "consumer" perspective [20]. In this regard, three different categories of GHG emissions exist, namely, the GHGs of Scope 1, Scope 2, and Scope 3. Scope 1 emissions are all direct GHG emissions generated from sources that are controlled or owned by an organization (e.g., emissions due to the combustion of fuels by furnaces, vehicles, boilers, etc.). Scope 2 and Scope 3 emissions are instead indirect GHGs occurring in places physically outside the organization's system boundary. However, while Scope 2 emissions refer to the purchase of energy flows such as electricity, steam, heat, etc., and are therefore accounted for in the organization's GHG inventory, Scope 3 emissions rely upon the activities (upstream and downstream) in the supply chain of the reporting organization. The main difference between the Consumer and the Producer perspective is that the former also accounts for Scope 3 emissions due to the very nature of the life cycle approach behind such a supply chain inventory.

Explicitly or implicitly or from a Producer or Consumer's perspective, life cycle thinking is, in all cases, always behind the estimation of the carbon footprint and climate neutrality. Regardless of the way carbon footprint is calculated, the items reported in Table 1 are deemed to guide practitioners into harmonized and comparable accounts of carbon emissions and removals. All these represent operating manuals for companies to conduct accurate carbon balances of their activities. This way, companies can develop their *ad hoc* carbon management strategies and demonstrate their efforts to fight climate change by becoming carbon neutral.

The goal of implementing a carbon management system at product or organizational scale is twofold. On one hand, companies can make improvements to life cycle processes in order to avoid or reduce their life cycle GHG emissions. On the other hand, the implementation of a carbon management plan can allow companies to make direct or indirect interventions on ecosystems to enhance carbon removals, which can then be accounted for to offset the residual emissions. Through such an impact mitigation and compensation program, companies can attempt to reach climate neutrality.

Alongside the development of standards and guidelines for carbon neutrality released by international standardization bodies, such as PAS 2060 developed by the BSI, companies can also adhere to climate neutrality certification schemes created in-house by consulting companies and not-for-profit organizations, such as the Climate Neutral Group (i.e., promoter of the Climate Neutral Certification Standard) [21], the Climate Neutral company (i.e., developer of the Climate Neutral Certified label) [22], Carbonfund.org (i.e., developer of the Carbonfree Product Certification) [23], or the Climate Impact Partners (i.e., developers of the CarbonNeutral® certification built around the Carbon Neutral Protocol) [24].

Those protocols not only guide companies but also individuals towards the measurement, reduction, and compensation of their carbon footprint to achieve net-zero GHG emissions over a defined time horizon. The most known and applied solution to offset residual emissions is to purchase carbon credits in the voluntary carbon markets. As further investigated in Secs. 4 and 5, however, other ways to mitigate the impact of climate change and achieve carbon neutrality are emerging. These may overcome the risk of carbon leakage due to the loss of traceability and other methodological shortcomings of the carbon-credits-based approach [25].

2.2. *Carbon management in territorial LCAs*

Next to product and organizational LCA, but not yet standardized, is positioned the so-called *territorial* LCA [25, 26]. This represents an additional extension of the LCA method conceptualized to encompass the complex spatial interactions of human-nature systems generating many outputs and activities. Literature studies on territorial LCA are relatively recent and mostly limited to multifunctional agrifood systems [27–29]. As it often happens with new environmental accounting approaches, the carbon footprint assessment has become quickly predominant in territorial LCA studies, with climate change taken as a reference impact category in pilot applications [30].

It is worth noticing that the most insightful developments of territorial LCA are proposed to improve land use planning in combination, or are inspired by, other methods such as urban metabolism assessments [31], optimization [32] or agent-based modeling [33]. These methods have longer histories than the LCA in addressing the complexities of energy and material inputs/outputs relationships in large territorial systems. The need to capture the spatial dimension of environmental stressors and impacts and their interaction through specific socio-ecological modeling is one prominent challenge of territorial LCA studies [34]. One promising method to tackle those challenges was the one proposed by Nitschelm and colleagues on the Spatialized Territorial LCA, which combines spatially explicit and territorial assessment frameworks to study land-use planning in an agricultural territory [35].

As an added value to conventional LCAs, carbon footprint assessments of territorial systems have the capability to account for natural capital attributes in the modeling framework and thus enrich the carbon balance in life cycle inventories with an evaluation of the carbon sequestration potential [36–40]. GHG emissions inventories at territorial scales, such as for urban [41], rural or mixed systems [42], up to country-scale systems [43], have been traditionally undertaken according to IPCC-based standards and guidelines, e.g., ISO 14064 [6, 44], and have demonstrated the feasibility to capture and incorporate the notion of ES provided by the local and distant ecosystems [45].

3. Opportunities from the Field of Ecosystem Services and Land Use Driven Impact Assessments

Ecosystem services (ES) are the benefits that people can get out of ecosystem functioning [46], which can be grouped into provisioning services, regulation & maintenance services, and recreational services [47]. Climate regulation through carbon removal and storage (i.e., carbon sequestration) is one well-known ES of the regulation & maintenance section. A vast LCA literature on agrifood products, construction biomaterials, and bioenergy and biofuel production, shows the importance of considering the carbon sequestered by biomass and soil when assessing climate change impacts (e.g., Refs. [48–56]). In one seminal study to account for carbon sequestration in the LCA, Brandão and colleagues built a framework to assess soil quality impacts and explicitly included soil organic carbon changes in the GHG balance across different energy crops [57].

Advances in the field of Life Cycle Inventory (LCI) database developments were also performed over time by incorporating biogenic carbon in datasets of crops and wooden materials as an environmental intervention flow in input to capture information about the carbon removed and stored in the biomass product [58]. Next to the importance of considering the contribution of "foregone" carbon sequestration in land use impact assessment [59], the temporally static approach of balancing anthropogenic versus biogenic carbon emissions was also questioned. In this regard, time in climate change impact assessment is key for determining the variability in the GHG emissions balance [60], which is an issue that should not be neglected in the LCA. For example, it was observed that forest management highly influences the value of characterization factors for CO_2 emitted through biomass combustion over different time horizons, questioning the climate neutrality associated with biogenic CO_2 emissions [61].

Several combinations between the LCA and modeling tools that could capture the dynamic behavior of biomass growth and the carbon natural cycle were effective in assessing carbon sequestration over time, such as, for example, in the coupling between LCA tools and BIOME-BGC [62], CO2FIX [63], InVEST-(LUCI)-LCA [64], CENTURY [65], iTree [66], or the Bern Carbon Cycle Model [67]. The Land-Use Change Improved (LUCI)-LCA model proposed by Chaplin-Kramer and colleagues [64], in particular, recommends integrated spatially explicit modeling of land use changes (LUC) and ES (see Fig. 2). This approach highlights the importance of specific information with regard to land use change effects relating to impacts on supply chain and innovation decisions. The authors emphasize that the strength of LUCI-LCA lies in highlighting the importance of location in predicting impacts of future land use rather than relying on regional averages of LUC.

The combination between the LCA and models like InVEST-(LUCI)-LCA, in particular, was proposed as an effective means for determining exogenous impact drivers for ES assessment in the LCA, such as in the case of land use and LUC [62, 64, 68]. The theme of ES accounting in the LCA through a land use impact assessment was extensively tackled by a working group of the UNEP/SETAC Life Cycle Initiative, which contributed to develop characterization factors for the assessment of land use

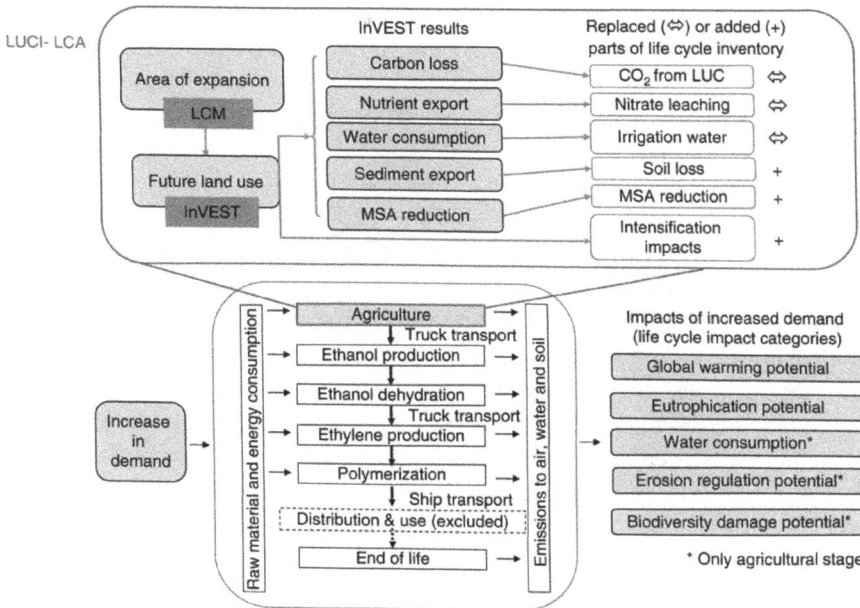

Fig. 2.   LUCI-LCA framework. (Reproduced from Ref. [65] via an Open Access license.)

(change) impacts on biodiversity and ES in the LCA [69]. Factors for several ES, such as the potential of biotic production, climate regulation, freshwater regulation, erosion regulation, and water purification, were calculated using a common impact characterization rationale compatible with existing LCI datasets. As a result, these characterization factors were made operational to assess "midpoint" impacts using physical units and can be translated into ES damage potential at the "endpoint" level. For example, using the elementary flows of land use and LUC as impact drivers, climate regulation potential can be determined by accounting for the CO_2 transfers between vegetation/soil and the atmosphere in the course of terrestrial release and re-storage of carbon [70]. Additional work was done to develop an impact characterization model for monetary valuations through social carbon cost accounting of carbon sequestration by soils [71].

Although the characterization factors developed under the UNE SETAC life cycle initiative still represent the most compatible and rea to-use instruments for conducting LCA-ES analyses, their implementa in LCA practice is very limited. One reason may be the inability of these indicators to link ES in bundles to the final value of natural capital and actual benefits to humans. Dynamic interactions over space and time occurring between land use and ecosystem functioning necessarily determine trade-offs and synergies in the provision of ES, which cannot be easily captured with existing linear and static LCA models.

For instance, while biofuels of first- and second-generation systems have, on average, lower carbon footprint than fossil fuels when LUC do not occur (as presented in Chap. 11), the reduction in GHG emissions from those biofuels are achieved at the expense of other impacts, such as acidification, eutrophication, water footprint, and biodiversity loss. In another example, provided by Jeswani and colleagues [72], it was highlighted how LUC played an important role in determining global warming impacts of first-generation biofuels (Figs. 3 and 4). As displayed in both graphs, LCA studies for the same feedstock-to-biofuel present contradictory results, ranging from favorable (without LUC) to unfavorable (with LUC).

Agriculture, forestry, and even urban systems are pertinent use cases for demonstrating the gaps, needs and opportunities to implement land-use-based impact assessments of ES, taking advantage of the knowledge gathered from the adoption of integrated models. For example, Othoniel and colleagues [62] built a dynamic (over time and space) land cover

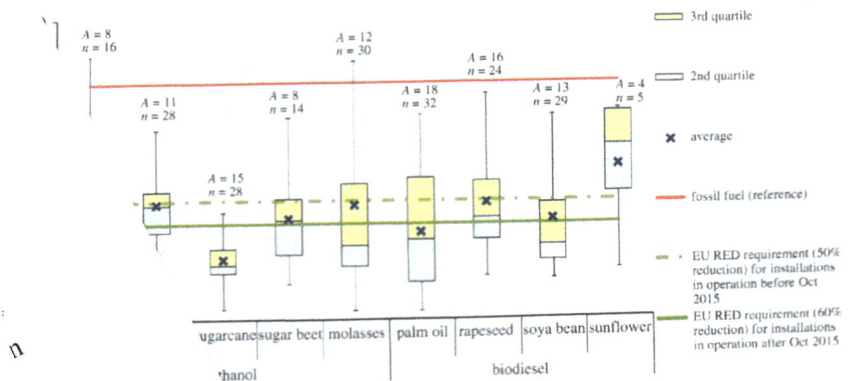

...eration biofuels without land-use change. (Reproduced from ... license.)

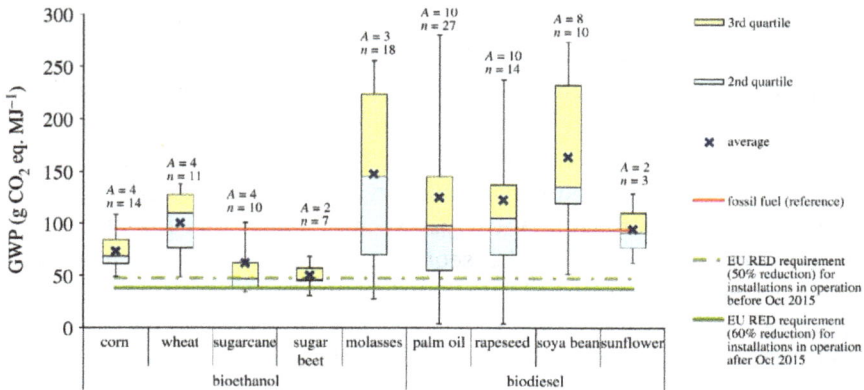

Fig. 4. GWP of first-generation biofuels with land-use change. (Reproduced from Ref. [72] via an Open Access license.)

change model to calculate characterization factors for two regulation & maintenance services (carbon sequestration and pollination). The model allowed one to spatially map the trade-offs and synergies between ES due to several alternatives of time-dependent land cover change on the scales of Luxembourg and its municipalities.

While this approach is operational only for conducting land-cover-change-based assessments in Luxembourg (possibly for use cases from the agricultural sector), the impact characterization model is replicable and ideally transferrable to any other place in the world. However, a certain level of sophistication and data adaptation for developing such a country-based integrated model of ES assessment would be needed. Research on the integration between socio-ecological modeling and the LCA is timely and a cutting edge for the improvement of knowledge on how to account for ES indicators at different spatial and temporal scales, as well as to track land use effects.

4. LCA-ESA-based Mitigation Hierarchy

A consensus exists about the need to implement decarbonization strategies to achieve climate neutrality, which would, however, follow preventive measures of impact reduction. In this regard, reduction measures have priority over compensation measures, as they are more tangible, time-independent and verifiable [73]. Moreover, there is a broad debate on the quality and effectiveness of different compensation mechanisms, projects and providers [74], which calls for more and more tools generating unquestionable quantitative evidence. The Mitigation Hierarchy developed in the field of biodiversity conservation represents one of those tools [75, 76]. As defined by the Cross Sector Biodiversity Initiative [77], the mitigation hierarchy is the sequence of actions to anticipate and *avoid*, and where avoidance is not possible, *minimize*, and, when impacts occur, *restore*, and where significant residual impacts remain, *offset* for biodiversity-related risks and impacts on affected communities and the environment.

The core principles of the mitigation hierarchy can be broadly applicable to domains outside the strict biodiversity-offsetting framework. It may become of inspiration to build a broader LCA-ES mitigation hierarchy structure. As the same mitigation hierarchy of "reduction before compensation" is also the current scientific consensus for carbon offsetting [73], designing a carbon management plan based on such an LCA-ES mitigation hierarchy is potentially feasible and suitable for organizations in any economic sector (primary, secondary, and tertiary). Nevertheless, such an LCA-ES mitigation hierarchy is not conceived to only design carbon management plans; more largely, an "environmental footprint plan" might be ideally designed in order to support the deployment of effective solutions to achieve environmental neutrality.

No LCA-related standard or guideline yet exists that explicitly considers the notion of ES. Nevertheless, a relatively high growth in research studies on ES accounting in the LCA was observed over the last ten years [78]. This suggests that there is a need to develop a Standards Application Protocol to guide organizations towards the implementation of LCA-ES mitigation hierarchy principles [79]. The rationale of applying the LCA in combination with an ES assessment is to address questions for carbon mitigation hierarchy. The LCA-ES framework assumes that the environmental impacts can be strongly mitigated with duly interventions on the life cycle system. Besides, the unavoidable impact can be mitigated and compensated with a certain time lag by an equally enhanced provision of ES. Transferred to the carbon management domain, a company first needs to measure and apply actions to avoid and reduce the carbon footprint of its production system, and then operate to implement, either directly or indirectly, specific sustainability management interventions on ecosystems with the aim to increase the value of carbon stock over a certain time, i.e., to reach a stable positive delta of carbon uptake.

As shown in Fig. 5, one can address the first two steps of the LCA-ES mitigation hierarchy through a rigorous application of the LCA method. Existing standards or guidelines for the quantification of environmental footprint can be applied, which can also serve to guide the company into the design of improvement actions to reduce and/or avoid impacts. According to an environmental footprint management plan (or "carbon management plan", if narrowed to address the single climate neutrality target), the company will identify natural or semi-natural ecosystems (e.g., forests, pasture lands, agricultural lands, etc.) that might necessitate a restoration, or where specific land management changes can be applied to increase the provision of one or more target ES. The more those ecosystems are physically close to the most impactful organization site(s), the better it is to avoid the risk of offsets displacement and loss of traceability of compensating ES. Target ES shall thus be quantified in physical terms at the first year of analysis, say "year zero". Those are ES potentially suitable to compensate the environmental issue in each corresponding impact category of reference, such as: carbon sequestration for mitigating climate change, accounted for in the mass of equivalent CO_2 that is absorbed and stored; air quality regulation for mitigating air pollution, quantified in the

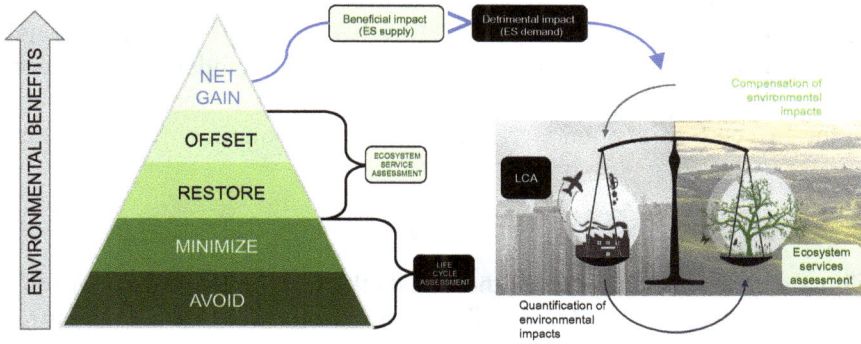

Fig. 5. LCA-ES Mitigation Hierarchy framework formulated to quantify, avoid and minimize environmental impacts through the Life Cycle Assessment and increase the provision of ecosystem services through restoration and offsetting.

mass of equivalent particulate matter removed from the air; freshwater provision for mitigating water depletion, in the volume of water supply; etc. [80]. In other words, the ES Demand generated by the product/organization life cycle (i.e., negative or detrimental environmental impacts) shall be neutralized by an equivalent production of ES Supply (i.e., positive or beneficial environmental impacts) sourced from the natural capital [81].

The very end point of such an LCA-ES mitigation hierarchy is to allow companies to offset their residual footprint through specific interventions on ecosystems, rather than by simply purchasing environmental credits in voluntary markets, such as carbon credits. For example, an agricultural company willing to achieve environmental impact neutrality may operate on its own land and implement long-term management changes to ensure stable increases in the provision of ES, compared to a benchmark quantified at "year zero". This may be regarded as a "direct" intervention producing an insetting result, where the impact is neutralized *in loco*. An organization from, e.g., the tertiary sector, which does not own any land, may invest in the implementation of restoration actions or nature-based solutions, which are sources of quantifiable ES. In this case, priority for land restoration should be given following a proximity principle, that is, investing preferentially on projects in the same or closest possible region or urban setting. Some anticipatory procedures on how to reach carbon neutrality in the agrifood sector according to a similar roadmap have been recently framed by Acampora and colleagues [82].

All this should bring organizations a net-zero impact or even gain, where the beneficial impact provided by ES overcomes the detrimental impact generated by the life cycle system (Fig. 2). A Standards Application Protocol that can keep recording and monitoring data over the five steps of the LCA-ES mitigation hierarchy shall be built on these requirements.

5. Roadmap for Implementation

Climate change is only one of the threats that can be evaluated with life cycle impact indicators. Trade-offs with many other environmental issues exist (e.g., eutrophication, acidification, air pollution, resource depletion, etc.), which all deserve reduction and compensation measures in parallel. Therefore, the implementation of an LCA-ES mitigation hierarchy approach that goes beyond the strict target of carbon neutrality is recommended.

LCA scholars have started to integrate the concept of mitigation hierarchy in the assessment of the environmental footprint of productive systems, either explicitly or implicitly. The rationale is to support companies and individuals towards the achievement of, more broadly and ambitiously speaking, an "environmental neutrality" state by enhancing the provision of ES. For example, the recent Circular Ecosystem Compensation approach proposes to compensate a broad set of environmental impacts in an existing ecosystem, which can be generated by products and services, organization and urban areas, and individuals, by renaturing degraded ecosystems [73]. In parallel, quantitative metrics to balance the environmental impacts in the form of ES demand with an increased ES supply associated with restoration actions are proposed in the so-called techno-ecological synergy (TES) framework [81].

While not directly referring to the TES, but following an analogous rationale, Babí Almenar and colleagues demonstrate that a tipping point in time may exist, after which urban nature-based solutions (NbS) start to generate a "net positive" environmental benefit that can compensate the environmental footprint generated by the implementation, management, and end-of-life activities of a deployed NbS [80]. In this regard, NbS can be considered the best available renaturation and restoration measures to implement in cities or across degraded landscapes to increase the supply of ES, such as carbon sequestration [83], and to provide offsetting solutions in the contexts of an LCA-ES mitigation hierarchy. Figure 6 graphically illustrates the relationships between ES supply and ES demand in terms of potential indicators. The notion of time and space is not considered in this example. However, in the investigated model, time and space play a key role, especially for what concerns global warming potential and carbon sequestration.

A large variety of investment options on NbS projects currently exist for producers, public authorities, and individuals willing to mitigate and compensate their environmental footprint, most often mediated by consulting firms operating in the finance market [84]. However, there is also very often huge uncertainty, unpredictability, and no transparency in the way organizations can offset their impacts or take advantage of the

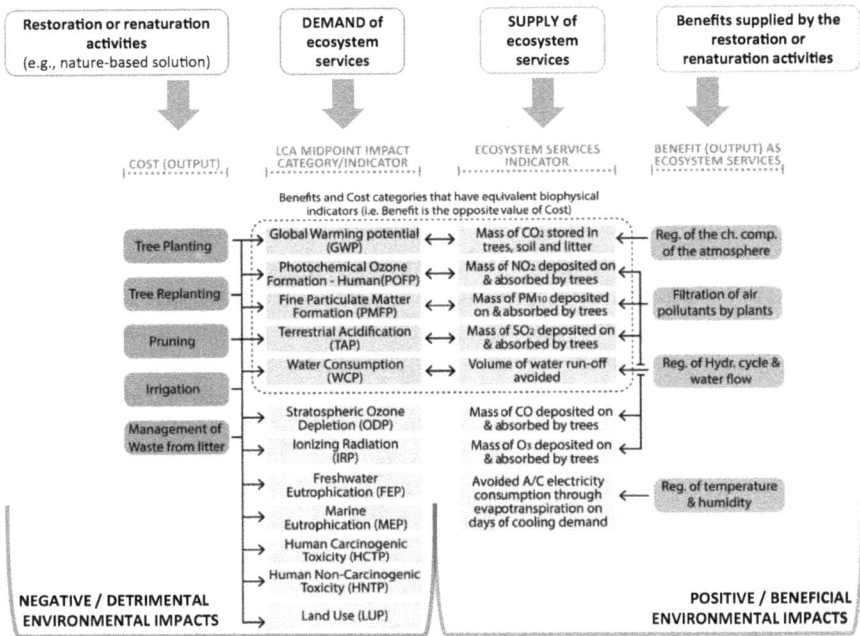

Fig. 6.   Relationship between ES demand and supply indicators in the case of a restoration/renaturation project implemented for environmental offsetting purposes. (Adapted from Ref. [90] with reproduction of the core image allowed via an Open Access license.)

benefits provided by investments in natural capital. To this end, several instruments have been created that may facilitate the access of private and public entities to quantifiable mitigation and compensation actions, such as the establishment of Payments for ES schemes [85] or the introduction of programs for the traceable adoption of trees planted in cities or rural contexts, e.g., Life-Terra [86], zeroCO2 [87], WOWnature [88], GoCarbonNeutral [89]. Such instruments are certainly not free of shortcomings and perfectible aspects, such as the potential lack of contractual transparency or the over- or under-estimation of environmental offset values. However, their introduction usually follows the bottom-up initiative of individuals and local communities and, for this reason, be considered a virtuous attempt to support the transition towards environmentally sustainable benchmarks.

Interestingly, certification schemes encouraging sustainable management of natural capital, such as forests, start to incorporate the notion of ES in their list of key performance indicators. This is, for example, the case of the well-known Forest Stewardship Council label (i.e., FSC®), which has been recognized for its commitment to responsible forestry with ES claims [91]. On one hand, FSC®-certified forest landowners or managers can quantitatively demonstrate their contribution to protecting biodiversity, storing carbon, purifying water, regenerating soil, and providing recreation. To this end, a so-called FSC® forest management plan is established that schedules periodic measurements, monitoring, and reporting of the forest ES supply. On the other hand, companies and individuals interested in demonstrating their environmental responsibility may financially support FSC® certified projects, and then duly communicate their commitment towards sustainability and environmental protection. Win-win agreements developed to such a scheme are already operational through platforms, such as WOWnature® [88], a bottom-up initiative that helps citizens, institutions, and companies grow new forests and protect existing ones.

The development and success of such collaboration schemes is guaranteed through the establishment of management plans and the adoption of certification programs, both covering the LCA and the ES pillars of the LCA-ES mitigation hierarchy framework. These may become integral components of the abovementioned Sustainability Standards Protocol. Focusing on climate change, fully-fledged and robust approaches to support carbon management at the product scale start to become operational, regularly upgraded, and used by consultants and companies. One of the most advanced approaches available so far is PAS 2060 [92], which explicitly targets carbon neutrality should a concrete carbon management be put in place. However, only carbon offsetting based on the purchase of carbon credits from the voluntary carbon markets are conceived in PAS 2060. Investments on NbS and land restoration projects with the possibility to track locally quantified ES, such as tons of sequestered carbon, should therefore become the next frontier of the standards. A concrete step to drive organizations to achieve net-zero GHGs has been undertaken with the International Workshop Agreement IWA 42 [93], see Table 1. This document provides an extensive set of recommendations and guiding principles to enable the implementation of an LCA-ES mitigation hierarchy framework with a focus on carbon emissions and carbon uptake.

Remarkably, such a guideline is getting followed by other promising initiatives, such as the ISO guidance on carbon neutrality — ISO/DIS 14068 (Greenhouse gas management and climate change management and related activities — Carbon neutrality), which will help organizations taking concrete actions.

6. Conclusions and Way Forward

Achieving climate neutrality by 2050 is the ultimate objective of the 194 Parties that signed the Paris Agreement entered into force on 4 November 2016. Climate neutrality is an extremely ambitious goal that requires a global effort of collaboration, transfer of knowledge, communication transparency, and behavioral change — all are actions that, without a clear top-down, unselfish, and shared political driving impulse, are far from being implemented.

Citizens and industries can play the role of game-changers only if they are economically stimulated; otherwise, they will passively follow the law, where enforced, or try to circumvent it. More and more compelling regulations are being created to limit GHG emissions. As a matter of fact, however, the share of renewable energy consumed annually at the global scale stays well below the 15% of the total primary energy consumption, with an increase of only 3% points over the last ten years [94]. This suggests that human society will remain far away from reaching the target of carbon neutrality if practical, innovative, and incentivizing measures are not deployed soon, which can offer business opportunities or cost reductions for people and organizations.

ES contribute to human well-being and economic welfare, and an increase in their supply can be fostered by taking actions for the sustainable management of natural capital. Both people and organizations can be involved in such a process. It is likely that the transition to a climate neutral standard of living should pass by a step where private organizations take leadership in enhancing the provision of ES, such as carbon sequestration. This would be timely, as organizations will be more and more solicited to commit with certificates of climate neutrality.

New standards, such as ISO 14068, are on the way to hopefully define clear and unbiased criteria for companies to mitigate their carbon footprint and find solutions to compensate for their residual impacts. Adopting a carbon credit-based offsetting approach is an alternative that already shows its shortcomings and will not solve the issue alone. Investments in traceable replantation projects and initiatives of sustainable management of existing forests is another option that appears to generate no negative rebound effects, even if renowned academics warn about the potential influence that large-scale reforestation projects may have on global circulation and remote climates [95]. In any case, avoiding deforestation and increasing reforestation is considered one of the most effective ways to achieve the Paris Agreement's targets [96, 97], especially if implemented as a complement (and not an alternative) to ambitious fossil fuel CO_2 emissions reductions [98].

Regardless the approach(es) that can be undertaken, to reach carbon neutrality, governments need to act immediately with policies that (i) incentivize bottom-up regionalized initiatives where citizens, local communities, cities, and SMEs can cooperate and co-create by making business with the sustainable management of natural capital, (ii) put in place serious environmental measures to reduce the impact produced by large-scale manufacturing, transportation and agrifood industrial groups, which are currently responsible for the highest shares of global GHGs emissions, and (iii) leverage on robust and transparent carbon footprint assessments in combination with proven, traceable and monitored actions to remove carbon from the atmosphere.

References

1. IEA, Announced net-zero CO2 or GHG emissions by 2050 reduction targets, International Energy Agency (2023). https://www.iea.org/data-and-statistics/charts/announced-net-zero-co2-or-ghg-emissions-by-2050-reduction-targets
2. ISO 14021:2016/Amd 1:2021 Environmental labels and declarations — Self-declared environmental claims (Type II environmental labelling) — Amendment 1: Carbon footprint, carbon neutral, Geneva, Switzerland (2016).
3. J. Janjusevic and N. Perovic, *Eco-certification in the Montenegrin Tourism as a Response on Climate Change*, Springer International Publishing, Cham, pp. 53–64 (2020).
4. F. André and J. Valenciano-Salazar, Becoming carbon neutral in Costa Rica to be more sustainable: An AHP approach, *Sustainability* **12**(2) (2020).

5. A. Birkenberg, M.E. Narjes, B. Weinmann and R. Birner, The potential of carbon neutral labeling to engage coffee consumers in climate change mitigation, *J. Clean. Prod.* **278** (2021).
6. S. Bastianoni, M. Marchi, D. Caro, P. Casprini and F.M. Pulselli, The connection between 2006 IPCC GHG inventory methodology and ISO 14064-1 certification standard — A reference point for the environmental policies at sub-national scale, *Environ. Sci. Policy* **44**, 97–107 (2014).
7. W. Badgery, B. Murphy, A. Cowie, S. Orgill, A. Rawson, A. Simmons, *et al.*, Soil carbon market-based instrument pilot-the sequestration of soil organic carbon for the purpose of obtaining carbon credits, *Soil Res.* **59**(1), 12–23 (2021).
8. C. Paul, B. Bartkowski, C. Donmez, A. Don, S. Mayer, M. Steffens, *et al.*, Carbon farming: Are soil carbon certificates a suitable tool for climate change mitigation? *J. Environ. Manage.* **330**, 117142 (2023).
9. L.J. Sonter, D.J. Barrett, C.J. Moran and B.S. Soares-Filho, Carbon emissions due to deforestation for the production of charcoal used in Brazil's steel industry, *Nat. Clim. Change.* **5**(4), 359–363 (2015).
10. ISO, ISO 14040:2006: Environmental management — Life cycle assessment — Principles and framework, International Organization for Standardization (ISO), Geneva, Switzerland (2006).
11. ISO 14044:2006: Environmental management — Life cycle assessment — Requirements and guidelines, Geneva, Switzerland (2006).
12. ISO 14067:2018 Greenhouse gases — Carbon footprint of products — Requirements and guidelines for quantification, Geneva, Switzerland (2018).
13. IPCC, 2006 IPCC guidelines for national greenhouse gas inventories, Institute for Global Environmental Strategies, Intergovernmental Panel on Climate Change (2006).
14. IPCC, 2019 Refinement to the 2006 IPCC guidelines for national greenhouse gas inventories, Task Force on National Greenhouse Gas Inventories, Intergovernmental Panel on Climate Change (2019).
15. N. Mirabella and K. Allacker, Urban GHG accounting: Discrepancies, constraints and opportunities, *Buildings and Cities* **2**(1) (2021).
16. ISO, ISO Survey of certifications to management system standards — International Organization for Standardization (ISO) Survey 2021 (2023). https://www.iso.org/the-iso-survey.html
17. A.M. Osorio, L.F. Úsuga, R.E. Vásquez, C. Nieto-Londoño, M.E. Rinaudo, J. A. Martínez, *et al.*, Towards carbon neutrality in higher education institutions: Case of two private universities in Colombia, *Sustainability* **14**(3) (2022).
18. K. Hergoualc'h, N. Mueller, M. Bernoux, Ä. Kasimir, T.J. van der Weerden and S.M. Ogle, Improved accuracy and reduced uncertainty in greenhouse gas inventories by refining the IPCC emission factor for direct N_2O

emissions from nitrogen inputs to managed soils, *Glob. Chang. Biol.* **27**(24), 6536–6550 (2021).

19. S.M. Ogle, L. Buendia, K. Butterbach-Bahl, F.J. Breidt, M. Hartman, K. Yagi, *et al.*, Advancing national greenhouse gas inventories for agriculture in developing countries: Improving activity data, emission factors and software technology, *Environ. Res. Lett.* **8**(1), 015030 (2013).
20. D. Caro, B. Rugani, F.M. Pulselli and E. Benetto, Implications of a consumer-based perspective for the estimation of GHG emissions. The illustrative case of Luxembourg, *Sci. Total Environ.* **508**, 67–75 (2015).
21. CNG, Climate Neutral Standard — Climate Neutral Certification Program (2020). https://www.climateneutralcertification.com/about/climate-neutral-standard-2021/
22. Climate Neutral, Climate Neutral Certification Standard (2023). https://www.climateneutral.org/standards
23. Carbonfund, Carbonfree certification (2023). https://carbonfund.org/
24. CarbonNeutral, The CarbonNeutral Protocol 2023 — The global standard for carbon neutral programmes, Climate Impact Partners (2023). https://carbon-neutral.com/the-carbonneutral-protocol
25. E. Loiseau, L. Aissani, S. Le Féon, F. Laurent, J. Cerceau, S. Sala, *et al.*, Territorial Life Cycle Assessment (LCA): What exactly is it about? A proposal towards using a common terminology and a research agenda, *J. Clean. Prod.* **176**, 474–485 (2018).
26. E. Loiseau, T. Salou and P. Roux, Territorial life cycle assessment, in *Assessing Progress Towards Sustainability — Frameworks, Tools and Case Studies*, 1st Edition, C. Teodosiu, S. Fiore and A. Hospido (eds.), Elsevier, pp. 161–188 (2022).
27. N. Borghino, M. Corson, L. Nitschelm, A. Wilfart, J. Fleuet, M. Moraine, *et al.*, Contribution of LCA to decision making: A scenario analysis in territorial agricultural production systems, *J. Environ. Manage.* **287**, 112288 (2021).
28. T. Ding, S. Bourrelly and W.M.J. Achten, Application of territorial emission factors with open-access data — a territorial LCA case study of land use for livestock production in Wallonia, *Int. J. Life Cycle Assess.* **26**(8), 1556–1569 (2021).
29. N. Rogy, P. Roux, T. Salou, C. Pradinaud, A. Sferratore, N. Géhéniau, *et al.*, Water supply scenarios of agricultural areas: Environmental performance through Territorial Life Cycle Assessment, *J. Clean. Prod.* **366** (2022).
30. P. De Toro and S. Iodice, Urban metabolism evaluation methods: Life cycle assessment and territorial regeneration, in *Regenerative Territories*, L. Amenta, M. Russo and A. Van Timmeren (eds.), **128**, pp. 213–230 (2022).
31. J. Sohn, G.C. Vega and M. Birkved, A methodology concept for territorial metabolism — Life cycle assessment: Challenges and opportunities in scaling from urban to territorial assessment, *Proc. CIRP.* **69**, 89–93 (2018).

32. T. Ding, B. Steubing and W.M.J. Achten, Coupling optimization with territorial LCA to support agricultural land-use planning, *J. Environ. Manage.* **328** (2023).

33. T. Ding and W.M.J. Achten, Coupling agent-based modeling with territorial LCA to support agricultural land-use planning, *J. Clean. Prod.* **380** (2022).

34. J.-B. Bahers, A. Athanassiadis, D. Perrotti and S. Kampelmann, The place of space in urban metabolism research: Towards a spatial turn? A review and future agenda, *Landscape Urban Plan.* **221**, 104376 (2022).

35. L. Nitschelm, J. Aubin, M.S. Corson, V. Viaud and C. Walter, Spatial differentiation in Life Cycle Assessment applied to an agricultural territory: Current practices and method development, *J. Clean. Prod.* **112**, 2472–2484 (2016).

36. C. Ingrao, R. Rana, C. Tricase and M. Lombardi, Application of carbon footprint to an agro-biogas supply chain in Southern Italy, *Appl. Energy* **149**, 75–88 (2015).

37. V. Timmermann and J. Dibdiakova, Greenhouse gas emissions from forestry in East Norway, *Int. J. Life Cycle Assess.* **19**(9), 1593–1606 (2014).

38. A.C. Demirkesen and F. Evrendilek, Integrating spatiotemporal dynamics of natural capital security and urban ecosystem carbon metabolism, *Environ. Dev. Sustain.* **20**(5), 2043–2063 (2018).

39. S. Saha, B. Bera, P.K. Shit, S. Bhattacharjee and N. Sengupta, Estimation of carbon budget through carbon emission-sequestration and valuation of ecosystem services in the extended part of Chota Nagpur Plateau (India), *J. Clean. Prod.* **380** (2022).

40. F. Meng, Q. Yuan, R.A. Bellezoni, J.A.P. de Oliveira, S. Cristiano, A. M. Shah, *et al.*, Quantification of the food-water-energy nexus in urban green and blue infrastructure: A synthesis of the literature, *Resour. Conserv. Recycl.* **188** (2023).

41. A. Zabeo, C. Bellio, L. Pizzol, E. Giubilato and E. Semenzin, Carbon footprint of municipal solid waste collection in the treviso area (Italy), *Environ. Eng. Manag. J.* **16**(8), 1781–1788 (2017).

42. M. Marchi, E. Neri, F.M. Pulselli and S. Bastianoni, CO_2 recovery from wine production: Possible implications on the carbon balance at territorial level, *J. CO_2 Util.* **28**, 137–144 (2018).

43. J. Wang, S. Jin, W. Bai, Y. Li and Y. Jin, Comparative analysis of the international carbon verification policies and systems, *Nat. Hazards* **84**, 381–397 (2016).

44. M. Marchi, F. M. Pulselli, S. Mangiavacchi, F. Menghetti, N. Marchettini and S. Bastianoni, The greenhouse gas inventory as a tool for planning integrated waste management systems: A case study in central Italy, *J. Clean. Prod.* **142**, 351–359 (2017).

45. T. Elliot, B. Goldstein, E. Gómez-Baggethun, V. Proença and B. Rugani, Ecosystem service deficits of European cities, *Sci. Total Environ.* **837**, 155875 (2022).
46. R. Costanza, R. de Groot, L. Braat, I. Kubiszewski, L. Fioramonti, P. Sutton, *et al.*, Twenty years of ecosystem services: How far have we come and how far do we still need to go? *Ecosyst. Serv.* **28**, 1–16 (2017).
47. R. Haines-Young and M. Potschin, Common International Classification of Ecosystem Services (CICES) V5.1 and guidance on the application of the revised structure, Fabis Consulting Ltd, UK (2018). www.cices.com
48. M.C. Heller, G.A. Keoleian and T.A. Volk, Life cycle assessment of a willow bioenergy cropping system, *Biomass Bioenergy* **25**(2), 147–165 (2003).
49. B.C. Renger, J.L. Birkeland and D.J. Midmore, Net-positive building carbon sequestration, *Build. Res. Inf.* **43**(1), 11–24 (2015).
50. E.I. Wiloso, R. Heijungs and G.R. De Snoo, LCA of second generation bio-ethanol: A review and some issues to be resolved for good LCA practice, *Renew. Sust. Energ. Rev.* **16**(7), 5295–5308 (2012).
51. P. Goglio, W.N. Smith, B.B. Grant, R.L. Desjardins, B.G. McConkey, C. A. Campbell, *et al.*, Accounting for soil carbon changes in agricultural life cycle assessment (LCA): A review, *J. Clean. Prod.* **104**, 23–39 (2015).
52. S.A. Archer, R.J. Murphy and R. Steinberger-Wilckens, Methodological analysis of palm oil biodiesel life cycle studies, *Renew. Sust. Energ. Rev.* **94**, 694–704 (2018).
53. M. Demertzi, J.A. Paulo, S.P. Faias, L. Arroja and A.C. Dias, Evaluating the carbon footprint of the cork sector with a dynamic approach including bio-genic carbon flows, *Int. J. Life Cycle Assess.* **23**(7), 1448–1459 (2018).
54. I. Batalla, M.T. Knudsen, L. Mogensen, Ó.D. Hierro, M. Pinto and J.E. Hermansen, Carbon footprint of milk from sheep farming systems in Northern Spain including soil carbon sequestration in grasslands, *J. Clean. Prod.* **104**, 121–129 (2015).
55. F. Cherubini and G. Jungmeier, LCA of a biorefinery concept producing bioethanol, bioenergy, and chemicals from switchgrass, *Int. J. Life Cycle Assess.* **15**(1), 53–66 (2010).
56. J. Gan, M. Chen, K. Semple, X. Liu, C. Dai and Q. Tu, Life cycle assessment of bamboo products: Review and harmonization, *Sci. Total Environ.* **849** (2022).
57. M. Brandão, L. Milà i Canals and R. Clift, Soil organic carbon changes in the cultivation of energy crops: Implications for GHG balances and soil quality for use in LCA, *Biomass Bioenergy* **35**(6), 2323–2336 (2011).
58. ecoinvent, The ecoinvent® v3 database, The Swiss Centre for Life Cycle Inventories, Dübendorf (2013). https://ecoinvent.org/

59. K. Koponen and S. Soimakallio, Foregone carbon sequestration due to land occupation — the case of agro-bioenergy in Finland, *Int. J. Life Cycle Assess.* **20**(11), 1544–1556 (2015).
60. A. Levasseur, P. Lesage, M. Margni, M. Brandão and R. Samson, Assessing temporary carbon sequestration and storage projects through land use, land-use change and forestry: Comparison of dynamic life cycle assessment with ton-year approaches, *Clim. Change* **115**(3–4), 759–776 (2012).
61. F. Cherubini, A.H. Strømman and E. Hertwich, Effects of boreal forest management practices on the climate impact of CO_2 emissions from bioenergy, *Ecol. Modell.* **223**(1), 59–66 (2011).
62. B. Othoniel, B. Rugani, R. Heijungs, M. Beyer, M. Machwitz and P. Post, An improved life cycle impact assessment principle for assessing the impact of land use on ecosystem services, *Sci. Total Environ.* **693**, 133374 (2019).
63. B. Rugani, K. Golkowska, I. Vázquez-Rowe, D. Koster, E. Benetto and P. Verdonckt, Simulation of environmental impact scores within the life cycle of mixed wood chips from alternative short rotation coppice systems in Flanders (Belgium), *Appl. Energy* **156**, 449–464 (2015).
64. R. Chaplin-Kramer, S. Sim, P. Hamel, B. Bryant, R. Noe, C. Mueller, *et al.*, Life cycle assessment needs predictive spatial modelling for biodiversity and ecosystem services, *Nat. Commun.* **8** (2017).
65. W. Liu, Y. Yan, D. Wang and W. Ma, Integrate carbon dynamics models for assessing the impact of land use intervention on carbon sequestration ecosystem service, *Ecol. Indic.* **91**, 268–277 (2018).
66. M. Oliveira, R. Santagata, S. Kaiser, Y. Liu, C. Vassillo, P. Ghisellini, *et al.*, Socioeconomic and environmental benefits of expanding urban green areas: A joint application of i-Tree and LCA approaches, *Land* **11**(12) (2022).
67. B.M. Petersen, M.T. Knudsen, J.E. Hermansen and N. Halberg, An approach to include soil carbon changes in life cycle assessments, *J. Clean. Prod.* **52**, 217–224 (2013).
68. J. Bare, Recommendation for land use impact assessment: First steps into framework, theory, and implementation, *Clean Technol. Environ. Policy* **13**(1), 7–18 (2011).
69. T. Koellner, L. De Baan, T. Beck, M. Brandão, B. Civit, M. Margni, *et al.*, UNEP-SETAC guideline on global land use impact assessment on biodiversity and ecosystem services in LCA, *Int. J. Life Cycle Assess.* **18**(6), 1188–1202 (2013).
70. R. Müller-Wenk and M. Brandão, Climatic impact of land use in LCA — carbon transfers between vegetation/soil and air, *Int. J. Life Cycle Assess.* **15**(2), 172–182 (2010).
71. V. Cao, M. Margni, B.D. Favis and L. Deschênes, Aggregated indicator to assess land use impacts in life cycle assessment (LCA) based on the economic value of ecosystem services, *J. Clean. Prod.* **94**, 56–66 (2015).

72. H.K. Jeswani, A. Chilvers and A. Azapagic, Environmental sustainability of biofuels: A review, *Proc. Math. Phys. Eng. Sci.* **476**(2243), 20200351 (2020).
73. D. Moore, V. Bach, M. Finkbeiner, T. Honkomp, H. Ahn, M. Sprenger, *et al.*, Offsetting environmental impacts beyond climate change: The Circular Ecosystem Compensation approach, *J. Environ. Manage.* **329**, 117068 (2023).
74. M. Finkbeiner and V. Bach, Life cycle assessment of decarbonization options — towards scientifically robust carbon neutrality, *Int. J. Life Cycle Assess.* **26**(4), 635–639 (2021).
75. G.M. Tucker, F. Quétier and W. Wende, Guidance on achieving no net loss or net gain of biodiversity and ecosystem services, Institute for European Environmental Policy, European Commission, DG Environment, Brussels (2020).
76. B.A. McKenney and J.M. Kiesecker, Policy development for biodiversity offsets: A review of offset frameworks, *Environ. Manage.* **45**(1), 165–176 (2010).
77. CSBI, A cross-sector guide for implementing the Mitigation Hierarchy, Cross Sector Biodiversity Initiative, The Biodiversity Consultancy, Cambridge, UK (2015). http://www.csbi.org.uk/our-work/mitigation-hierarchy-guide/
78. C.P. VanderWilde and J.P. Newell, Ecosystem services and life cycle assessment: A bibliometric review, *Resour. Conserv. Recycl.* **169**, 105461 (2021).
79. B. Rugani and M. Allocco, A Standards Application Protocol to establish environmental footprint neutrality based on combined LCA-ES assessment procedures, Heraklion, Greece (2022). https://www.espconference.org/europe22/wiki/754946/session-overview-book-of-abstracts#Thematic%20 sessions
80. J. Babí Almenar, C. Petucco, G. Sonnemann, D. Geneletti, T. Elliot and B. Rugani, Modelling the net environmental and economic impacts of urban nature-based solutions by combining ecosystem services, system dynamics and life cycle thinking: An application to urban forests, *Ecosyst. Serv.* **60**, 101506 (2023).
81. Y. Xue and B.R. Bakshi, Metrics for a nature-positive world: A multiscale approach for absolute environmental sustainability assessment, *Sci. Tot. Environ.* **846**, 157373 (2022).
82. A. Acampora, L. Ruini, G. Mattia, C.A. Pratesi and M.C. Lucchetti, Towards carbon neutrality in the agri-food sector: Drivers and barriers, *Resour. Conserv. Recycl.* **189** (2023).
83. E. Kavehei, G.A. Jenkins, M.F. Adame and C. Lemckert, Carbon sequestration potential for mitigating the carbon footprint of green stormwater infrastructure, *Renew. Sust. Energ. Rev.* **94**, 1179–1191 (2018).
84. EIB, Investing in nature: Financing conservation and nature-based solutions. A practical Guide for Europe — Including how to access support from the

European Investment Bank's dedicated Natural Capital Financing Facility (2019).

85. U. Aslam, M. Termansen and L. Fleskens, Investigating farmers' preferences for alternative PES schemes for carbon sequestration in UK agroecosystems, *Ecosyst. Serv.* **27**, 103–112 (2017).

86. Life-Terra, Life Terra project, co-financed by the European Commission through the LIFE Programme (LIFE19 CCM/NL/001200) (2021). https://www.lifeterra.eu/

87. zeroCO2, Plant, adopt, or gift a tree (2023). https://zeroco2.eco/it/

88. WOWnature, WOWnature® project (2023). https://www.wownature.eu/

89. A. Mora Rollo, A. Rollo and C. Mora, The tree-lined path to carbon neutrality, *Nat. Rev. Earth Environ.* **1**(7), 332–332 (2020).

90. J. Babí Almenar, C. Petucco, T. Navarrete Gutiérrez, L. Chion and B. Rugani, Assessing net environmental and economic impacts of urban forests: An online decision support tool, *Land* **12**(1), 70 (2023).

91. Ecosystem services procedure: Impact demonstration and market tools — FSC-PRO-30-006 V1-2 EN, Bonn, Germany (2021).

92. PAS 2060:2014 — Specification for the demonstration of carbon neutrality, London, United Kingdom (2014).

93. Net Zero Guidelines — Accelerating the transition to net zero. International Workshop Agreement IWA 42:2022(E) (2022).

94. OurWorldinData.org, Global primary energy consumption by source (2023). https://ourworldindata.org/grapher/global-energy-substitution

95. R. Portmann, U. Beyerle, E. Davin, E. M. Fischer, S. De Hertog and S. Schemm, Global forestation and deforestation affect remote climate via adjusted atmosphere and ocean circulation, *Nat. Commun.* **13**(1), 5569 (2022).

96. J. Busch, J. Engelmann, S.C. Cook-Patton, B.W. Griscom, T. Kroeger, H. Possingham, *et al.*, Potential for low-cost carbon dioxide removal through tropical reforestation, *Nat. Clim. Change* **9**(6), 463–466 (2019).

97. J.-F. Bastin, Y. Finegold, C. Garcia, D. Mollicone, M. Rezende, D. Routh, *et al.*, The global tree restoration potential, *Science* **365**(6448), 76–79 (2019).

98. H.D. Matthews, K. Zickfeld, M. Dickau, A.J. MacIsaac, S. Mathesius, C.-M. Nzotungicimpaye, *et al.*, Temporary nature-based carbon removal can lower peak warming in a well-below 2°C scenario, *Commun. Earth Environ.* **3**(1), 65 (2022).

Chapter 13

Multi-Criteria Decision-Making Framework For Sustainability Assessment

Dhanush MAJJI and Arnab DUTTA*

*Department of Chemical Engineering,
Birla Institute of Technology and Science (BITS) Pilani,
Hyderabad Campus
Jawahar Nagar, Kapra Mandal, Hyderabad 500 078,
Telangana, India*

**arnabdutta@hyderabad.bits-pilani.ac.in*

Owing to increased concerns pertaining to global warming and its impact on climate change, alternate fuels as a replacement for conventional transportation fuels have emerged as a potential greener solution to reduce carbon emissions. However, it is important to assess the sustainability of any alternate option. In this context, mathematical tools like multi-criteria decision-making are important techniques that can aid in assessing different options under conflicting sustainability criteria. In this chapter, the evaluation of the following fuels: conventional gasoline (CG), conventional diesel (CD), compressed natural gas (CNG), a blend of 15% gasoline and 85% methanol by volume (M85), a blend of 85% ethanol and 15% gasoline by volume (E85), and pure ethanol (E100) are carried out. Based on the results obtained from two different multi-criteria decision-making methods, i.e., AHP-TOPSIS and AHP-PROMETHEE, it is evident that compressed natural gas is observed to be a better choice as opposed to conventional fuels for alternative fuel vehicles.

1. Background

The rise in population and improved standards of living in the last two decades of the 20th century has created great demand for the usage of transportation vehicles [1]. As a result, the consumption of fossil fuels has seen a significant rise in carbon emissions [2]. Combustion of current fossil fuels like petrol and diesel emit a lot of greenhouse gases (GHGs), leading to global warming, a serious environmental concern [3]. Efforts for developing alternative fuels for transportation have increased to reduce harmful emissions. Rising carbon dioxide (CO_2) emissions worldwide (Fig. 1(a) and 1(b)) make the development of alternative fuel vehicles (AFVs), which run on unconventional fuels, i.e., other than petrol or diesel, potential options to address rising climate concerns globally [4]. This will not only help to improve the urban air quality but also make nations more sustainable.

Several factors — technical, economic, and environmental — should be considered to assess the development and deployment of AFVs. Economic factors include vehicle costs as well as operational and

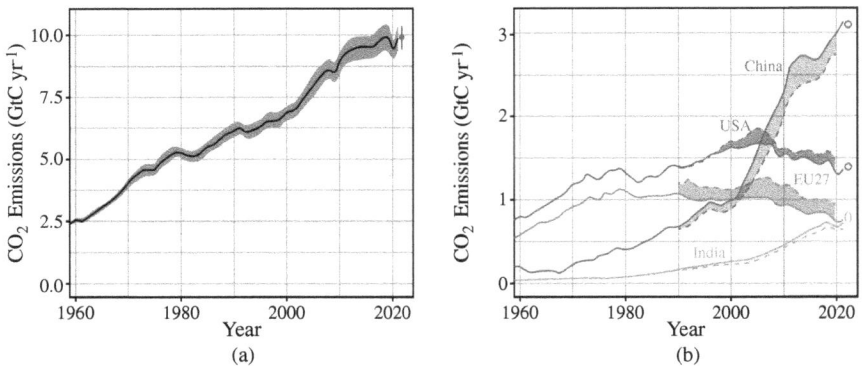

Fig. 1.   Fossil CO_2 emissions for: (a) the globe, including an uncertainty of $\pm 5\%$ (gray shading) and a projection through the year 2022 (red dot and uncertainty range), and (b) territorial (solid lines) and consumption (dashed lines) emissions for the top three country emitters (US, China, India) and the European Union (EU27). (Reproduced from Ref. [4] under the Creative Commons license.)

maintenance costs. Environmental factors include emissions of GHGs and various other atmospheric pollutants, whereas technical factors include the availability of technologies for the development of the vehicle, infrastructure, government taxes, etc. Based on all these factors, an alternative fuel, which can be a substitute for current fossil fuels, is to be chosen for the AFVs. In this context, multi-criterion decision-making (MCDM) is indeed an effective tool that can be used for the selection of the best alternative under several conflicting criteria. Analytical Hierarchy Process (AHP), Technique for Order of Preference by Similarity to Ideal Solution (TOPSIS), and PROMETHEE (Preference Ranking Organization METHod of Enrichment Evaluation) are some of the most commonly used methods to formulate an MCDM framework [5–8]. Figure 2 illustrates a general MCDM framework [9]. Uliasz-Misiak and colleagues [10] used the MCDM technique to select sites for integrated CO_2 storage and geothermal energy recovery based on different geological criteria. Strantzali and colleagues [11] developed a decision support framework using the MCDM technique, which can evaluate different liquefied natural gas supply options for Greece. Puig-Gamero and colleagues [12] used an MCDM technique to assess 15 different alternatives and obtain the optimum blend in the co-gasification of olive pomace, almond shells and petcoke, taking into account four criteria: H_2/CO ratio, gasification reactivity, CO_2 burden, and raw material price. Osorio-Tejada and colleagues [13] used the MCDM technique to assess the impact of LNG in comparison to hydrotreated vegetable oil and diesel oil as transport fuels for freight in Spain. Hansson and colleagues [14] used the MCDM methodology to rank several alternative marine fuels using economic, environmental, technical, and social aspects. Streimikiene and colleagues [15] used the MCDM framework to assess the environmental technologies in road transport. Zhou and colleagues [16] assessed the performance of conventional fuels and renewable energies that can be used for AFVs. Several fuels like conventional gasoline (CG), conventional diesel (CD), compressed natural gas (CNG), a blend of 15% gasoline and 85% methanol by volume (M85), a blend of 85% ethanol and 15% gasoline by volume (E85), and pure ethanol (E100) were assessed based on several factors. Life cycle cost (LCC), global warming potential (GWP), net energy yield (NE), and non-renewable resource depletion potential (NRDP) were taken into account in order to evaluate the performance of these fuels.

Thus, it is evident that MCDM can undoubtedly be an important mathematical framework to assess different fuel options for AFVs. In the

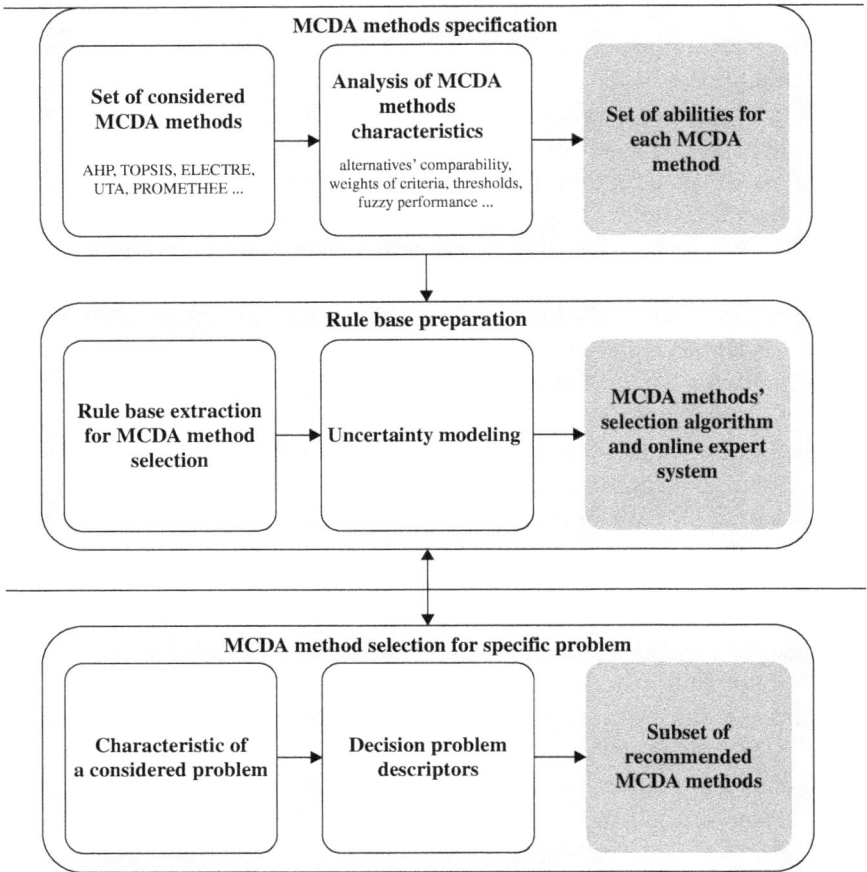

Fig. 2.   General MCDM framework. (Reproduced from Ref. [9] under the Creative Commons license.)

next section, different methods pertaining to the MCDM framework are presented in details. This is then followed by a case study related to the selection of fuels for AFVs.

2. MCDM Methodology

2.1. *AHP method*

AHP is a technique to find the weights for each criterion [17]. The following steps elucidate the steps involved in the AHP method.

Step 1: The relative importance of each criterion with respect to others are obtained from the information available in the open literature. Based on this information, a pairwise comparison matrix is computed, which gives the relative importance of each criterion with respect to others.

Step 2: The pairwise normalized matrix is obtained by normalizing the comparison matrix using Eq. (1)

$$a'_{pj} = \frac{a_{pj}}{\Sigma^n_{p=1} a_{pj}} \forall p=1, 2, ...,n \quad \text{and} \quad j=1, 2, ...,n \tag{1}$$

where n represents the number of criteria.

Step 3: Criterion weights (w_j) are obtained from the pairwise normalized matrix using Eq. (2).

$$w_j = \frac{\Sigma^n_{j=1} a'_{pj}}{n}. \tag{2}$$

Step 4: The pairwise comparison matrix is multiplied with criterion weights to obtain the weighted pairwise comparison matrix using Eq. (3).

$$W_{pj} = w_j * a_{pj}. \tag{3}$$

Step 5: Weighted sum values are calculated from the weighted pairwise comparison matrix using Eq. (4).

$$\textbf{weighted sum value}_p = \sum_{j=1}^{n} W_{pj}. \tag{4}$$

Step 6: The value of λ_{max} is obtained using the criteria weights and weighted sum values using Eq. (5).

$$\lambda_{max} = \frac{\Sigma^n_{p=1} \frac{(\text{weighted sum value})_p}{w_p}}{n}. \tag{5}$$

Step 7: The values of the consistency index (CI) and the consistency ratio (CR) are evaluated using Eqs. (6) and (7).

$$CI = \frac{\lambda_{max} - n}{n - 1}, \tag{6}$$

$$CR = \frac{CI}{\text{Random index}}. \tag{7}$$

Table 1.   Random index (RI) values corresponding to the number of criteria (n).

N	1	2	3	4	5	6	7	8	9	10
RI	0	0	0.58	0.9	1.12	1.24	1.32	1.41	1.45	1.49

The value of the random index (RI) depends on the number of criteria involved in the problem under investigation [17]. Table 1 denotes the value of RI corresponding to the value of n.

If the consistency ratio is <0.10, then the degree of consistency is considered to be satisfactory, and the criteria weights obtained from the AHP method can be used within the MCDM framework.

2.2. *TOPSIS method*

TOPSIS is a distance-based MCDM method [18]. This approach is predicated on the notion that the selected alternative is closest to the ideal value and farthest from the negative ideal value. The steps involved in the TOPSIS method are given below.

Step 1: The decision matrix is used to obtain the values of the ideal best (X_j^+) and the ideal worst (X_j^-) for each criterion. For the beneficial criteria, the maximum and minimum values correspond to X_j^+ and X_j^-, respectively. However, for the non-beneficial criteria, the minimum and maximum values correspond to X_j^+ and X_j^-, respectively. Based on the values of X_j^+ and X_j^- for each criterion, the decision matrix is normalized using Eq. (8).

$$b_{ij} = \frac{X_{ij} - X_j^-}{X_j^+ - X_j^-}. \tag{8}$$

Step 2: The decision matrix is normalized using Eq. (8), and the weighted normalized matrix (b_{ij}') is obtained by multiplying criteria weights from the AHP method with the normalized matrix using Eq. (9).

$$b_{ij}' = w_j * b_{ij}. \tag{9}$$

Step 3: The ideal (f_j^*) and the negative ideal (f_j^{**}) values for each criterion as given by Eqs. (10) and (11), respectively, are computed using the weighted normalized matrix.

$$\mathbf{f_j^*} = \max_i(\mathbf{b_{ij}'}), \tag{10}$$

$$\mathbf{f_j^{**}} = \min_i(\mathbf{b_{ij}'}). \tag{11}$$

<u>*Step 4*</u>: The values of separation measures (D_i^+ and D_i^-) are calculated using the concept of Euclidean distance, as given in Eqs. (12) and (13).

$$\mathbf{D_i^+} = \sqrt{\sum_{j=1}^{n} (\mathbf{b_{ij}'} - \mathbf{f_j^*})^2}, \tag{12}$$

$$\mathbf{D_i^-} = \sqrt{\sum_{j=1}^{n} (\mathbf{b_{ij}'} - \mathbf{f_j^{**}})^2}. \tag{13}$$

<u>*Step 5*</u>: The relative closeness (C_i) values for each alternative are calculated using the values of separation measures, as per Eq. (14).

$$\mathbf{C_i} = \frac{\mathbf{D_i^-}}{\mathbf{D_i^- + D_i^+}}. \tag{14}$$

<u>*Step 6*</u>: The alternatives are ranked according to the descending order of the relative closeness values.

2.3. *PROMETHEE method*

PROMETHEE is one of the outranking-based MCDM methods developed by Brans in 1982 and is based on the preference function approach [19]. In this method, it is assumed that the alternatives are evaluated according to the beneficial and non-beneficial criteria [20]. The following steps describe the PROMETHEE method.

<u>*Step 1*</u>: The decision matrix is normalized using Eq. (8), and the pairwise differences $d_j(x, y)$ between the alternatives x and y for criterion j are calculated.

<u>*Step 2*</u>: The preference function value $P_j(x, y)$ matrix is calculated using Eq. (15)

$$\mathbf{P_j(x, y)} = \begin{cases} \mathbf{0,} & \mathbf{d_j(x, y) \le 0} \\ \mathbf{d_j,} & \mathbf{d_j(x, y) > 0} \end{cases} \tag{15}$$

<u>*Step 3*</u>: The multicriterion preference index $\pi(x, y)$ for all criteria is calculated using Eq. (16)

$$\pi(x, y) = \frac{\sum_{j=1}^{n} w_j * P_j(x, y)}{\sum_{J=1}^{n} w_j}.$$ (16)

Step 4: The values of the leaving $\phi^+(i)$ and the entering $\phi^-(i)$ outranking flows are calculated using Eqs. (17) and (18), respectively.

$$\phi^+(i) = \frac{\sum \pi(x, y)}{m-1},$$ (17)

$$\phi^-(i) = \frac{\sum \pi(y, x)}{m-1}.$$ (18)

Step 5: The net outranking flow ϕ_i, which is the difference between the leaving $\phi^+(i)$ and the entering $\phi^-(i)$ outranking flows for each alternative, is calculated using Eq. (19).

$$\phi_i = \phi^+(i) - \phi^-(i).$$ (19)

Step 6: The alternatives are ranked based on the descending order of the net outranking flow values ϕ_i.

3. Case Study

In this case study, the performances of conventional fuels and alternative renewable fuels were compared based on the conflicting environmental factors. Six alternative fuels, i.e., conventional gasoline (CG), conventional diesel (CD), compressed natural gas (CNG), a blend of 15% gasoline and 85% methanol by volume (M85), a blend of 85% ethanol and 15% gasoline by volume (E85), and pure ethanol (E100), were chosen for performing the comparative analysis [16]. These six alternative fuels were assessed based on the following criteria:

1. LCC (Yuan/km)
2. GWP (gCO$_2$/km)
3. NE yield (MJ/MJ)
4. NRDP (MJ/km)

LCC, GWP and NRDP were considered as the non-beneficial criteria (NB), whereas NE is considered as the beneficial criteria (B). The decision matrix used for the MCDM framework is shown in Table 2.

Table 2. Decision matrix for six fuels [16].

Sl. No.	Sample	LCC (Yuan/km)	GWP (gCO$_2$/km)	NE (MJ/MJ)	NRDP (MJ/km)
1	CG	0.396	514	0.763	9.010
2	CD	0.402	378	0.804	7.123
3	CNG	0.183	472	0.791	4.6
4	M85	0.366	491	0.487	5.469
5	E85	0.299	378	0.441	3.632
6	E100	0.292	362	0.301	1.298

4. Results and Discussion

Six alternative fuels (CG, CD, CNG, M85, E85, and E100) are used for the selection of the best alternative fuel for transportation. Four different criteria (LCC (Yuan/km), GWP (gCO$_2$/km), NE (MJ/MJ), and NRDP (MJ/km)) are considered for the MCDM framework. LCC, GWP, and NRDP are considered as the non-beneficial criteria (NB), whereas NE is considered as the beneficial criteria (B). The results obtained from the proposed MCDM framework are presented in the following sections.

4.1. *AHP*

The relative importance of each criterion from the perspective of an alternate renewable fuel to conventional fuel was extracted from Zhou and colleagues [16]. The pair-wise comparison matrix, as shown in Table 3,

Table 3. Pairwise comparison matrix used for fuel ranking.[#]

Criteria	LCC	GWP	NE	NRDP
Type	NB	NB	B	NB
LCC	1	0.5	1	1
GWP	2	1	1	1
NE	1	1	1	2
NRDP	1	1	0.5	1

[#]B: Beneficial, NB: Non-beneficial

Table 4.   Normalized pairwise comparison matrix and criteria weights obtained using the AHP method.

Criteria	LCC	GWP	NE	NRDP	Weights
LCC	0.2	0.1429	0.2857	0.2	0.2071
GWP	0.4	0.2857	0.2857	0.2	0.2929
NE	0.2	0.2857	0.2857	0.4	0.2929
NRDP	0.2	0.2857	0.1429	0.2	0.2071

indicates the relative importance of each criterion with respect to other criterions.

The pairwise comparison matrix was then normalized using Eq. (1), and weights were then applied to each criterion. The normalized pairwise comparison matrix, along with weights for individual criteria, is given in Table 4.

The criteria weights obtained using the AHP method were further checked to assess the consistency index and the consistency ratio. The consistency index was found to be 0.0404, and the consistency ratio was found to be 0.0449 for a random index value of 1.25. Since the consistency ratio obtained in AHP was less than 0.1, the weights can be used in TOPSIS and PROMETHEE methods to rank all the fuels.

4.2. *Rankings based on the AHP-TOPSIS method*

The best alternative in TOPSIS was chosen based on the shortest Euclidean distance from the ideal solution. The performance and emission criteria of the alternative fuels were normalized using Eq. (8). The weighted normalized matrix (Table 5) was obtained by multiplying the normalized matrix with criteria weights obtained from the AHP method. The ideal and negative ideal values shown in Table 6 were obtained from the weighted normalized matrix. Based on these values, the separation measures and relative closeness for each alternative were calculated, as given in Table 7. The fuel samples were sorted in descending order of their relative closeness values and ranked accordingly.

Figure 3 illustrates the ranking of all the fuel samples obtained using the AHP-TOPSIS method. CNG was ranked as the best alternative, followed by CD. E85 and E100 were ranked 3 and 4, respectively. CG and M85 were considered the worst alternatives, with rankings of 5 and 6, respectively.

Table 5. Weighted normalized decision matrix.

S. No.	Sample	LCC	GWP	NE	NRDP
1	CG	0.0057	0.0000	0.2690	0.0000
2	CD	0.0000	0.2620	0.2929	0.0507
3	CNG	0.2071	0.0809	0.2853	0.1185
4	M85	0.0341	0.0443	0.1083	0.0951
5	E85	0.0974	0.2620	0.0815	0.1445
6	E100	0.1040	0.2929	0.0000	0.2071

Table 6. Ideal and negative ideal values for each criterion.

	LCC	GWP	NE	NRDP
Ideal	0.2071 (CNG)	0.2929 (E100)	0.2929 (CD)	0.2071 (E100)
Negative ideal	0 (CD)	0 (CG)	0 (E100)	0 (CG)

Table 7. Values of separation measures and relative closeness in the AHP-TOPSIS method.

Sl. No.	Sample	Separation Measure w.r.t Ideal Values	Separation Measure w.r.t Negative-Ideal Values	Relative Closeness
1	CG	0.4121	0.2690	0.3950
2	CD	0.2614	0.3962	0.6025
3	CNG	0.2299	0.3806	0.6235
4	M85	0.3720	0.1546	0.2936
5	E85	0.2482	0.3251	0.5671
6	E100	0.3105	0.3735	0.5461

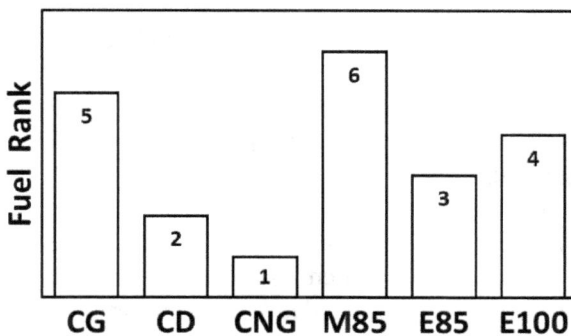

Fig. 3. Ranking of fuels based on the AHP-TOPSIS method.

4.3. *Rankings based on the AHP-PROMETHEE method*

The values of pairwise differences between the alternatives are given in Table 8. Using these values, the preference function values were evaluated, as shown in Table 9. Based on criteria weights from the AHP method, the preference index values (Table 10) were computed and used to calculate the leaving, entering, and net outranking flows of the fuel samples. The values of leaving, entering, and net outranking flows and the ranking of fuels for different loads are given in Table 11.

Figure 4 illustrates the rankings of all fuel samples. CNG was considered the best alternative fuel followed by CD. E100 and E85 were ranked 3 and 4, respectively. M85 and CG were considered the worst alternatives, with rankings of 5 and 6, respectively.

5. Conclusion

Sustainability performance evaluation of the following fuels — conventional gasoline (CG), conventional diesel (CD), compressed natural gas (CNG), a blend of 15% gasoline and 85% methanol by volume (M85), a blend of 85% ethanol and 15% gasoline by volume (E85), and pure ethanol (E100) — were assessed on several factors. The MCDM approach was applied consisting of methods like AHP-TOPSIS and AHP-PROMETHEE, which were successfully used to rank all the fuel alternatives. Based on the above results it was observed that:

- The criteria weights obtained using the AHP method were found to be satisfactory, with a consistency ratio of 0.0449, which is less than 0.1. Thus, these weights can be used reliably for AHP-TOPSIS and AHP-PROMETHEE methods.
- It can be clearly observed from the AHP method that GWP and NE have more weightage.
- It is evident from the rankings obtained from the AHP-TOPSIS and AHP-PROMETHEE methods that CNG is the best substitute for the conventional fuels.

CNG can be a potential alternative fuel that can be used in the engines of AFVs with low levels of GHG emissions. This case study demonstrates how the MCDM framework can indeed be an important methodology for a sustainability assessment of different unconventional options and aid in its adoption for a greener future.

Table 8.   Pairwise differences matrix.

LCC						
Sample	CG	CD	CNG	M85	E85	E100
CG	0.0000	0.0274	−0.9726	−0.1370	−0.4429	−0.4749
CD	−0.0274	0.0000	−1.0000	−0.1644	−0.4703	−0.5023
CNG	0.9726	1.0000	0.0000	0.8356	0.5297	0.4977
M85	0.1370	0.1644	−0.8356	0.0000	−0.3059	−0.3379
E85	0.4429	0.4703	−0.5297	0.3059	0.0000	−0.0320
E100	0.4749	0.5023	−0.4977	0.3379	0.0320	0.0000

GWP						
Sample	CG	CD	CNG	M85	E85	E100
CG	0.0000	−0.8947	−0.2763	−0.1513	−0.8947	−1.0000
CD	0.8947	0.0000	0.6184	0.7434	0.0000	−0.1053
CNG	0.2763	−0.6184	0.0000	0.1250	−0.6184	−0.7237
M85	0.1513	−0.7434	−0.1250	0.0000	−0.7434	−0.8487
E85	0.8947	0.0000	0.6184	0.7434	0.0000	−0.1053
E100	1.0000	0.1053	0.7237	0.8487	0.1053	0.0000

NE						
Sample	CG	CD	CNG	M85	E85	E100
CG	0.0000	−0.0815	−0.0557	0.5487	0.6402	0.9185
CD	0.0815	0.0000	0.0258	0.6302	0.7217	1.0000
CNG	0.0557	−0.0258	0.0000	0.6044	0.6958	0.9742
M85	−0.5487	−0.6302	−0.6044	0.0000	0.0915	0.3698
E85	−0.6402	−0.7217	−0.6958	−0.0915	0.0000	0.2783
E100	−0.9185	−1.0000	−0.9742	−0.3698	−0.2783	0.0000

NRDP						
Sample	CG	CD	CNG	M85	E85	E100
CG	0.0000	−0.2447	−0.5718	−0.4592	−0.6974	−1.0000
CD	0.2447	0.0000	−0.3272	−0.2145	−0.4527	−0.7533
CNG	0.5718	0.3272	0.0000	0.1127	−0.1255	0.4282
M85	0.4592	0.2145	−0.1127	0.0000	−0.2382	−0.5408
E85	0.6974	0.4527	0.1255	0.2382	0.0000	−0.3026
E100	1.0000	0.7553	0.4282	0.5408	0.3026	0.0000

Table 9.   Preference function values.

LCC						
Sample	CG	CD	CNG	M85	E85	E100
CG	0.0000	0.0274	0.0000	0.0000	0.0000	0.0000
CD	0.0000	0.0000	0.0000	0.0000	0.0000	0.0000
CNG	0.9726	1.0000	0.0000	0.8356	0.5297	0.4977
M85	0.1370	0.1644	0.0000	0.0000	0.0000	0.0000
E85	0.4429	0.4703	0.0000	0.3059	0.0000	0.0000
E100	0.4749	0.5023	0.0000	0.3379	0.0320	0.0000

GWP						
Sample	CG	CD	CNG	M85	E85	E100
CG	0.0000	0.0000	0.0000	0.0000	0.0000	0.0000
CD	0.8947	0.0000	0.6184	0.7434	0.0000	0.0000
CNG	0.2763	0.0000	0.0000	0.1250	0.0000	0.0000
M85	0.1513	0.0000	0.0000	0.0000	0.0000	0.0000
E85	0.8947	0.0000	0.6184	0.7434	0.0000	0.0000
E100	1.0000	0.1053	0.7237	0.8487	0.1053	0.0000

NE						
Sample	CG	CD	CNG	M85	E85	E100
CG	0.0000	0.0000	0.0000	0.5487	0.6402	0.9185
CD	0.0815	0.0000	0.0258	0.6302	0.7217	1.0000
CNG	0.0557	0.0000	0.0000	0.6044	0.6958	0.9742
M85	0.0000	0.0000	0.0000	0.0000	0.0915	0.3698
E85	0.0000	0.0000	0.0000	0.0000	0.0000	0.2783
E100	0.0000	0.0000	0.0000	0.0000	0.0000	0.0000

NRDP						
Sample	CG	CD	CNG	M85	E85	E100
CG	0.0000	0.0000	0.0000	0.0000	0.0000	0.0000
CD	0.2447	0.0000	0.0000	0.0000	0.0000	0.0000
CNG	0.5718	0.3272	0.0000	0.1127	0.0000	0.0000
M85	0.4592	0.2145	0.0000	0.0000	0.0000	0.0000
E85	0.6974	0.4527	0.1255	0.2382	0.0000	0.0000
E100	1.0000	0.7553	0.4282	0.5408	0.3026	0.0000

Table 10.   Preference index values.

Sample	CG	CD	CNG	M85	E85	E100
CG	0.0000	0.0057	0.0000	0.1607	0.1875	0.2690
CD	0.3366	0.0000	0.1887	0.4023	0.2113	0.2929
CNG	0.4171	0.2749	0.0000	0.4100	0.3135	0.3884
M85	0.1678	0.0785	0.0000	0.0000	0.0268	0.1083
E85	0.4982	0.1912	0.2071	0.3304	0.0000	0.0815
E100	0.5984	0.2913	0.3006	0.4306	0.1001	0.0000

Table 11.   Values of leaving, entering, and net outranking flows in the AHP-PROMETHEE method.

Sl. No.	Sample	Leaving Outranking Flow	Entering Outranking Flow	Net Outranking Flow
1	CG	0.1246	0.4036	−0.2791
2	CD	0.2863	0.1683	0.1180
3	CNG	0.3608	0.1393	0.2215
4	M85	0.0763	0.3468	−0.2705
5	E85	0.2617	0.1678	0.0938
6	E100	0.3442	0.2280	0.1162

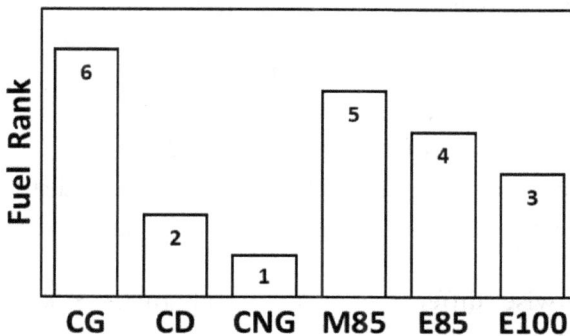

Fig. 4.   Ranking of fuels based on the AHP-PROMETHEE method.

Nomenclature

a_{pj} = element of pairwise comparison matrix measured between criteria p and j

a'_{pj} = element of normalized pairwise comparison matrix measured between criteria p and j

n = number of criteria

w_j = weight of criterion j obtained from AHP method

w_{pj} = element of weighted pairwise comparison matrix measured between criteria p and j

w_p = weight of criterion p obtained from AHP method

(weighted sum value)$_p$ = weighted sum value obtained for criterion p

λ_{max} = maximum eigen value

CI = Consistency Index

CR = Consistency Ratio

RI = Random Index

X_{ij} = element of decision matrix for alternative i and criterion j

X_j^+ = ideal best value w.r.t criterion j

X_j^- = ideal worst value w.r.t criterion j

b_{ij} = element of normalized decision matrix for alternative i and criterion j

v = maximum group utility

m = number of alternatives

b'_{ij} = element of weighted normalized decision matrix for alternative i and criterion j

f_j^* = ideal alternative value w.r.t. criterion j

f_j^{**} = negative ideal alternative value w.r.t. criterion j

D_j^+ = separation measure of alternative i w.r.t ideal value

D_j^- = separation measure of alternative i w.r.t negative ideal value

C_i = relative closeness of alternative i w.r.t negative ideal separation measure

d_j (x, y) = pairwise differences between alternatives x and y w.r.t. criterion j

P_j (x, y) = preference function value measured between alternatives x and y w.r.t criterion j

π (x, y) = multicriterion preference index calculated between alternatives x and y

$\phi^+(i)$ = leaving outranking flow of alternative i

$\phi^-(i)$ = entering outranking flow of alternative i

ϕ_i = net outranking flow of alternative i

References

1. B.D. McNicol, D.A.J. Rand and K.R. Williams, Fuel cells for road transportation purposes — Yes or no? *J. Power Sources* **100**, 47–59 (2001). https://doi.org/10.1016/S0378-7753(01)00882-5

2. A. Dutta, S. Farooq, I.A. Karimi and S.A. Khan, Assessing the potential of CO_2 utilization with an integrated framework for producing power and chemicals. *J. CO_2 Util.* **19**, 49–57 (2017). https://doi.org/10.1016/j.jcou.2017.03.005

3. A. Dutta, Life cycle assessment strategies for carbon capture and utilization processes, in *Life Cycle Assessment: New Developments and Multi-Disciplinary Applications*, World Scientific Publishing, Singapore, pp. 55–75 (2022). https://doi.org/10.1142/9789811245800_0004

4. P. Friedlingstein, M.W. Jones, M. O'Sullivan, R.M. Andrew, D.C.E. Bakker, J. Hauck, *et al.* Global Carbon Budget 2021. *Earth Syst. Sci. Data* **14**, 1917–2005 (2022). https://doi.org/10.5194/essd-14-1917-2022

5. Z. Wang, Y. Wang, G. Xu and J. Ren, Sustainable desalination process selection: Decision support framework under hybrid information. *Desalination* **465**, 44–57 (2019). https://doi.org/10.1016/j.desal.2019.04.022

6. S.S. Bohra, A comprehensive review on applications of multicriteria decision-making methods in power and energy systems, *Int. J. Energy Res.* **46**(4), 4088–4118 (2022). https://doi.org/10.1002/er.7517

7. Y. Tang and F. You, Multicriteria environmental and economic analysis of municipal solid waste incineration power plant with carbon capture and separation from the life-cycle perspective, *ACS Sustain. Chem. Eng.* **6**(1), 937–956 (2018). https://doi.org/10.1021/acssuschemeng.7b03283

8. A. Kumar, B. Sah, A.R. Singh, Y. Deng, X. He and P. Kumar, A review of multi criteria decision making (MCDM) towards sustainable renewable energy development. *Renew. Sustain. Energy Rev.* **69**, 596–609. https://doi.org/10.1016/j.rser.2016.11.191.

9. J. Wątróbski, J. Jankowski, P. Ziemba, A. Karczmarczyk and M. Ziolo, Generalised framework for multi-criteria method selection. Omega (United Kingdom) **86**, 107–124. https://doi.org/10.1016/j.omega.2018.07.004

10. B. Uliasz-Misiak, J. Lewandowska-Śmierzchalska and R. Matuła, Criteria for selecting sites for integrated CO_2 storage and geothermal energy recovery, **285** (2021). https://doi.org/10.1016/j.jclepro.2020.124822.

11. E. Strantzali, K. Aravossis, G.A. Livanos and C. Nikoloudis, A decision support approach for evaluating liquefied natural gas supply options: Implementation on Greek case stud, *J. Clean Prod.* **222**, 414–423 (2019). https://doi.org/10.1016/j.jclepro.2019.03.031

12. M. Puig-Gamero, L. Sanchez-Silva, F. Dorado and P. Sánchez, Multi-criteria analysis for selecting the optimum blend in the co-gasification process, *Comput. Chem. Eng.* **141**, 106983 (2020). https://doi.org/10.1016/j.compchemeng.2020.106983

13. J.L. Osorio-Tejada, E. Llera-Sastresa and S. Scarpellini, A multi-criteria sustainability assessment for biodiesel and liquefied natural gas as alternative fuels in transport systems, *J. Nat. Gas Sci. Eng.* **42**, 169–186 (2017). https://doi.org/10.1016/j.jngse.2017.02.046

14. J. Hansson, S. Månsson, S. Brynolf and M. Grahn, Alternative marine fuels: Prospects based on multi-criteria decision analysis involving Swedish stakeholders, *Biomass Bioenergy* **126**, 159–173 (2019). https://doi.org/10.1016/j.biombioe.2019.05.008

15. D. Streimikiene, T. Baležentis and L. Baležentiene, Comparative assessment of road transport technologies, *Renew. Sustain. Energ. Rev.* **20**, 611–618 (2013). https://doi.org/10.1016/j.rser.2012.12.021

16. Z. Zhou, H. Jiang and L. Qin, Life cycle sustainability assessment of fuels, *Fuel* **86**, 256–263 (2007). https://doi.org/10.1016/j.fuel.2006.06.004

17. R.W. Saaty, The analytic hierarchy process — what it is and how it is used, *Math. Model.* **9**, 161–176 (1987). https://doi.org/10.1016/0270-0255(87)90473-8

18. I. Ertuğrul and N. Karakaşoğlu, Performance evaluation of Turkish cement firms with fuzzy analytic hierarchy process and TOPSIS methods, *Expert Syst. Appl.* **36**, 702–715 (2009). https://doi.org/10.1016/j.eswa.2007.10.014

19. J.P Brans, P. Vincke and B. Mareschal, How to select and how to rank projects: The PROMETHEE method, *Eur. J. Oper. Res.* **24**, 228–238 (1986).

20. M. Behzadian, R.B. Kazemzadeh, A. Albadvi and M. Aghdasi, PROMETHEE: A comprehensive literature review on methodologies and applications, *Eur. J. Oper. Res.* **200**, 198–215 (2010). https://doi.org/10.1016/j.ejor.2009.01.021

Chapter 14

Environmental and Economic Impacts of Food Waste Management — A Focus on Sustainability and Life Cycle Assessment

Matthew FRANCHETTI*,**, Alex SPIVAK*,
Shrijith Ashok KUMAR* and
Lakshika KURUPPUARACHCHI*

*Department of Mechanical, Industrial,
and Manufacturing Engineering,
The University of Toledo,
2801 W. Bancroft St., Ohio, 43606, USA*

**matthew.franchetti@utoledo.edu

The food sector accounts for nearly 30% of the world's total energy consumption and around 22% of the total greenhouse gas (GHG) emissions. To address global food waste concerns, the comparison of different food waste to energy technologies from an economic, energy and emissions standpoint using the Life Cycle Assessment (LCA) is presented in this chapter. The integration of multi-objective optimization and LCA framework, which aims to minimize GHG emissions and energy consumption over the life cycle of food waste management options, will also be reviewed. A total of three prominent waste-to-energy techniques are presented: (1) Combustion Plants — Using a boiler to capture and convert the released heat into electricity and steam, (2) Gasification and

Pyrolysis — Generating electricity by heating the fuel without allowing enough oxygen for total combustion, and (3) Anaerobic Digestion — Using microorganisms to convert the organic wastes into biogas, which can be utilized to generate electricity.

1. Introduction

One third of the food waste that is being disposed of worldwide is a growing concern in terms of environmental, social and economic sustainability. The waste management industry faces its greatest challenges in diverting food waste from landfills. The improper and inefficient strategies in the disposal of food waste can lead to grave environmental, economic and social consequences. Furthermore, one of the most notable barriers to effective proper food waste management is the high economic cost associated with their treatment facilities. This chapter discusses the use of a Life Cycle Assessment (LCA) and optimization modeling techniques for a sustainable food waste management system. Also included in this chapter are a discussion of sustainable goals, food waste to energy conversion techniques, and LCA modeling of food waste management.

2. Food Waste Management: Developing Sustainable Development Goals

2.1. *Introduction*

A total of nearly 1.4 billion tons of food get wasted every year, as per the records of the Food and Agricultural Organization (FAO) of the United Nations (UN) [1]. Countries all over the world are striving to ensure that the amount of food waste is reduced substantially to address the needs of the people globally. Human behavior influences the amount of food waste through all phases — planning, storage, preparation, and consumption of food [2]. Additionally, the lifespan of raw materials, like fruits and vegetables, is not predictable, and when they get purchased in abundance, there is a high possibility that they can get spoiled and damaged, resulting in them getting disposed of or unconsumed.

The focus is primarily on food waste management, as the reduction in food waste ensures a reduction in the loss of food, reduced greenhouse gas (GHG) emissions, and reduced disposal in landfill sites, which reflects in reduced climatic changes.

2.2. Food waste and its management

Generally, disposed food is a combination of food waste and food loss. Research analysis has shown that food waste occurs at the consumer level while food loss occurs at the post-harvest level [3]. Additionally, developed and developing countries have different stages for food waste generation. Developing countries face this difficulty in the early stages of the food supply chain due to various issues during the harvesting, storing and cooling stages of the materials.

Developed countries, on the other hand, face these issues in the final stages of the food supply chain caused by a customer's careless behavior, careless spending, ineffective management of the stocks, etc. [1]. To eliminate the issues faced with food waste, various management techniques are suggested, which include animal feeding, composting, anaerobic digestion, incineration, and landfills. It is highlighted herein that environmental issues are faced when food waste is disposed of at landfills. The food waste gets converted into methane and GHGs, which then contaminates the groundwater level and poisons neighboring ecosystems.

To avoid environmental concerns related to food waste, a focus on anaerobic digestion (AD) and composting is more favorable, as the byproducts of this process generate useful products such as biogas and organic fertilizers [4–6]. Applications of the LCA for food waste management will be presented in Sec. 4. AD and composting is found to have a more significant and positive impact on the ecosystem and do not cause any damage [6].

2.3. Impact of food waste

Food waste causes an adverse impact on different life cycle stages that may include production, transportation or distribution. Apart from negative effects on the environment, useful resources like water, energy, soil and money have gone to waste. The impact is significant and classified into environmental, economic, and social impacts. The decomposition of food waste, which is disposed of in landfills, releases GHGs, with methane being the primary byproduct. Methane is one of the most poisonous GHGs and is 21 times stronger than carbon dioxide. Gas emissions from food disposal contribute to 7% of total GHG emissions, and the main source of this is the improper disposal of food waste.

Food waste impacts the balance of the food ecosystem, and as a result, nearly a billion people all over the world suffer from malnutrition and hunger. The food generated is usually in surplus, and it gets disposed of even though it is in good condition. The consumers and producers are part of the current economic system, and they interact.

Consumer preference influences food producer behavior and the generation of waste. Food waste affects the system and results in price hikes in food stocks. This results in people with minimum incomes struggling to meet their basic needs. To combat the impacts of food waste, the United Nations (UN) has proposed the Sustainable Development Goals, which aims to address and eliminate the impact on the food ecosystem [7].

2.4. *Sustainable Development Goals*

The Sustainable Development Goals (SDGs) were established by the Member States of the UN in 2015. There are 17 goals that focus on creating sustainable development by 2030, as displayed in Fig. 1 [4]. As listed among the 17 goals, SDG 12, which states: "Responsible Consumption and Production", addresses the issues related to food waste.

Fig. 1.   United Nations Sustainable Development Goals.

2.5. *Methods to attain the Sustainable Development Goals*

Countries work to attain SDG 12 to address the issues related to Food Waste, and this is undertaken by primarily focusing on AD and composting of the organic wastes available in Municipal Solid Waste (MSW).

Composting is a biological process where organic waste gets decomposed within an aerobic or anaerobic environment. These processes are sustainable in terms of economic aspects as they involve lower operation costs and produce excellent agricultural soil amendments compared to other waste management options. There are three important factors for composting: microorganisms, moisture control, and aerobic [2].

3. Conversion of Food Waste to Energy: Net-Zero Carbon Initiatives

3.1. *Introduction*

According to UN reports, nearly 1 billion people go undernourished, and another 795 million people go hungry. Besides, 2 billion people consume excessive amounts of food, resulting in them being overweight or obese, leading to the detriment of human health and the environmental ecosystem's balance. Reports show that the food sector accounts for nearly 30% of the world's total energy consumption and around 22% of the total GHG emissions. This issue, combined with poor food supply chains in developed and developing countries, increases the percentage of the total MSW collection across the globe [5].

Waste-to-energy (WTE) is a technologically advanced means of waste disposal that converts non-recyclable waste materials into usable heat, electricity or fuel through different processes like combustion, gasification, AD, and landfill gas recovery [2]. Net-zero carbon is a balance achieved between carbon emissions sent into the atmosphere and what is removed from them. The goal is to obtain carbon neutrality so that the emissions are not higher than the amount that can be removed.

3.2. *Net-Zero Carbon*

Net-zero carbon refers to the global process of adapting energy technologies to minimize GHG emissions. The net-zero goal is not to stop human emissions but to balance the ecosystem by removing an equal amount of

GHG emissions emitted on a regular basis. The process of achieving a balance between the emissions released and emissions that are taken out of the atmosphere is the goal of net-zero carbon and is referred to as carbon neutrality [6].

Various reports indicate that there are seven barriers to achieving net-zero carbon. They are: Economic Barriers, Legislative Barriers, Professional/Technical Barriers, Socio-Cultural Barriers, Technological Barriers, Market Barriers, and Geographical Barriers. Among these barriers, economic barriers are the most significant, while the geographical barrier is the least significant [9]. To overcome these barriers, various steps to be undertaken have been identified. This includes utilizing materials from renewable resources like bamboo, reclaimed wood, timber, and recycled steel, and renewable energy sources like solar water heaters, wind turbines, solar photovoltaics, and Combined Heat and Power (CHP). Utilizing these components reduces carbon emissions and the amount of waste disposed of in landfills, thus having a positive impact on the carbon emission rate.

3.3. *Food waste to energy — conversions*

WTE is the process of converting waste materials into usable energy through technological processes to reduce the emissions of GHGs. With increased waste production, it is essential to utilize and convert the waste generated into a source of energy, as this provides a solution to the energy problem and helps reduce the emission of GHGs that would be released in landfills. Of the total food produced, one-third gets wasted, and they are disposed of in landfills, which, when decomposed, releases a GHG mixture of methane and carbon dioxide.

To combat and eliminate this issue, there are three prominent WTE techniques involved. They are: (1) Combustion Plants — Use of a boiler to capture and convert the released heat into electricity and steam, (2) Gasification and Pyrolysis — Generate electricity by heating the fuel without allowing enough oxygen for total combustion, and (3) AD — Use of microorganisms to convert the organic wastes into biogas [6], which can be utilized to generate electricity [6]. Figure 2 provides a process map for AD [6].

Energy is recovered from the Combustion Plants where the waste from landfills are collected in a bunker and then burned in a boiler.

Anaerobic Digester Facility

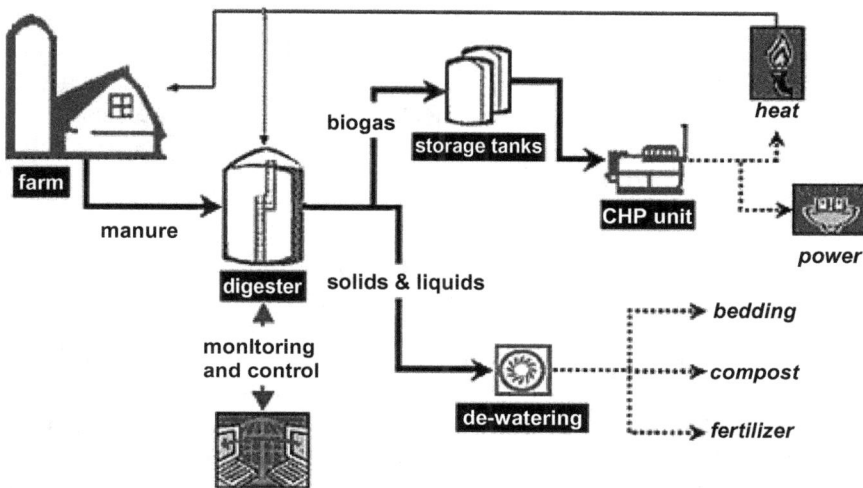

Fig. 2.   Anaerobic Digestion process map.

The heat emitted from combustion is then used to convert water into steam, which is later used to operate a turbine generator to produce electricity. It is made of three components, namely:

- Mass Burn Facilities, which use waste as fuel for burning with excess air. They utilize a sloping, moving grate that vibrates or moves to agitate the waste and mix it with air to be burned properly. They require minimal pre-processing before burning commences, and they process between 25 and 3,000 tons per day, depending on the size and demand.
- Modular Systems, which use various combinations of systems of incineration to burn unprocessed and mixed waste. They are smaller in size and are portable from site to site. They are usually prefabricated units with small to medium capacities of 5 to 120 tons of solid waste per day. Facilities have between 1 and 4 units for a total plant capacity of between 15 and 400 tons per day. The majority output produced is steam as the sole energy product. It is made up of two combustion chambers, which ensures more complete combustion and serves as a primary means of pollution control.

- Refuse Derived Fuel (RDF) Systems, which use mechanical methods to shred incoming waste, separate non-combustible materials, and produce a mixture that can be utilized as fuel. The fuel produced is then used to burn pretreated municipal solid waste and it harnesses the energy to produce steam and electricity. There are two processing wet and dry processing systems. Wet RDF processing is used primarily in the pulp and paper industries, while Dry RDF is used to process solid waste through dry processing into different products like fluff RDF, densified RDF, and powdered RDF.

3.4. *Interaction of net-zero carbon and energy*

WTE technologies, especially those integrated with carbon capture [8], are one among the many solutions to reach the goal of net-zero carbon emissions. Apart from that, utilizing food waste to generate energy can potentially help reduce GHG emissions. Landfills are a source of GHG emissions that can be combated by collecting waste and using them to be converted into energy [6]. According to a 2018 EPA report, over 63 million tons of food waste was generated from commercial, institutional and residential sectors, where only 4% were composted. A total of 24% of the waste in MSW constitutes food waste [9]. In order to realize the net-zero carbon goals, food waste present in the MSW must be converted into energy, and this can be achieved by using various methods. There are different technologies currently in use that focus on energy generation from food waste. They are: (1) Biological Technology — AD, (2) Ethanol Fermentation, and (3) Thermal and Thermochemical Technology — Incineration, Pyrolysis and Gasification, and Hydrothermal Carbonization (HTC).

AD is a process done in the absence of oxygen. This takes place in a sealed vessel called the reactor, where there is no oxygen involved in the process of converting organic matter like food wastes, wastewater biosolids, and animal manure into biogas. The byproducts derived from this process, besides biogas, are liquid digestate and solid digestate. The majority of the food waste collected (90–95%) can be utilized for AD, but high salt concentrations in food waste inhibit the process. This can be prevented by the co-digestion of food waste with low nitrogen, lipid content waste, sewage sludge, and municipal waste, while it helps to increase the CH_4 yield in biogas by 40–50%.

An advantage of AD is that food waste can be co-digested with different substrates, which improve the waste management and production of the desired products. A drawback is in the difference in the types of reactors, composition and quality of the co-substrates, process control factors (temperature, pH, OLR, HRT), and microbial dynamics that need to be optimized to achieve maximum benefits [10]. A case study of Matthew Tomich of Energy Vision and Marianne Mintz of Argonne National Laboratory explores the production and use of renewable compressed natural gas (R-CNG), derived from the AD of organic waste, which can be used as a fuel source for heavy-duty refuse vehicles and other natural gas vehicles in Sacramento, California. This study was conducted in 2017, with the approval and sponsorship of the US Department of Energy's Clean Cities Program [11].

Biogas can be used as vehicle fuel or processed further into an alternative for natural gas, energy products, or advanced biochemicals and bioproducts. Digestate is composed of liquid and solid portions, and they have multiple uses like animal bedding (solids), nutrient-rich fertilizers (solids and liquids), organic-rich compost (solids), and soil amendments (solids), which are used in farms as fertilizers [10]. Alternatively, Ethanol Fermentation involves different pretreatment methods such as acid, alkali, thermal and enzymatic processes that increase the cellulose digestibility. The common method used is Enzymatic Hydrolysis. It is a lucrative approach as it avoids competition from food crops, reduces food waste, and lowers the carbon footprint. It can be used for different types of food waste collected and is a more attractive option for energy conversion. It utilizes a variety of treatment methods to increase its cellulose digestibility, and as a result, the overall economic viability is still an issue that needs to be addressed.

Incineration, or WTE systems, is a process where waste materials get combusted and are converted to heat and energy. This can be used to operate the steam turbines for energy production or as heat exchangers to heat up process steam in industries. One advantage of this process is that it reduces the volume of the waste by 80–85%. However, apart from air pollution concerns, research shows that direct energy recovery from food waste by incineration is a grueling process. This is due to the high moisture content and non-combustible components present in the food waste. When food waste alone is incinerated, the large amount of water content makes it difficult to gather the predicted energy levels for recovery. When they are mixed with MSW, a positive impact is seen, and this is the solution considered. In order to combat GHG emissions, WTE combined with CCS has been proposed as a solution to achieve net-zero carbon emissions [8].

Pyrolysis and Gasification are both thermal processes. Pyrolysis converts food waste in an oxygen-free environment into bio-oil as the major product, along with syngas ($CO + H_2$) and solid biochar. Gasification converts food waste into a combustible gas mixture by partially oxidizing food waste at high temperatures, typically in the range of 800–900°C. This is used as an alternate source to incineration to combat food waste. Gasification and Pyrolysis is a complex process that is made up of a number of physical and chemical interactions that occur at a temperature greater than 600°C, and they depend on the reactor type and waste characteristics, namely, ash softening and melting temperatures. Since there is no process solely dependent on food waste, it is better to combine food waste with solid waste to get a significant result. The cost associated with gasification and pyrolysis is significantly lesser than the incineration plants, and these processes also mitigate the emissions of air pollutants by using no or low oxygen. Since they produce syngas composed mainly of CO and H_2 (85%) with a small proportion of CO_2 and CH_4, it is suitable for food wastes. The results of treatment of the food waste with pre-processed waste or feedstocks have shown that they have good potential for solid waste thermal treatment with an aim for power generation.

HTC is one of the thermal conversion technologies preferred by researchers, especially for waste streams with high moisture content (80–90%). This wet process converts food wastes into valuable, energy-rich resources under autogenous pressures and relatively low temperatures (180–350°C). The advantages of this process can enable reduced overall carbon footprints, greater waste volume reductions, and no process-related odors. It takes only a few hours for the biological processes to succeed, and the high process temperature helps to eliminate the pathogens and inactivate other potential organic contaminants [12].

3.5. *Barriers to food waste to energy production*

The process to achieve energy from food waste faces different barriers. They are: (1) social, (2) technical, (3) economic, and (4) institutional.

- Social barriers involve a lack of public awareness upon the impact of food waste being converted into energy; thus, the energy sector has a great disadvantage. The lack of public awareness is also seen in the increase in the dumping of food waste. Thus, it is essential to create awareness amongst the public about food waste dumping and instill awareness about the benefits of recycling.

- Technical barriers involve the lack of understanding of the working of equipment. Thus, it is essential to employ workers with experience or train new recruits to bring down the failure rate and gain a more sustained output.
- Economic barriers involve reduced investment cost, a lack of competitiveness, and a lack of formal markets, for the co-products are the factors that affect the economic sector. A standardized investment for the process as the initial set-up helps, and a steady inflow of cash to be generated.
- Institutional barriers involving policy changes, disconnectivity from stakeholders, and a lack of communication between management and workers are the main issues faced by the institutions. Steps must be taken by the upper management to interact with workers, provide them with updates, and educate them on the changes in the policy, while providing monetary benefits so that they will feel the need to interact and provide inputs that will help develop the company and gain a stronghold in the competitive market [13].

In order to overcome these barriers, it is essential to understand the aspects that cause these barriers. We highlight in this chapter that proper research approaches be conducted for LCA applications, especially in food waste management. Identifying the proper segments and segregating the policies suitable for the company are steps that must be undertaken.

4. The Use of LCA Models in Food Waste Management

4.1. *Introduction*

LCA modeling of food management is used to analyze the current system and make predictions about the environmental outcomes of proposed models. The scope of the LCA varies depending on the customer's needs or research goals. Organizational assessments may range in scope from the analysis of food waste from specific facilities, such as restaurants, factories and schools to regional, national or global LCAs encompassing food waste from chains of restaurants, global organizations, or school systems. In one noteworthy example, the LCA case study of centralized and decentralized AD with different biogas applications was carried out in Ref. [5]. The overall scope and system boundary of the sample LCA case study is illustrated in Fig. 3.

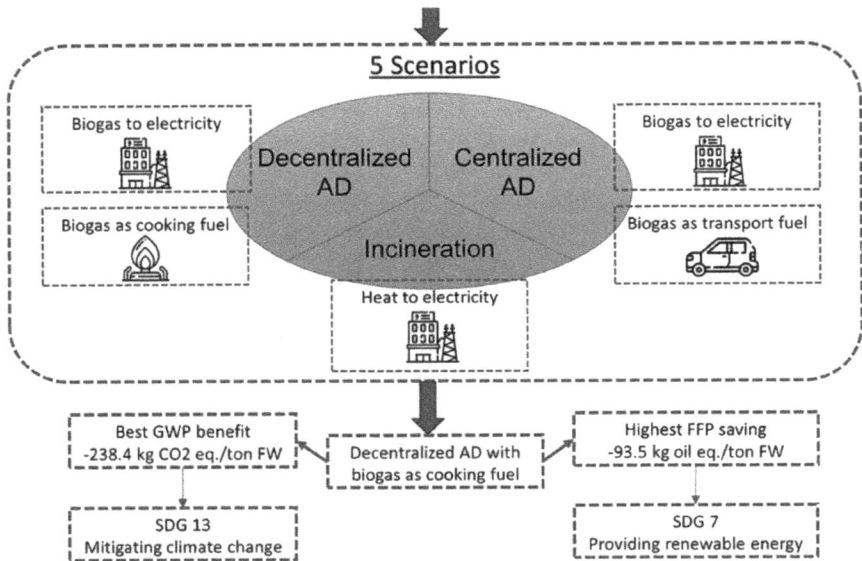

Fig. 3. Life Cycle Assessment system of centralized vs. decentralized Anaerobic Digestion. (Reproduced from Ref. [6] with permission from Elsevier.)

Alternatively, instead of studying all food waste products in one entity, product-based LCAs may be limited to specific products, such as beef waste or tomato waste, on the same or greater scale. Other versions of the product-based LCA may be a study, such as vegetable waste in specific organizations or on the national level. Assessments of the food waste generated by consumers may also be conducted on any level. These LCAs are often more difficult to conduct without associated LCI data relating to the scope of the food waste LCA geographical boundary. Finally, food waste may be analyzed on a national level or, with much greater uncertainty, estimated on a global level.

The general goal of the food waste LCA is to demonstrate how unconsumed food affects environmental and economic factors at the studied level. Such analysis is limited to the part of the food that is not consumed by intended customers. The typical goal of such an LCA is to trace such products across their entire life cycle, typically from seed production, soil, and livestock management to incineration or other end-of-cycle processes.

4.2. *LCA modeling*

Traditionally, LCA models are subdivided into input-output, process-based and hybrid. The latter model encompasses components of the first two models. Another division is between attributional and consequential LCAs.

Generally, attributional assessments give an estimate of what part of the global environmental burdens belong to the study object. In attributional life cycle analysis of the food waste, the average environmental burdens (e.g., carbon generation) due to specific categories of food waste are partitioned between the life cycles considered in the study. This method allows one to estimate the share of the global environmental burdens belonging to a product [14]. It provides information about the impacts of the processes used to produce a product but does not consider indirect effects arising from changes in its output [15, 16]. Attributional LCA aims to describe the environmentally relevant physical flows to and from a life cycle and its subsystems [14, 17]. Such subsystems may include allocated environmental impact of production of farm machinery, crop management, farm stock feed (this process may be cyclical since food waste may be used as wet or dry pig feed), etc.

Consequential assessments give an estimate of how the production and use of the study object affect global environmental burdens [14]. In consequential food waste life cycle analysis, specific sub-cycle data, such as the impact of crop production from a specific location, is not distributed across the entire studied system. Such an approach generally increases the accuracy of the analysis when compared to the attributional method. Consequential LCA provides information about the consequences of changes in the level of output (and consumption and disposal) of a product, including effects both inside and outside the life cycle of the product [15].

In food waste LCA, such analysis allows one to estimate how changes in production, management and consumption practice will impact the direct effect of food waste on the environment, and to some degree, how reduction in, for example, packaging and transportation affect parallel industries. Consequential LCA aims to describe how environmentally relevant flows will change in response to possible decisions [14, 17, 18], such as the reduction of portions in hospitals and restaurants, discounting of food near its expiration date, education of consumers about proper eating habits, etc. These and other management decisions are discussed later in the chapter.

Generally, LCAs are globally accepted and recognized scientific environmental management tools that are evolving to include new sets of environmental indicators to encompass ecosystem services. Some studies are now considering animal welfare, food security, long-term values of organic farming, and other factors. Such a need in food waste LCA is generated by, for example, cases where environmental impacts associated with providing more living space to each chicken is considered. The animal welfare factor is used to compensate for the need of larger facilities. Consequential LCAs with extended input-output are evolving to include long-term effects of organic farming or other practices, such as a reduction in the use of antibiotics in animal farming. Even though the authors of this book are not aware of any studies of long-term environmental consequences due to the improved health of consumers of organic products, such an LCA analysis is conceivable. However, such an approach may not be well suited for food waste analysis.

Another direction in the development of food waste LCA may be towards predictions of consequences of changes in technology or the introduction of new technologies, such as laboratory-produced meat and protein-based meat substitutes. Such studies have been conducted both as predictive studies (not common) and pilot studies (more common).

4.3. *LCA in food waste management options*

LCAs in food waste management include areas such as nearly expired and expired food management, consumer behavior, optimization of packaging and deliveries, and shelf life extension methodologies. The upstream processes include food production, processing, distribution, sales (retail and wholesale), consumption, and waste treatment. The system is often circular since, for example, waste may be used as animal food, fertilizers, and/or energy. Both types of LCA are used in food management analysis, and both methods have advantages and disadvantages. LCAs are used to investigate food management of various scales. Some of the approaches and associated data requirements in LCA applications for waste disposal and recycling options have been discussed in the literature [18]. In our work concerning food waste management, we present the following: (1) Conducted national or regional food waste LCAs, (2) Food waste prevention consequences on a food waste management system, (3) Household food waste minimization, (4) Perishable food waste in the supermarkets, (4) School cafeteria food waste analysis, and (5) Fresh fruit and vegetables waste analysis.

Other outcomes of food waste studies that showed the environmental impact due to a reduction in food waste have been categorized. Some studies considered food waste as part of the study but were not necessarily focused on this topic. Their authors extrapolated some results based on available data. Food production alternatives: (1) Meat/protein-based alternatives may in the future reduce the environmental impact of food waste due to the lower impact of production and there being no bones produced [19, 20], (2) Extrudate vegetable meat alternatives consisting of protein combined with amaranth or buckwheat flour and a vegetable milk alternative made from lentil proteins will ultimately reduce food waste due to the initially lower environmental costs of production, (3) Reduction on production quantities, and (4) Reduction and management of food portions in schools, hospitals, catering, and other food services result in lesser food waste [21, 22, 23]. In the example presented by Mistretta and colleagues [21], the LCA covering the scope of cradle-to-gate is performed for the amount of food production, consumption, and waste in an institutional food catering supply chain. The LCA flow diagram is displayed in Fig. 4. Their work concluded that meat and fish production stages resulted in the highest environmental impacts.

Food waste diversion involves the reduction of environmental impacts due to alternative food waste utilization strategies [24, 25]. Management improvements, such as the management of nearly expired food waste,

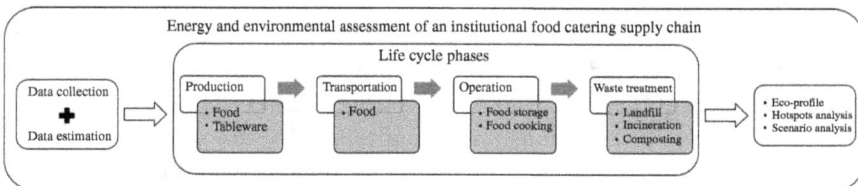

Fig. 4.   Cradle-to-gate LCA of the amount of food consumed for an institutional food catering supply chain. (Reproduced from Ref. [21] with permission from Elsevier.)

result in a significant reduction of the environmental impact [26–30]. Technological comparisons of food waste methods can be found in various types of LCA application case studies [5, 24, 31]. Education and enforcement on household food waste may be significantly reduced as well [32, 33].

4.4. *Concluding remarks: Food waste minimization from an LCA perspective*

The United Nations' Sustainable Development Goal (SDG) 12.3 sets to reduce the per capita global food waste (FW) at the retail and consumer levels by 50% by 2030. Food losses contribute 1.4 kilograms (kg) of carbon dioxide equivalents (CO_2-eq) capita^{-1}day^{-1} (28%) to the overall carbon footprint of the average diet in the United States; in total, this is equivalent to the emissions from 33 million passenger vehicles annually [34]. Currently, one-third of global food production is lost or wasted during the various phases of the food value chain, from farm to final consumption [35, 36]. This causes an economic loss of about $7,500 [37, 38]. Nearly 1.3 billion tonnes (metric tons) of food produced for human consumption is lost or wasted globally each year, with fruits and vegetables accounting for 40–50% [39, 40].

A reasonable definition for food waste is food that is originally produced for human consumption but is then directed to either a non-food use (including animal feed) or waste disposal [41]. In 2017, the United States Environmental Protection Agency (US EPA) estimated that only 6.3% of waste food was diverted from landfills and incinerators for composting. Food accounts for 22% of our municipal solid waste, more than any other type of waste in the US's daily garbage [40, 42]. AD appears to be the most successful technology to convert non-edible food waste into a resource [5, 43]. Food waste minimization is one of the important components of the effort to improve the sustainability of food generation. However, environmental outcomes of the methods discussed earlier in the chapter are affected by initial conditions such as local, regional and global biodiversity, climate, cultural norms, and economic conditions. In return, the listed conditions are also affected by, among other factors, food waste minimization. In other words, there is a co-dependence of factors on each other. The LCA is often a "snapshot" of the current or proposed conditions.

References

1. M. Grosso and L. Falasconi, Addressing food wastage in the framework of UN Sustainable Development Goals, *Waste Manag. Res.* **36**(2), 97–98 (2018).
2. N.H. Nordin, N. Kaida, N.A. Othman, F.N.M. Akhir and H. Hara, Reducing food waste: Strategies for household waste management to minimize the impact of climate change and contribute to Malaysia's sustainable development, *IOP Conf. Ser.: Earth Environ. Sci.* **479**, 012035 (2020).
3. A. Lemaire and S. Limbourg, How can food loss and waste management achieve sustainable development goals? *J. Clean. Prod.* **234**, 1221–1234 (2019).
4. H. Saleh, B. Surya and H. Hamsina, Implementation of sustainable development goals to Makassar zero waste and energy source, *Int. J. Energy Econ. Policy.* **10**(4), 530–538 (2020).
5. H. Tian, X. Wang, E. Y. Lim, J. T.E. Lee, A.W.L. Ee, J. Zhang, *et al.*, Life cycle assessment of food waste to energy and resources: Centralized and decentralized anaerobic digestion with different downstream biogas utilization, *Renew. Sust. Energ. Rev.* **150**, 111489 (2021).
6. E. Vahle, What is net zero carbon waste-to-energy? *Energy Link* (2020). https://goenergylink.com/blog/net-0-carbon-waste-to-energy/
7. A. Seberini, Economic, social and environmental world impacts of food waste on society and zero waste as a global approach to their elimination, *SHS Web Conf.* **74**, 03010 (2020). https://www.shs- conferences.org/articles/shsconf/pdf/2020/02/shsconf_glob2020_03010.pdf [7]
8. P. Wienchol, A. Szlęk and M. Ditaranto, Waste-to-energy technology integrated with carbon capture — Challenges and opportunities, *Energy.* **198**, 117352 (2020).
9. E. Ohene, A.P.C. Chan and A. Darko, Prioritizing barriers and developing mitigation strategies toward net-zero carbon building sector, *Build. Environ.* **223**, 109437 (2022).
10. US EPA, AgStar — Learn about biogas discovery — How does anaerobic digestion work? United States Environmental Protection Agency (2022) https://www.epa.gov/agstar/how-does-anaerobic-digestion-work
11. M. Tomich and M. Mintz, Waste-to-fuel: A case study of converting food waste to renewable natural gas as a transportation fuel, Report/Paper No.

ANL/ESD-17/9, National Academy of Sciences, US (2017). https://trid.trb. org/view/1483159

12. T.P.T. Pham, R. Kaushik, G.K. Parshetti, R. Mahmood and R. Balasubramanian, Food waste-to-energy conversion technologies: Current status and future directions, *Waste Manag.* **38**, 399–408 (2015).

13. J. Ludlow, F. Jalil-Vega, X.S. Rivera, R A. Garrido, A. Hawkes, I. Staffell, *et al.*, Organic waste to energy: Resource potential and barriers to uptake in Chile, *Sustain. Prod. Consum.* **28**, 1522–1537 (2021).

14. T. Ekvall, Attributional and consequential life cycle assessment, in *Sustainability Assessment at the 21st Century*, M.J. Bastante-Ceca, J.L. Fuentes-Bargues, L. Hufnagel, F.-C. Mihai and C. Iatu (eds.), IntechOpen (2020).

15. C.M. Braguglia, A. Gallipoli, A. Gianico and P. Pagliaccia, Anaerobic bio-conversion of food waste into energy: A critical review. *Bioresour. Technol.* **248**, 37–56 (2018).

16. H. Zeng, Y. Yan, F. Liberti, B. Pietro and F. Fantozzi, Technical and eco-nomic feasibility analysis of an anaerobic digestion plant fed with canteen food waste, *Energy Convers. Manag.* **180**, 938–948 (2019).

17. G. Finnveden, M.Z. Hauschild, T. Ekvall, J. Guinée, R. Heijungs, S. Hellweg, *et al*. Recent developments in life cycle assessment, *J. Environ. Manage.* **91**, 1–21 (2009).

18. V. Bisinella, R. Götze, K. Conradsen, A. Damgaard, T.H. Christensen and T.F. Astrup, Importance of waste composition for Life Cycle Assessment of waste management solutions, *J. Clean. Prod.* **164**, 1180–1191 (2017).

19. C. Varela-Ortega, I. Blanco-Gutiérrez, R. Manners and A. Detzel, Life cycle assessment of animal-based foods and plant-based protein-rich alternatives: A socio-economic perspective, *J. Sci. Food Agric.* **102**(12), 5111–5120 (2022).

20. A. Detzel, M. Krüger, M. Busch, I. Blanco-Gutiérrez, C. Varela, R. Manners, *et al.*, Life cycle assessment of animal-based foods and plant-based protein-rich alternatives: An environmental perspective, *J. Sci. Food Agric.* **102**(12), 5098–5110 (2022).

21. M. Mistretta, P. Caputo, M. Cellura and M. A. Cusenza, Energy and environ-mental life cycle assessment of an institutional catering service: An Italian case study, *Sci. Total Environ.* **657**, 1150–1160 (2019).

22. R. Salemdeeb, D. F. Vivanco, A. Al-Tabbaa and E.K.H.J. Zu Ermgassen, A holistic approach to the environmental evaluation of food waste prevention, *Waste Manag.* **59**, 442–450 (2017).

23. D. E. Amodeo and C. Klimas, A life cycle assessment of hospital food waste: A model for large scale commercial impact reduction, *DePaul Discoveries* **10**(1) (2021). https://via.library.depaul.edu/depaul-disc/vol10/iss1/5

24. I.S. Arvanitoyannis and A. Kassaveti, Fish industry waste: Uses, treatments, environmental impacts, current and potential, *Int. J. Food Sci. Technol.* **43**, 726–745 (2008).

25. R. Salemdeeb, E.K.H.J. Zu Ermgassen, M.H. Kim, A. Balmford and A. Al-Tabbaa, Environmental and health impacts of using food waste as animal feed: A comparative analysis of food waste management options, *J. Clean. Prod.* **140**, 871–880 (2017).

26. J. Laso, M. Margallo, I. Garcia-Herrero, P. Fullana, A. Bala, C. Gazulla, A. Polettini, *et al.*, Combined application of Life Cycle Assessment and linear programming to evaluate food waste-to-food strategies: Seeking for answers in the nexus approach, *Waste Manag.* **80**, 186–197 (2018).

27. G. Mondello, R. Salomone, G. Ioppolo, G. Saija, S Sparacia and M.C. Lucchetti, Comparative LCA of alternative scenarios for waste treatment: The case of food waste production by the mass-retail sector, *Sustainability* **9**, 827 (2017).

28. P. Brancoli, K. Rousta and K. Bolton, Life cycle assessment of supermarket food waste, *Resour. Conserv. Recycl.* **118**, 39–46 (2017).

29. K. Scholz, M. Eriksson and I. Strid, Carbon footprint of supermarket food waste. *Resour. Conserv. Recycl.* **94**, 56–65 (2015).

30. E.C. Gentil, D. Gallo and T.H. Christensen, Environmental evaluation of municipal waste prevention, *Waste Manag.* **31**, 2371–2379 (2011).

31. H.H. Khoo, T.Z. Lim and R.B.H. Tan, Food waste conversion options in Singapore: Environmental impacts based on an LCA perspective, *Sci. Total Environ.* **408**, 1367–1373 (2010).

32. L. Gruber, C. Brandstetter, U. Bos, J. Lindner and S. Albrecht, LCA study of unconsumed food and the influence of consumer behavior, *Int. J. Life Cycle Assess.* **21**(5), 773–784 (2016).

33. A.B.S. Schott and T. Andersson, Food waste minimization from a lifecycle perspective, *J. Environ. Manage.* **147**, 219–226 (2015).

34. M.C. Heller and G.A. Keoleian, Greenhouse gas emission estimates of U.S. dietary choices and food loss, *J. Ind. Ecol.* **19**(3), 391–401 (2015).

35. M. Kummu, H. de Moel, M. Porkka, S. Siebert, O. Varis and P.J. Ward, Lost food, wasted resources: Global food supply chain losses and their impacts on freshwater, cropland, and fertilizer use, *Sci. Total Environ.* **438**, 477–489 (2012).

36. S. Scherhaufer, G. Moates, H. Hartikainen, K. Waldron and G. Obersteiner, Environmental impacts of food waste in Europe, *Waste Manag.* **77**, 98–113 (2018).

37. P. Bartocci, M. Zampilli, F. Liberti, V. Pistolesi, S. Massoli, G. Bidini, *et al.*, LCA analysis of food waste co-digestion, *Sci. Total Environ.* **709**, 136187 (2020).

38. Committee on World Food Security, Food losses and waste in the context of sustainable food systems, The Committee on World Food Security, 41st Session (2014). https://www.fao.org/3/av037e/av037e.pdf
39. J. Gustavsson, C. Cederberg, U. Sonesson, R. van Otterdijk and A. Meybeck, Global food losses and food waste, Food and Agriculture Organization, Rome (2011). https://www.fao.org/3/mb060e/mb060e.pdf
40. Y. Omolayo, B.J. Feingold, R.A. Neff and X.X. Romeikoa, Review life cycle assessment of food loss and waste in the food supply chain, *Resour. Conserv. Recycl.* **164**, 105–119 (2021).
41. C. Beretta, M. Stucki and S. Hellweg, Environmental impacts and hotspots of food losses: Value chain analysis of Swiss food consumption, *Environ. Sci. Technol.* **51**, 11165–11173 (2017).
42. US EPA, Sustainable management of food basics, United States Environmental Protection Agency (2023). https://www.epa.gov/sustainable-management-food/sustainable-management-food-basics
43. H. Zhou, Q. Yang, E. Gul, M. Shi, J. Li, M. Yang, *et al.*, Decarbonizing university campuses through the production of biogas from food waste: An LCA analysis, *Renew. Energ.* **176**, 565–578 (2021).

Chapter 15

The Net-Zero Carbon Dioxide Framework for Life Cycle Assessment of Carbon Dioxide Capture, Utilization and Storage Techniques

Arezoo AZIMI[a], Li SHEN[b] and Mijndert VAN DER SPEK*

[a]*School of Engineering and Physical Sciences,*
EH14 4AS, Edinburgh, UK
[b]*Copernicus Institute of Sustainable Development,*
Utrecht University, 3584 CB Utrecht, The Netherlands

*mv103@hw.ac.uk

Carbon dioxide (CO_2) Capture, Utilization and Storage (CCUS) is a technology that can substantially limit CO_2 emissions from point source emitters, though it is understood that point-source CCUS does not mitigate all emissions from value chains using fossil feedstock. To allow CCUS technologies and systems to operate in future settings, they will need to abide by a strict net-zero $CO_{2\text{-eq}}$-emissions constraint, meaning additional measures like CO_2 removal may be needed to compensate for residual greenhouse gas (GHG) emissions. To help design truly net-zero CCUS systems, the net-zero $CO_{2\text{-eq}}$-emissions framework was developed. This chapter exemplifies this framework by using two CCUS case studies — seasonal energy storage for intermitted renewable energy supply (iRES) and iron and steelmaking. Life Cycle Assessments (LSAs) are reported on the whole value chain of these two systems for climate change and other environmental impacts to understand the potential trade-offs, while the net-zero framework is used to design for zero-GHG emissions.

1. Introduction

It is understood that climate change is one of the largest threats to our planet and humanity. The IPCC Sixth Assessment Report [1] and the IPCC (special) report on the impacts of global warming of 1.5°C above pre-industrial levels [2] show that greenhouse gas (GHG) emissions must reach net zero in the second half of the century or by 2050 to fulfil the Paris climate targets of two, respectively, one-point-five degrees temperature change compared to pre-industrial eras. After net zero is reached, humankind must effectively remove carbon dioxide (CO_2) from the atmosphere to offset legacy emissions. Subsequent IPCC assessment reports have also shown that a portfolio of clean energy and clean industrial technologies needs to be deployed to pave credible pathways towards net zero by the aforementioned target dates (e.g., see Refs. 1 and 3). CO_2 Capture and Storage and CO_2 Capture and Utilization (CCS and CCU but often conjugated to CCUS for convenience) are expected to play a pivotal role in our journey to net-zero GHG emissions and beyond [1].

CCS is a technology where CO_2 is filtered from (mostly) point-source CO_2 emitters, such as the smokestacks of power and industrial plants, and then permanently stored in underground storage reservoirs or in stable mineral form above ground (see Fig. 1) [4]. CCU involves the capture of CO_2 from point sources or the air and the subsequent conversion of the

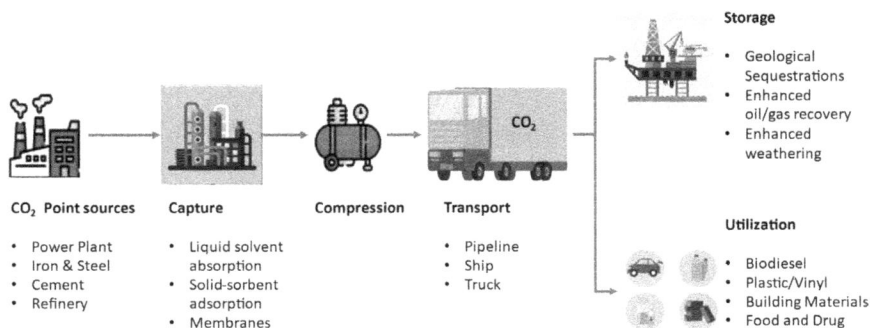

Fig. 1.   Schematic representation of possible CCUS supply chains.

CO_2 to useful products, e.g., fuels, chemicals, plastics, or construction materials (Fig. 1) [5]. Where CCS has the sole purpose of reducing CO_2 emissions to the atmosphere, CCU's main merit is that it provides a source of carbon for hydrocarbon materials in a world where fossil fuels are no longer used. CCU's role in the abatement of CO_2 emissions is, however, debated because CO_2-based hydrocarbon materials are typically combusted at their end-of-life (e.g., a jet fuel is combusted in jet engines, plastics are mostly combusted in waste incinerators, even in a 2050 scenario); thereby the CO_2 molecule is still released to the atmosphere, contributing to climate change [6–8].

The Life Cycle Assessment (LCA) body of knowledge shows that CCS and CCU alone do not reach net-zero GHG emissions over their life cycle (see Sec. 2). Meanwhile, many economies have committed to reach net zero between 2045 and 2060, in line with Paris climate targets, which are only one or two investment cycles away. Meanwhile, production plants have typical lifetimes of greater than 20 years, while developing and constructing a CCUS project takes many years. This suggests that projects developed during this decade and the next will certainly "live to see" a net-zero society and thus have to function in a net-zero system. Moreover, there has been intense scientific and political debate over the GHG reduction potential of CCU technologies, where the answer differs based on assumed system boundaries, data inputs, and allocation assumptions to LCA studies, for example. Finally, it is understood that CCS and CCU technologies may also have environmental impacts beyond climate change, such as increased particulate matter formation or eutrophication of water bodies, mostly because they increase the energy used per unit of main product produced [9, 10]. These observations have signaled the need to not use the LCA as an ex-post method to assess the climate impacts and other environmental burdens of CCUS technologies but rather as an ex-ante method to help design technology solutions that abide to the net-zero constraint, while minimizing environmental burdens other than climate change. As a response, Sutter and colleagues [8] have postulated the net-zero framework for CCUS systems analysis, while Wevers and colleagues [11] have progressed this method to be used in the LCA.

This chapter discusses the net-zero framework as a useful approach to design and analyze net-zero GHGs (here onwards, net-zero CO_2-eq-emission) CCUS systems. It aims to provide a guide to the use of the net-zero framework in a few simple steps to exemplify its use and highlight further development needs.

The structure of the chapter is as follows. Section 2 provides a concise overview of the LCA literature on CCU and CCS technologies and highlights current issues with the LCA of CCUS systems. Section 3 introduces the net-zero framework for CCUS LCAs as a response to the observed challenges. Subsequently, Sec. 4 exemplifies the use of the net-zero framework using two case studies: the storage of excess renewable electricity in chemical energy carriers, among which methane produced from CO_2 and hydrogen, and the full decarbonization of large-scale steel production using biomass, CCUS, and CO_2 removal technologies. Section 5 discusses the framework's current limitations and areas for further development to optimize its utility.

2. Concise Overview of LCA Studies on CCU and CCS Systems

Before introducing the net-zero framework, we first provide a concise overview of existing LCA studies on CCS and CCU, applied to selected industries (cement, steel, and petroleum refining) and power production from fossil fuels. These studies highlight issues with the current approaches used to assess the environmental impacts of CCS and CCU, listed at the end of this section. The literature sweep was done using Google Scholar, Web of Science, and Scopus, with specific search terms such as "life cycle assessment", "carbon capture storage", "carbon capture utilization", "LCA", "CCU", "CCUS", "iron and steel making", "power plant", "cement plant", and "refinery". We applied two criteria for selection: (1) there is an LCA on the selected industries or power production plant, and (2) it should include CCS and/or CCU as a form of carbon management. Based on these two criteria, 32 studies were included. This section presents the results of this review per industry type, including system boundaries reported, selected functional units (FUs), handling of multifunctionalities, environmental impact categories presented, and climate change reduction potentials of CCS and CCU technologies.

2.1. *Cement industry*

Cement is one of the main constituents of a building. Cement production consumes large amounts of energy and raw materials. Cement manufacturing is responsible for approximately 7% of global anthropogenic $CO_{2\text{-eq}}$

emissions and is considered one of the major consumers of natural resources [12, 13]. We found seven LCA studies conducted on cement production, including CCS [14–20] and one study, including CCU [21].

System boundaries. All the reviewed papers in this study used a cradle-to-gate system boundary. The CCS studies differed such that some studies [15, 18, 19] only included the CO_2 capture stage but not the transport and storage, while others [14, 16, 17] considered the complete value chain from capture to storage. Chauvy and colleagues [21] considered on-site capture and conversion of CO_2 into methane using renewable hydrogen, which is produced through water electrolysis. The produced raw synthetic natural gas was then used either for on-site industrial symbiosis or other uses like grid injection. All CCU processes happened within the cement plant gate; therefore, there were no environmental impacts related to CO_2 transport and storage.

Functional unit. LCA studies estimate the environmental impacts of a product or process system based on its function. For example, they report the climate change impact per kilogram of useful product produced. In the cement industry, most studies select 1 metric tonne of cement [14, 17, 18] or 1 metric tonne of clinker [15, 16, 19, 20] as the FU. In the CCU study [21], 1 GJ of synthetic natural gas produced was chosen as the FU.

Multifunctionality. The reviewed studies mostly included CCS, and the main product of the plants was cement or clinker. Therefore, no useful manufactured co-product exists, and no allocation or expansion was applied. The CCU study did not mention multifunctionality either, though this would have been needed, given that the plant moved from one (cement) to two (cement and methane) useful products for the case where the methane was injected to the grid.

Environmental impact categories. The studies used different software, such as SimaPro and Gabi, with various databases, such as ecoinvent, and have attempted to include as many impact categories as possible. The main impact category studied was climate change (CC), and the most used impact category assessment method was ReCiPe, using both the midpoint [17, 19] and endpoint [20, 21] analysis methods. Cavalett and colleagues [16] and An and colleagues [14] used integrated assessment methods (IAMs) and the CML2001 method, respectively.

Climate change reduction of studied CCS and CCU options. Schakel and colleagues [20] reported that the addition of CO_2-capture plants and subsequent storage reduced the net climate change of the base cement plant by 76% and 84%, respectively, when retrofitted with coal and using natural-gas-fired calcium-looping CO_2-capture technology. Cavalett and colleagues [16] stated that the configuration reached negative emissions between −24 and −169 gCO_2-eq per $kg_{clinker}$ by the additional use of bio-mass and alternative fuel coupled with oxyfuel CO_2 capture. Rolfe and colleagues [19] integrated calcium carbonate looping (CCL) and oxy-fuel combustion processes into a cement plant and discovered that the CCL unit has less environmental impacts than the oxy-fuel combustion process because of the additional benefit of electricity generation through the heat recovery system. Chauvy and colleagues [21] stated that integrating CCU in the cement plant reduced the net CO_2 emissions from 81.61 kg CO_2-eq to 27.94 kg CO_2-eq (a 66% reduction), while fossil and mineral resource depletion will also be lower.

2.2. Steel industry

The second-largest industry in terms of production volume is the iron and steel industry. It is also the first largest direct CO_2 emitter [22, 23] responsible for 6% of global CO_2 emissions and 8% of energy-related emission (including power consumption by steel industry) [24]. There are two main routes to steel production: integrated steel mills, which are based on the blast furnace-basic oxygen furnace (BF-BOF) configuration and mini mills based on the electric arc furnace process (EAF). We reviewed both routes, including CCS and/or CCU, for which we found five studies.

System boundaries. Both the CCS and CCU studies have mostly selected cradle-to-gate as the system boundary, except for Tanzer and colleagues [25], who chose to expand the system under consideration to include the end-of-life phase of steel and monoethanolamine, thus using a cradle-to-grave system boundary.

Functional unit. The studies investigating CCS [25–27] mostly chose 1 metric ton of hot rolled coil (HRC) as the FU, while CCU studies [28, 29] used the amount of byproduct produced as the FU. For example, Rigamonti and Brivio [28] adopted the treatment of 1,000 kg of process gases from steel production as the FU, and Thonemann and colleagues [29] used 1 kg of methanol produced from steel mill gases.

Multifunctionality. Thonemann and colleagues [29] tried to solve the multifunctionality issue created by the production of the methanol byproduct with system expansion, allocation, and substitution approaches. They stated that the method of handling multifunctionality strongly influences the resulting climate change. The substitution method was useful for calculating the environmental burdens of only one target product; in this case, either steel or methanol. The system expansion worked best for an unbiased approach and should be used to avoid allocation. Rigamonti and Brivio [28] applied the system expansion with the substitution method to solve the multifunctionality issue, meaning that they expanded the system boundary to also consider the avoided impacts of displacing alternative technologies, such as methanol production from natural gas, which is the pathway responsible for producing 65% of global methanol production.

Environmental impact categories. Three studies [26–28] assessed midpoint environmental impact categories, and the studies showed good performance on some impact categories like climate change (CC) with some environmental trade-offs in others. For example, Tanzer and colleagues [25] reduced life-cycle CO_2 emissions by deploying biomass with CCS in iron and steelmaking while showing this would put a large burden on land use — they cautioned that the delay in the carbon reuptake by newly grown biomass, along with other impacts of biomass production, may increase the climate change potential of biogenic CO_2, potentially negating part of the emissions avoided in the steel plant.

Climate change reduction of studied CCS and CCU options. Rigamonti and Brivio [28] stated that when the produced methanol was used as a marine transport fuel instead of heavy fuel oil, the impact on climate change decreased from 504 kg $CO_{2\text{-eq}}$ to 491 kg $CO_{2\text{-eq}}$ per 1,000 kg of the process gases. Tanzer and colleagues [25] reported that the use of CCS alone resulted in a higher net CO_2 reduction than the use of bioenergy alone, while their combination resulted in a higher net CO_2 reduction compared to the sum of separate interventions and was able to reach net-negative $CO_{2\text{-eq}}$ emissions.

2.3. *Oil refining*

Three LCA studies were found for oil refining, of which two investigated CCU attached to a hydrogen manufacturing unit [9, 30], producing polyols and/or dimethyl ether (DME) from the captured CO_2. The third study investigated CCS [31] in a US refinery.

System boundaries. All studies selected a cradle-to-gate system boundary.

Functional unit. In the system with a multi-product CCU [30], the CO_2 is captured at the refinery and converted to DME and polyols, either simultaneously or in two consecutive cycles. In a first step, 1 MJ of H_2 produced was selected as the FU. Then, to create a harmonized basket of products (H_2, DME and polyol), they calculated the amount of captured CO_2 per MJ of H_2 produced and estimated how much CO_2-based DME and polyols could be produced from the captured CO_2. This led to an overall FU of 1 MJ H2 + 0.78 MJ DME + 0.04 kg polyol. While for the single product CCU study [9], the FU selected was the production of 1 MJ H_2 (LHV base), 0.03 kg polyols, and 0.187 kg low pressure steam. For the CCS study [31], 1 kg of $CO_{2\text{-eq}}$ abated was defined as the FU.

Multifunctionality. Fernandez-Dacosta and colleagues [9, 30] selected a system expansion for both the single and multi-product CCU systems. They [30] defined a harmonized basket of products (H_2, DME and polyol, respectively, H_2, polyol and LP steam), which are produced in the same amounts in all systems.

Environmental impact categories. Young and colleagues [31] used the Tool for Reduction and Assessment of Chemicals and other environmental Impacts (TRACI) v.2.1 to assess the impacts. They studied six impact categories: climate change potential (CCP), particulate matter formation potential (PMFP), photochemical oxidant formation potential (POFP), freshwater eutrophication potential (FEP), acidification potential (AP), and water consumption (WC). Fernandez-Dacosta and colleagues [9, 30] used the ReCiPe impact characterization method for both the single and multi-product CCU study. They chose seven environmental indicators for the single product CCUS study (including CCP, terrestrial acidification potential (TAP), FEP, PMFP, POFP, human toxicity (HT), and fossil resources scarcity (FRS)) and just two impact categories (CCP and FRS) for the multi-product CCUS system due to the complexity of the system of study.

Climate change reduction of studied CCS and CCU options. The single product CCU study [9] revealed that implementing CCUS in the hydrogen unit combining with the storage of the remaining CO_2 decreased the climate change impact category by 23% compared to the reference case (where no CO_2 was captured). The results of the multi-product CCU plant showed that the largest direct CO_2 emission reductions are achieved with CCS without utilization (–70%) but at the expense of higher total costs (+7%). They reported that a combination of multi-product CCU with storage of the remaining CO_2 is an economically attractive pathway for reducing climate change (–18% comparing to the reference case) while still being economically feasible. Young and colleagues [59] reported that adding a capture unit captures CO_2 from the refinery while also adding CO_2 emissions due to its energy; therefore, additional CO_2 needs be captured to achieve a net CO_2 avoided of 1 kg $CO_{2\text{-eq}}$.

2.4. *Power production*

In 2021, 500 coal-fired power plants were reported to be under construction worldwide, and there are plans for another 100. Power plants are an important source of CO_2 emissions if unabated but can be retrofitted with CCUS for emissions reduction. Here, we reviewed 16 LCA studies [32–47] conducted on coal, natural gas, or other fossil fuel-fired power plants.

System boundaries. Four of the reviewed studies [36, 38, 39, 44] chose cradle-to-gate, with CO_2 capture and compression as the system boundary, while others [32, 40, 43] also included CO_2 transport and storage. Among the reviewed studies, there were five studies [33, 35, 37, 46] that stated cradle-to-grave as their system boundary, although four were, in fact, cradle-to-gate with CO_2 transport and storage (thus using incorrect terminology).

Functional unit. The selected FUs for power plants with CCS was 1 kWh or 1 MWh of electrical energy produced. In the case of CCU, Von der Assen and Bardow [47] chose 1 kg polyols and 0.36 kWh of grid electricity as the FUs as they stated that their system had two main functions of producing polyols for polyurethane production and the supply of electricity to the German electricity grid.

Multifunctionality. Von der Assen and Bardow [47] solved their multifunctionality using allocation with two options — first (as the worst-case allocation scenario for polyols), they assigned the impacts of the entire CCU system to polyols (assuming grid electricity has no impacts at all), then they assigned a credit to the polyethercarbonate polyols in the CCU system for the supply of grid electricity.

This grid electricity is assumed to substitute the electricity from a conventional power plant without CCS. This way, the environmental benefits or burdens are fully assigned to CO_2-based polyol production.

Environmental impact categories. In the case of power plants, researchers chose diverse impact assessment methods like ReCiPe [40, 42, 47], EDIP 97 [41], CML [33, 35, 38, 45, 46], and IPCC [32], studying midpoint environmental impact indicators. Matin and colleagues [44] used four different methods of IPCC, AWARE, ReCiPe Midpoint (H), and TRACI for their LCA. They studied the effects of CO_2-capture efficiency via amine or ammonia-based absorption of flue gases on climate change. The results showed that CCS plants have trade-offs on other impact categories like water consumption (WC), stratospheric ozone depletion (ODP), ionizing radiation potential (IRP), freshwater eutrophication potential (FEP), photochemical oxidant formation potential (POFP), and fossil resources scarcity (FRS). The performance of MEA-based CCS plants is better compared to the ammonia-based plants in terms of CCP and FEP.

Climate change reduction of studied CCS and CCU options. None of the studies reached net-zero CO_2 emissions. Matin and Flanagan [44] stated that adding a CCS plant would reduce CCP; however, there are environmental trade-offs, and other impact categories, such as acidification (AP and TAP), eutrophication (FEP and MEP), and HT, will show higher values compared to the plants without CCS. The highest reduction in life cycle GHG emissions of 68–92% per FU was reported by Volkart and colleagues [32]. Von der Assen and Bardow [47] stated that producing polyols with 20 wt% CO_2 in the polymer chains reduces the GHG emissions to 2.65–2.86 kg $CO_{2\text{-eq}}$ per 1 kg of polyols (11–19% reduction compared to conventional polyether polyols), so it cannot be considered as a GHG sink. Increasing the CO_2-based carbon content in polyols will increase the CCP reduction.

2.5. General observations

Reviewing these studies elicited several methodological issues relating to the LCA of CCUS systems coupled with point source emitters. They are listed as follows.

1. **Functional unit choice:** Many different FUs have been used for the assessment of CCU or CCS systems, especially when the captured CO_2 molecule is converted into a useful product (i.e., CCU). This makes intercomparison of studies very difficult, if not impossible, plus it bears the risk of conveying incorrect messages.

2. **System boundaries:** There is a wide variety in the system boundaries applied and how specific boundaries are named, despite clear guidelines [48] on how these should be reported. For example, some authors stated that their studies were cradle-to-grave studies, but they did not include the waste disposal or end-of-life stage, or they confused cradle-to-grave with cradle-to-gate plus CO_2 storage.

3. **Life cycle impact assessment:** Many studies did not include all the midpoint or endpoint impact categories. Some of them focused on CCP only while excluding other environmental trade-offs.

4. **Less than 100% reduction in climate change:** When including the use and/or end-of-life phase(s) of the captured carbon molecules, the net climate change reduction is very often well less than 100%, especially if the CO_2 molecules are converted to short-lived products that are combusted end-of-life. This suggests that many of the studied systems are incompatible with the Paris Agreement's goal of 1.5-to-2-degree temperature increase (roughly translated into net-zero GHGs by 2050 or 2070, respectively).

Finally, on a positive note, more recent studies tend to adhere much better to LCA guidelines [48] and provide more detailed reporting of selected system boundaries and other scoping elements.

3. The Net-Zero CO_2-emissions Framework

The literature review in Sec. 2 highlighted the following: (1) different FU choices are made when evaluating the life cycle of CCU and CCS systems, (2) the nomenclature used for system boundary selection is not always clear, (3) often, systems are assessed on climate change potential only while neglecting other environmental impacts, and (4) CCS and CCU configurations do not render net-zero CO_2 (or net-zero $CO_{2\text{-eq}}$) production systems. The latter is the focus of this chapter but strongly correlates with the other problems identified.

The gap to net-zero $CO_2(_{-eq})$ is especially large when CO_2 with a fossil origin is produced and transformed into a short-lived product like a fuel or a chemical, and this fossil CO_2 molecule is emitted into the air at the disposal phase of the short-lived product. This implies that when a fossil CO_2 molecule is reused once in a product that replaces a fossil product fuel or chemical (as in Fig. 2), the theoretical maximum CO_2 emission reduction is 50% (in Fig. 2, compare the two CO_2 units emitted under schemes 1(a) and 1(b) with the single CO_2 unit emitted by scheme 2), while often, the real CO_{2-eq} reduction is much smaller than 50%, which was nicely illustrated in recent work by De Kleijne and colleagues [49]. This poses a problem for CCUS, as many countries have by now pledged to reach net-zero emissions around mid-century, and the technology developed and deployed today will need to operate in such net-zero systems, given that there are only one or two investment rounds left until 2050. Note that, historically, there has also been much debate on the true climate change mitigation value of short-lived CCU products (e.g., Ref. [50]) (less so on CCS applications, although there has been some; compare, e.g., Refs. [51] and [52]), where early work used narrow system boundaries and suggested — incorrectly — that CO_2 utilization could have significant climate benefits (e.g., Refs. [53] and [54]), while the true

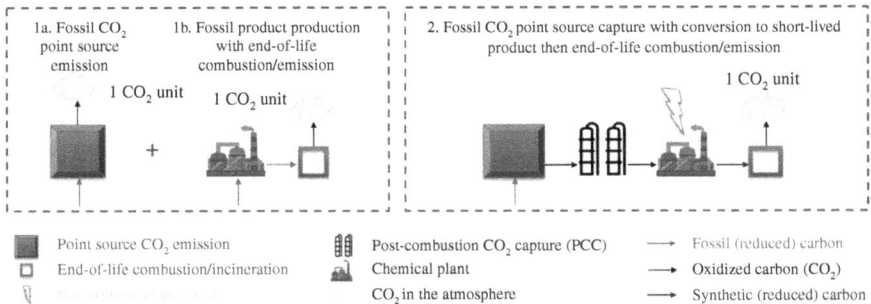

Fig. 2.   Once-through fuel combustion and industrial production (1(a) and 1(b)) versus a situation where a CO_2 molecule is reused once. In both cases, fossil CO_2 is eventually emitted into the atmosphere. 1(a) depicts a situation where a fossil resource is used, emitting fossil CO_2, while in 1(b), the fossil resource is first converted into a short-lived product and then emitted at its end-of-life through combustion/incineration/degradation. The right panel (2) depicts a situation where a fossil resource is used while the emitted CO_2 molecule is captured and converted into a short-lived product and then combusted/incinerated/degraded into a fossil CO_2 molecule end-of-life.

climate impacts can only be fully appreciated when including the complete carbon value chain from mining to the end of life (e.g., Refs. [49] and [55]).

To address the issue of CCU climate change potential and converge the debate on this, Mazzotti and colleagues [8] postulated the net-CO_2 emissions framework, where they used life cycle thinking to design systems that are inherently net-zero CO_2-emissions, contrary to classical LCA practice where a system is designed and then assessed on its climate change and other environmental merits. The idea is that if a comparative assessment between technologies and systems is done, and any of these systems is constrained to be net zero, then their merits can be assessed based on other indicators than climate change, e.g., land use, energy efficiency, etc., leading to a more robust and like-for-like assessment of climate change mitigation technologies. Given that all systems are net zero by design, the practitioner can focus on solutions that minimize other environmental burdens (or the burden of the most relevance to a specific situation). Sutter and colleagues [8] used a case study on chemical energy carriers to design systems that are net-zero CO_2, where the energy carrier was produced using renewable electricity and was consumed either in an industrial setting or in transport. Wevers and colleagues [11] expanded on this by undertaking a full LCA of net-zero $CO_{2\text{-eq}}$ renewable energy storage in chemical energy carriers, with subsequent re-electrification of these energy carriers in power production plants [11]. Meanwhile, Gabrielli and colleagues [6] followed up with a net-zero assessment of CCS, CCU, and biomass use in the chemical industry, where Becattini and colleagues did the same for net-zero CO_2-emissions aviation [56].

Here, we break down the net-zero framework for the LCA, as published by Wevers and colleagues [11] into six distinct steps.

1. **Define the goal and scope, functional unit, and system boundaries.** We advise using FUs typical for a non-decarbonized production system, e.g., 1 GWh of electricity production or 1 tonne of methanol or steel production, i.e., use a FU that reflects the main product of the incumbent system. The system boundaries can either be cradle-to-grave or cradle-to-gate but including the CO_2 end-of-life/disposal/fate and depending on the key product produced. If the key product does not contain carbon (e.g., hydrogen) or emit carbon at the end-of-life (e.g., steel), then a cradle-to-gate boundary can be selected for the main product's value chain, while any CO_2 captured from that value

chain and its fate should be included in the system boundary. If the main product emits carbon at the end-of-life, a cradle-to-grave approach is needed (see Fig. 7).

2. **Build life cycle inventory.** This should include the inputs and outputs of all the blocks within the system boundary, including the background systems (e.g., the mining and transport of raw materials, usually obtained from life cycle databases like ecoinvent).

3. **Identify $CO_{2\text{-eq}}$ emissions in the base production system (CO_2 positive).** This encompasses scrutinizing the whole value chain inside the boundaries for GHG emissions across the boundaries. This can, e.g., result from a hotspot analysis or from assessing the climate change potential of the individual blocks. Ideally, the source of the emissions is identified (e.g., a transport modality or other economic activity) so that adequate mitigation measures can be designed for these emissions in step 4.

4. **Introduce measures/technologies to return $CO_{2\text{-eq}}$-emissions to the system.** For the GHG emission sources identified in step 3, appropriate mitigation measures need to be added. Given our focus on CCUS, the main decarbonization technologies implemented may be CCU or CCS, but there will be remaining emissions across the value chain, as was discussed. The mitigation measures for these can span the full range of decarbonization technologies available and will normally depend on the type of GHG emission, its size, and the scope of the study. Examples include but are not limited to electrification, fuel switching, or the use of CO_2 removal technologies (for emissions that are dispersed or otherwise hard/expensive to abate). The resulting system will have net-zero CO_2 or $CO_{2\text{-eq}}$ emissions, depending on the scope of the study (and exemplified in Sec. 4, case study 4.1).

5. **Calculate performance indicators, e.g., LC mid-point indicators.** When the systems under evaluation are designed to the net-zero constraint, the next step is to calculate performance indicators typical for LCA, e.g., using the usual life cycle impact assessment methods.

6. **Assess/compare systems on calculated performance indicators.** The final step involves the comparative assessment of the systems studied on the impact categories other than climate change (CC) (since each system was designed net-zero $CO_{2\text{-eq}}$). Where desired, the practitioner or their clients can suggest environmental burdens of the most importance and thereby steer which environmental impacts to give the most weight to.

The six steps of the net-zero framework are illustrated in Sec. 4 using two case studies.

4. Framework Exemplification Using Case Studies

4.1. *Intermittent renewable energy systems (iRES) using chemicals as an energy storage buffer*

The first case study investigates the use of chemical energy carriers to store intermittent renewable electricity, especially between seasons. The transition towards a very low or zero-emissions energy system will strongly depend on the availability of renewable electricity sourced from intermittent solar and wind [57], but there is a significant challenge to balance the large differences between intermittent Renewable Energy Supply (iRES) and continuous end-user demand [58, 59].

In Europe, natural gas-fired power plants (so-called peaking plants) are used to balance this asynchronized demand and supply. For future energy supply based on a very high share of iRES, the grid-scale, long-term storage of electricity is, however, a necessity.

In this case study, three iRES systems with three types of chemical energy carriers as the storage media are designed based on the net-zero constraint, and their life cycle environmental performances are compared with two reference systems — natural gas combined cycle power supply (NGCC) and an NGCC system with CCS and a direct air capture and storage plant (DACCS). A simplified scheme of the comparisons is shown in Fig. 3.

4.1.1. *Define the goal and scope, functional unit, and system boundaries*

The aim of this LCA is to understand whether using chemicals as energy storage media is an environmentally feasible option as a part of the future iRES to achieve net zero. The goal is also to quantify the benefits (e.g., net-zero CO_2) and the trade-offs (e.g., resources depletion to build the iRES). The FU is defined as *1 kWh of dispatchable electricity production*. Three iRES systems with three chemical storage media, namely, hydrogen, methane, and ammonia, were designed and analyzed. This storage is only used if the intermittent supply does not match the demand profile from the grid

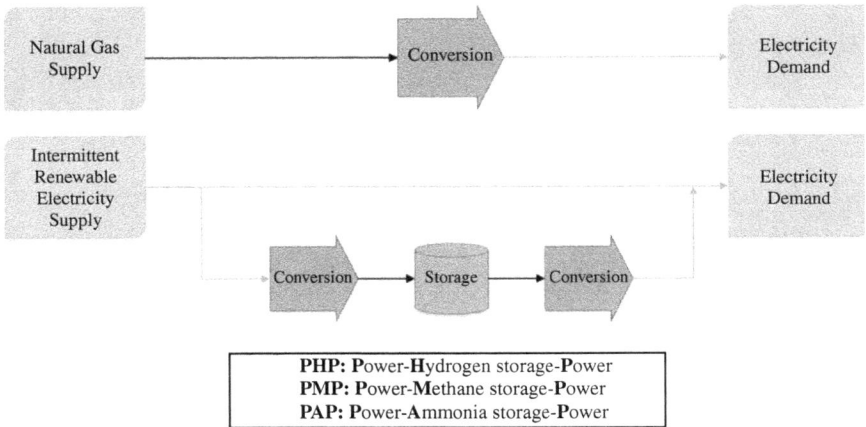

Fig. 3. The P-X-P systems analyzed (bottom) compared with the reference case (top). Yellow lines indicate electricity; black lines indicate chemical energy carriers, modified from Ref. [11].

at a point in time. The temporal scope of the study is 2030. The focus geographical scope was Central West Europe (CWE) because it has a large interconnection capacity between the EU member states, and it adopts a flow-based market coupling, an advanced tool to optimize the grid capacity. The supply chains of the manufacturing of the iRES systems were global.

Six environmental impact categories were reported to capture both environmental benefits and trade-offs based on the available data. The six categories are climate change (CC), photochemical ozone formation potential (POFP, sometimes also called smog), particulate matter formation potential (PMFP), terrestrial acidification potential (TAP), mineral eutrophication potential (MEP), mineral resources scarcity (MRS), and fossil resources scarcity (FRS). The detailed impact assessment models adopted can be found in Ref. [11].

4.1.2. *Design the iRES: the net-zero constraint and system size definition*

The systems under consideration were configured such that their direct CO_2 emissions to the atmosphere were set to zero: either the system does not emit direct CO_2 by nature (e.g., renewable power-to-hydrogen-to-power) or the system was expected to take up the same amount of

CO_2 emitted during its activities, for instance, through CO_2 capture from a combustion process and/or the air. The systems thus have net-zero direct CO_2 emissions.

To determine the system dimension with optimal capacities, the energy supply should meet the demand. One of the critical parameters is the percentage of iRES that is to be stored to achieve the desired function. An optimization model written in Python was developed using electricity supply and demand data at hourly and daily resolutions (see Fig. 4) to determine the sizes of the storage needed for peak shaving, as well as the sizes of other components in the studied iRES. The maximum capacity of the four iRES + storage systems is assumed to be 1 GW. The wind and solar profiles of the CWE regions were combined to obtain the iRES supply profile.

The results of the system optimization are shown in Table 1. These data are also the key inputs as the foreground data for the LCA. The background data was obtained from ecoinvent Database and validated wherever possible via independent literature. Country-specific and technology-specific background data were prioritized. Some of them were

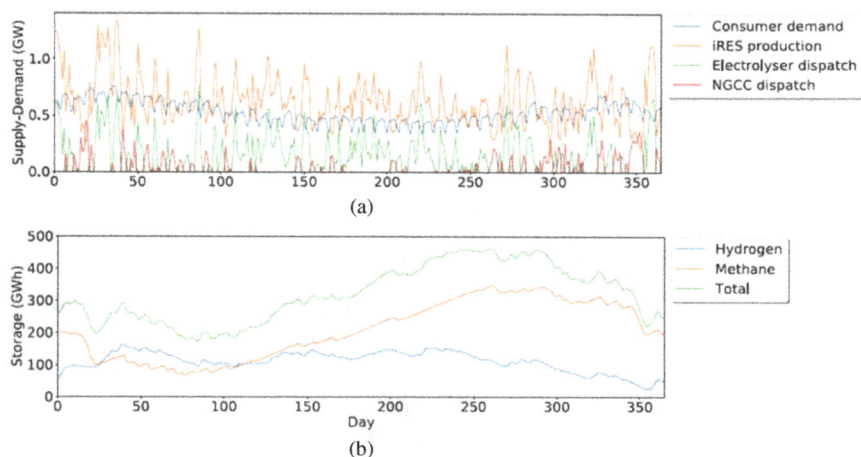

Fig. 4. (a) Daily fluctuations in the projected demand and supply for the P-M-P system (shown as lines "Consumer demand" and "iRES production") using 2015 meteorological data, and the resulting dispatch capacity required for the electrolyzer and natural gas combined cycle power supply (GW). (b) Hydrogen and methane storage requirements (GWh) and the two curves combined. Day 1 begins on 1 January, and the small peak in hydrogen production around day 45 represents a peak in wind power production in February 2015 [11].

Table 1.   Summary of the system dimensions and optimized capacities of the P-X-P systems aligned with the wind and solar supply and demand profiles, reconstructed based on Table 3 of Ref. [11].

Product Systems	P-H-P	P-M-P	P-A-P
Estimated electricity demand (GWh/year)		4,628	
Calculated iRES production (GWh/year)	5,500	5,647	5,628
Optimized capacity and system parameters			
Electrolyzer size (MW)	881	859	913
Capacity factor electrolyzer	16.6%	17.3%	17.2%
Power plant/fuel cell size (MW)	553	548	549
Methane/Ammonia plant (MW)	N.A.	110	110
Hydrogen storage required (GWh)	1,140	347	330
Methane/ammonia storage (GWh)	N.A.	859	763
Roundtrip efficiency	31.8%	26.8%	27.4%

modified based on the latest technology developments. If country-specific data was not available, generic data was used.

4.1.3. *Results and conclusions*

Figure 5 shows the LCA results of five energy supply systems. Compared to the two NGCC systems, the P-X-P systems offer an 80% reduction in CC (climate change), a 40–50% POFP (photochemical oxidant formation potential), and a 90% reduction in FRS (fossil resources scarcity). The benefits are strongly influenced by the reduction of fossil fuel extraction and combustion compared to the NGCC systems.

However, the P-X-P systems have high impacts on PMFP (particulate matter formation potential), MEP (marine eutrophication potential), and MRSP (mineral resource scarcity potential) compared to the NGCC systems. For TAP (terrestrial acidification potential), the NGCC case has the lowest impact, whereas the P-A-P system has the highest impact. For the P-X-P systems, the impacts from the production of renewable electricity dominate (90–98%) for all seven impact categories. A significant contributor is the manufacturing of PV panels and wind turbines for which aluminum and copper are extracted, and the manufacturing processes are currently still relying on fossil fuel energy systems. The impacts of other

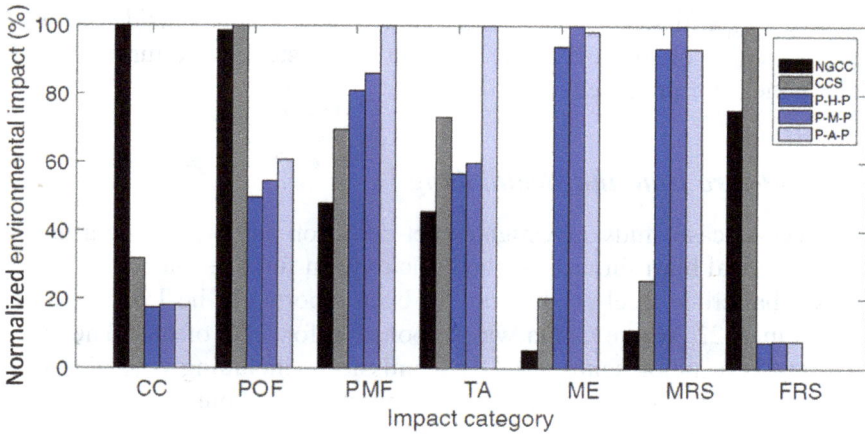

Fig. 5. Cradle-to-factory gate environmental impacts of 1 kWh of dispatchable electricity for three P-X-P systems using chemicals as storage media, compared with NGCC and NGCC with CCS.

components, such as the electrolyzer, storage tanks, fuel cells, and combustion facilities of the P-X-P systems are relatively low. They only contribute about 1–5% of the total impacts except for mineral resources scarcity, which accounts for a contribution of 9–12%.

Fully abating the direct CO_2 emissions of the NGCC system ("NGCC+CCS" in Fig. 5) reduces its CC impact by around 70% compared to the unabated NGCC system, but it leads to an increased impact on all other impact categories. The noticeable increase (approximately 25%) in fossil resource scarcity of the NGCC+CCS system compared to the NGCC is largely due to the reduced efficiency of the NGCC and, thus, the increased use of natural gas, which was also observed by the studies reviewed in Sec. 2. Infrastructure impacts are the highest for the CCS system due to the CO_2 pipeline and storage tanks.

This case study addressed the system design, sizing optimization, and the LCA of the P-X-P systems using chemical energy storage to balance the intermittent renewable electricity supply in the future. A key observation is that a net-zero direct CO_2 emissions system does not necessarily imply a net-life-cycle-zero-$CO_{2\text{-eq}}$ system, especially if there are still fossil CO_2 emissions in the background production systems, e.g., solar panel production. Another key observation from this case study was that many of the $CO_{2\text{-eq}}$ emissions of the background system happen outside the geographic region of study, e.g., the location where steel, wind turbines, and solar panels are produced.

4.2. *Net-zero iron and steelmaking*

The second case study investigates net-zero iron and steelmaking using the traditional blast furnace — the basic oxygen furnace route.

Global crude steel production has been reported to be 1,885 million tonnes in 2022, where China was responsible for 54% of this. The steel industry is a major source of GHG emissions, including 6% of global CO_2 emissions. The most dominant method of steelmaking is the blast furnace-basic oxygen furnace route (BF/BOF), which is responsible for 71% of global steel production and releases approximately 2,000 to up to 4,000 kg $CO_{2\text{-eq}}$ per tonne of steel produced. This process uses mostly coal and coal derivatives for energy input and iron ore reduction.

The iron and steelmaking value chain starts from mining and extraction of raw materials like iron ore, coal, and limestone. These are transported to the steel plant and processed to produce sinter, coke, and lime, respectively, for primary ironmaking in the blast furnace. An overview of an integrated BF/BOF steelmaking plant is presented in Fig. 6.

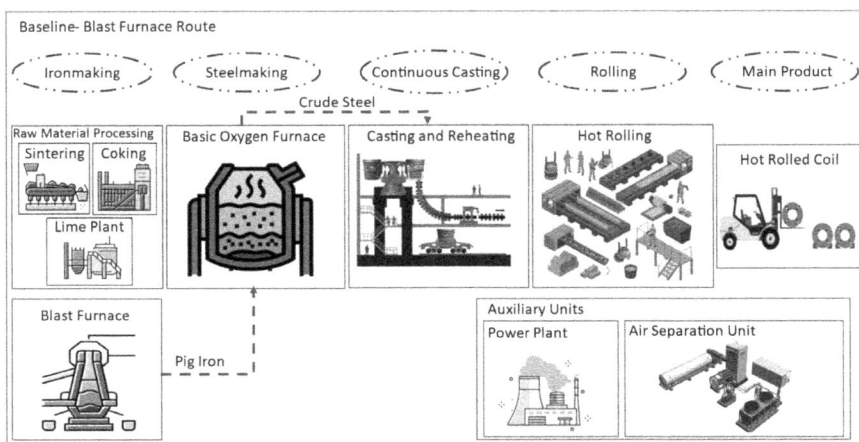

Fig. 6.   Graphic overview of the iron and steelmaking process using the blast furnace — basic oxygen furnace route.

In this study, we aim to design measures to fully decarbonize the integrated BF/BOF steelmaking process by using the LCA and the net-zero framework. The challenge of decarbonizing a steelmaking plant pivots around the reduction of iron ore in the blast furnace and providing the electricity and heat needed for running the plant, as these operations are the largest emitters of CO_2. To exemplify the net-zero framework for steelmaking, we selected two primary emission reduction case studies: (1) CCS and (2) using biomass as the source of heating energy needed (and as a reducing agent to a small extent). In a third case study, we also combined these two options. The case study results below show that neither fuel substitution with biomass nor CCS alone is sufficient to fully decarbonize the steel plant and, therefore, we applied a DACCS to compensate for any residual emissions.

The application of the net-zero framework to the iron and steelmaking process is presented in the next sections.

4.2.1. *Define the goal and scope, functional unit, and system boundaries*

The goal of this study was to evaluate how the steelmaking value chain could be decarbonized fully and how different decarbonization options compare on environmental indicators other than climate change. The FU selected for this study is 1 tonne of hot rolled coil (HRC) produced, which is the basic output product of integrated iron and steelmaking. According to Tanzer and Ramirez [25], the system boundary selected includes all the emissions from resource extraction, energy and material use, the processing plants inside the steel plant factory gate, and all the mitigating technologies introduced (but excluding the end-of-life phase); that is, in line with the net-zero guidelines laid down in Sec. 3, a cradle-to-gate boundary was applied, extended to the CO_2 storage and biomass supply for the case studies that made use of this (Fig. 7). The LCA undertaken was a process-based attributional LCA with a cut-off, while the temporal scope was 2023 and the location was the United Kingdom.

4.2.2. *Build the life cycle inventory*

The mass and energy balance of a typical integrated BF/BOF steelmaking plant were extracted from a 2013 IEAGHG report [60]. Some steel plants

Cradle-to-gate with biomass and CO_2 storage

Cradle-to-gate with storage of CO_2

Biomass processing

CO_2 CO_2 CO_2

Gate-to-gate

Post-combustion CCS

Compression

Transport

CO_2

Resource extraction, upstream energy and material use

Direct air capture

Ambient air CO_2-filtered air

Storage

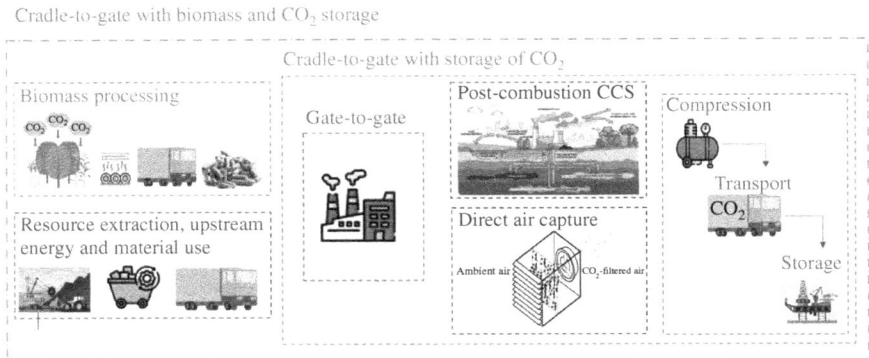

Fig. 7.   Iron and steelmaking system and system boundaries. The figure shows different system boundary options, ranging from gate-to-gate to cradle-to-gate, with carbon neutral biomass and CO_2 storage. Here, a cradle-to-gate boundary was selected to capture all emissions from the steelmaking value chain, complemented with CO_2 storage and biomass supply where these were used.

may purchase some of the intermediates from outside the fence rather than producing them on-site and/or purchasing the needed electricity from the grid. But in our model, all unit operations and the power plant are included to understand their impact. The CO_2 capture configuration was also designed based on the data reported in the IEAGHG report [60], capturing CO_2 from the blast furnace hot stoves, steam generation plant, cokes oven batteries, and lime kiln. Post-combustion capture using MEA as an absorbent and with a 90% capture efficiency was assumed, while the heat required for CO_2 capture was provided by an additional steam generation plant (the heat availability in the base case steel plant is insufficient to cover the CO_2 capture needs). The biomass was assumed to be wood chips imported from Quebec, Canada, and subsequently torrefied on the steelmaking site. For a DACCS, we assumed an alkaline absorbent-based plant following the design reported in Qiu and colleagues [61], from which the life cycle inventory (LCI) was also taken. This type of DACCS uses natural gas as the main energy input, complemented with some electricity. An overview of the mass and energy balances for the steel mill with and without the capture plant is provided in Table A.1; also, two figures depicting the internal flow of gases in the steel mill and mass and energy input of the plant is provided in the Appendix. Finally, all inventories were implemented in SimaPro v.9.5.0, and the data needed for the background system was taken from ecoinvent v.3.9.1.

4.2.3. *Identify the CO_{2-eq} emissions in the steelmaking base case*

The emissions to air from each processing plant inside the factory gate were calculated from the flue gas compositions and flowrates. The details of the emissions from each plant are also reported in Table A.1, while a block diagram with gas flows in the steel plant is presented in Figs. A.1 and A.2. From the above assumptions and inventories, the unabated integrated BF/BOF steelmaking plant was calculated to emit 2,278 kg CO_{2-eq} per tonne of hot rolled coil (HRC) (Fig. 8, leftmost bar). The power plant is

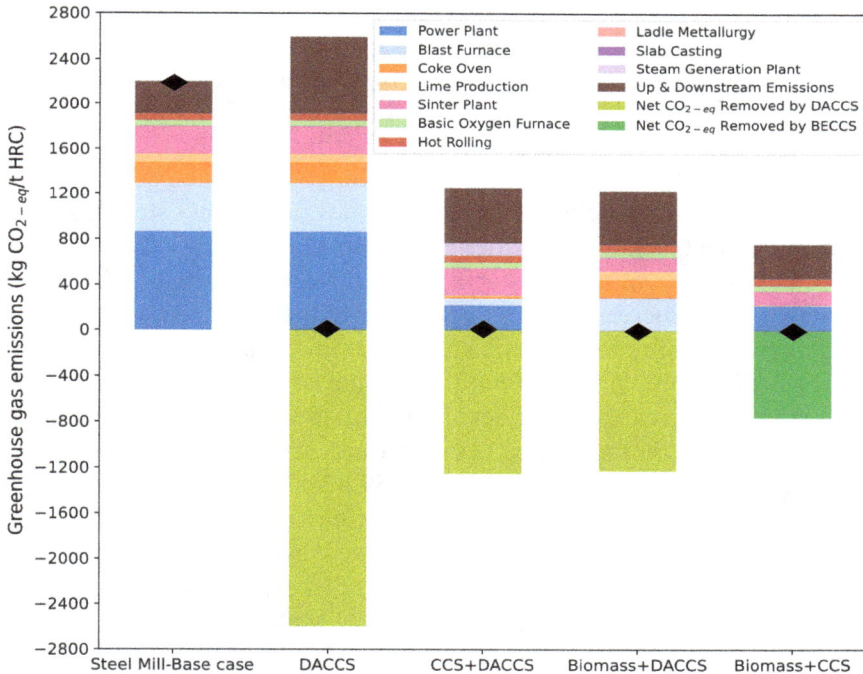

Fig. 8. CO_{2-eq} emissions for the steelmaking base case (leftmost bar) and for all net-zero cases, broken down into the origins of the emission. Note that neither the CO_2 capture nor feed substitution with biomass feeding renders full decarbonization, and as a result, a DACCS is needed as an offset.

the largest emitter, producing 865.6 kg CO_{2-eq} per FU, followed by the blast furnace, responsible for 426.9 6 kg CO_{2-eq} per FU. Meanwhile, the cokes ovens and sinter plant also show high CO_{2-eq} emissions (Fig. 8). Finally, there are also emissions outside the factory gate, including upstream emission like mining and transportation of raw materials and downstream emissions like waste treatment.

4.2.4. *Introduction of CO_2 mitigation technologies*

After estimating the total CO_{2-eq} emissions of the integrated BF/BOF steelmaking plant and their origins, measures were implemented to reduce them, namely, CCS and partial substitution of the coal feed with biomass. As mentioned, where these emission mitigation measures were insufficient to reach net-zero CO_{2-eq}, they were complemented with DACCS.

CCS: Given our focus on CCUS, the first decarbonization technology investigated was post-combustion CO_2 capture with MEA as a solvent and subsequent compression to pipeline specifications for the final geological storage. As in the IEAGHG report [60], the capture plant removes 90% of the CO_2 from selected flue gases: blast furnace hot stoves, steam generation plant, cokes oven batteries, and lime kiln. The addition of CCS to the base steel plant requires an additional steam generation plant, and the power plant increases in size compared to the base case, as the electricity consumption of the steel plant will go up (from 400.1 kWh/t HRC to 621.7 kWh/t HRC). Therefore, the amount of natural gas used in the steel mill also increases. Other details are provided in Table A.2, while the case study results are presented in Fig. 8 (middle bar): the CCS case sees a reduction of the total emissions to 1,059.8 kg CO_{2-eq} per FU — more than 50% but clearly not reaching zero. The chief remaining contributors are sintering with 246.1 kg CO_{2-eq}, followed by the power plant with 219.2 kg CO_{2-eq} per FU. The emissions of other plants like cokes ovens, blast furnace, and lime production are largely abated.

Biomass: Our study assumed wood chips were imported from Quebec, Canada, to the United Kingdom, with subsequent torrefaction on the steelmaking site. In a torrefaction process, the biomass is heated slowly in an inert or oxygen-deficient environment between 200 and 300°C. The

CO_2 emissions of this process were reported to be 0.20 gr per 1 gr of woodchips. In general, the biomass before felling is assumed to be carbon neutral, and the CO_2 emissions of combustion are labeled as biogenic, meaning they do not contribute to climate change in the LCA calculations. The emissions of biomass production, harvesting, and transport were accounted for in the Life Cycle Inventory (LCI). Here, we substituted 5% of the coking coal feed to the coke ovens (with a biomass/coking coal substitution ratio equal to their respective carbon contents), 50% of the cokes used in sintering (using carbon content ratio), and 100% of the Pulverized Coal Injection (PCI) coal in the blast furnace (with a biomass/ PCI coal substitution ratio equal to their respective lower heating values), by torrefied wood pellets. The total emission of the integrated blast furnace-basic oxygen furnaces (BF/BOF) steelmaking plant reduced to 1,037.99 kg $CO_{2\text{-eq}}$ per FU. The blast furnace and cokemaking plant remained the most emitting plants, with 271 and 167.45 kg $CO_{2\text{-eq}}$ per FU, respectively (Fig. 8, second right bar).

BECCS: Given that neither CCS nor partial coal substitution with biomass renders net-zero $CO_{2\text{-eq}}$ emissions over the steel production value chain, a logical option is to combine the two. In this case study, the natural gas fed into the steam generation plant (needed for CO_2 capture) was substituted with torrefied wood pellets (with the substitution ratio equal to their respective lower heating values). This intervention caused the total net life-cycle emissions from steelmaking to drop to zero, given that the captured and stored biogenic CO_2 emissions are considered net-negative emissions (Fig. 8, dark green shaded area, rightmost bar). The power plant and sintering plant are still the largest emitters of fossil CO_2 emissions in this case. Sintering is positive (emitting 123.06 kg $CO_{2\text{-eq}}$) despite the partial biomass feed because no emissions are captured from this plant. The power plant in the BECCS case is bigger than in the biomass without the CCS case study because of the capture and compression electricity needs, and is therefore emitting more CO_2. Conversely, the lime plant is emitting less because its emissions are captured, while the two largest emitters (blast furnace and steam generation plant) are fully decarbonized.

DACCS: Finally, the second left bar in Fig. 8 shows the breakdown of GHG emissions in case all positive emissions are offset with DACCS. The figure shows that the total positive emissions increase, resulting from the natural gas and grid electricity use of the direct air capture (DAC) plant and the grid electricity use for CO_2 transport and storage (up and downstream emissions).

4.2.5. *Analysis of environmental burdens other than climate change*

Figure 9 allows the study of a selection of environmental impact categories beyond climate change (CC); namely, land use (LU), water consumption (WC), mineral resource scarcity (MRS), and fossil resource scarcity (FRS). These midpoint environmental impact categories were calculated using the ReCiPe 2016 Midpoint (E) v.1.07. Understandably, LU is high in cases where biomass is used and is the highest in the BECCS case study, given that the steam generation plant here is also fed with biomass. MRS is roughly the same across the cases, which could change if the impacts of infrastructure are incorporated. Interestingly, WC and FRS are highest in the case where all steel plant emissions are offset with DACCS,

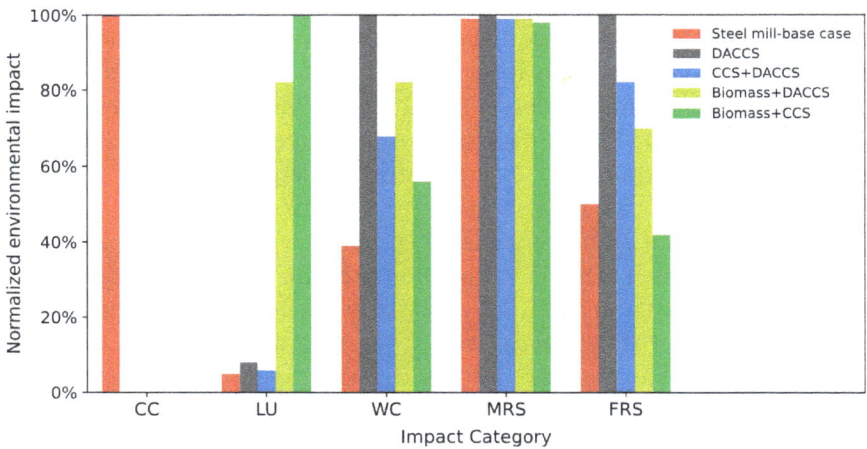

Fig. 9.   Comparative environmental impacts across five midpoint indicators.

resulting from the high water losses in this type of (alkaline solvent) DACCS plant and from the use of natural gas and grid electricity to power the DACCS plant. The WC in the CCS and biomass cases is comparable and is higher than in the reference case. FRS is lowest in the BECCS case, while it is approximately the same for the CCS and biomass case, owing to the necessity to apply DACCS in the biomass case.

5. Limitations, Questions Arising, and Future Research Needs

Although the net-zero framework is seen as a step forward in the assessment of CCUS technologies and systems, a clear limitation was highlighted, while the use of the framework has also surfaced several questions.

The key limitation is that the studies using this framework have thus far used the current technology status and "current" or "past" LCI databases to generate the environmental emissions of the foreground, respectively, background systems, while net zero needs to be reached in the *future*, not in the past or current time. This means it would be more appropriate to use future technology performances and future background databases, where available.

The recent publication of, e.g., Premise, a streamlined approach to producing prospective LCA databases enables the latter. Using Premise, ecoinvent databases can be adjusted to reflect environmental emissions corresponding to different socio-techno-economic development pathways, specifically the Shared Socio-economic Pathways (SSPs) and the Representative Concentration Pathways (RCPs), with the latter corresponding to certain temperature increase targets (e.g., 1.5°C, 2°C). This way, the LCA practitioner can evaluate their system against different possible futures to understand how the gap to net zero — and corresponding need for additional decarbonization measures — may vary. This may prove a useful combination with the net-zero LCA framework.

Thonemann and colleagues [62] provided a nice review of a prospective LCA, and discussed how technology performance can be scaled from low-technology readiness levels (corresponding to, e.g., lab scale) to high-technology readiness levels (corresponding to commercial scale) *and* how technology improvement over time may play a role, e.g., through technological learning.

Other questions that surfaced while using the net-zero framework include how to deal with emissions outside the geographical scope of the study and who should be responsible for these? Should they also be made net zero by the problem owner, or is this the responsibility of the country where the products are produced? Such questions are very timely given the global nature of supply chains and recent policy initiatives to start taxing the embodied carbon of products at regional or country boundaries [63].

Future research needs to be centered around the above-discussed issues. It is imperative good quality databases are (further) developed and curated with background data for different future scenarios and regions, while the use of such databases should become commonplace in prospective LCA studies, including these using the net-zero framework. Meanwhile, it is critical to develop and/or improve the methods to project future technology performance to allow like-for-like comparison of alternative technologies.

6. Summarizing Statements

In this chapter, we introduced the net-zero LCA framework for the assessment of CCUS as a CO_2-emission mitigation option, following the notion that neither CCS nor CCU will/may reduce a system's emissions to zero $CO_{2(-eq)}$. The chapter introduced a step-by-step approach for executing the net-zero method and applied this to two case studies. In both cases, CCUS alone (and in the second case, the partial replacement of coal with biomass) was insufficient to reach net zero, and therefore, the remaining emissions were compensated for with direct air CCS. Both cases show that different options to the full decarbonization of a production system show trade-offs on other environmental indicators, whose importance can vary per case study (e.g., in some countries, water consumption may be limiting, while in others, land use or eutrophication). In future work, it is imperative to combine the net-zero approach with a prospective LCA to model the production system and its environment in a manner more representative of when net-zero needs can be reached.

Appendix A.

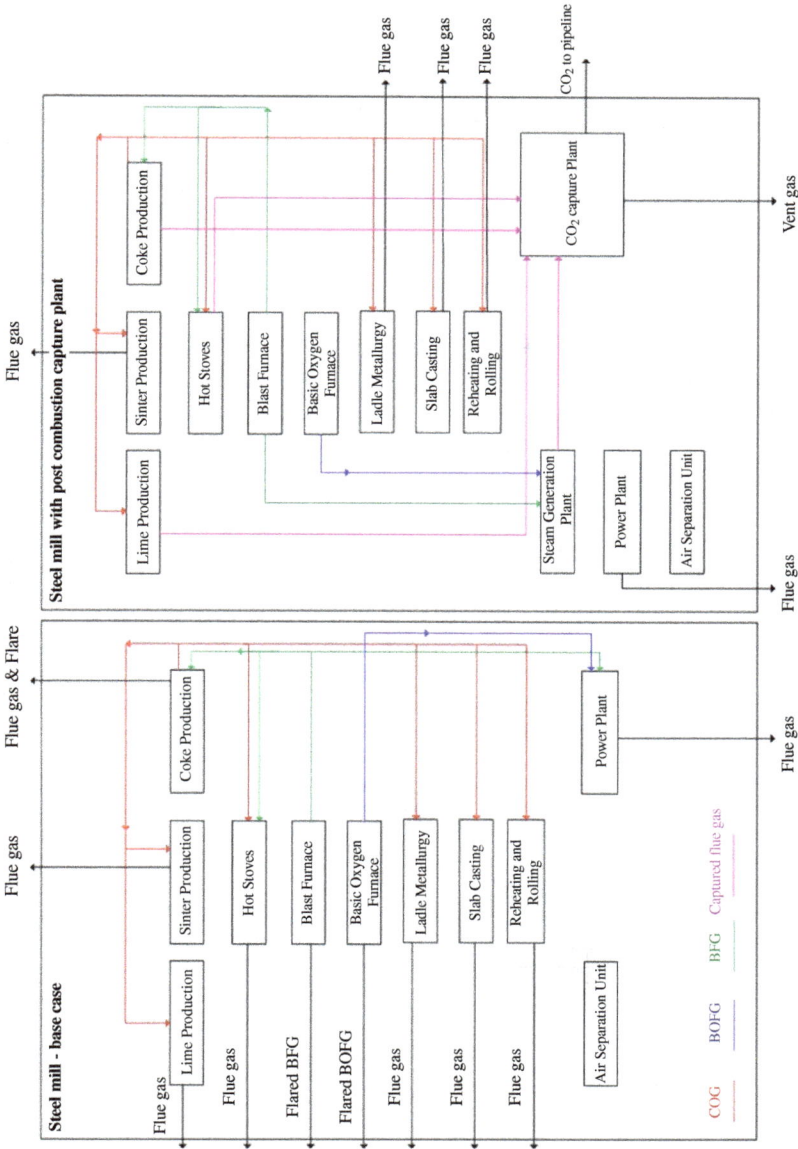

Fig. A.1.   Internal gas flow of a steel mill (coke oven gas, basic oxygen furnace gas, blast furnace gas, and the captured emissions) and flue gases with and without the capture plant.

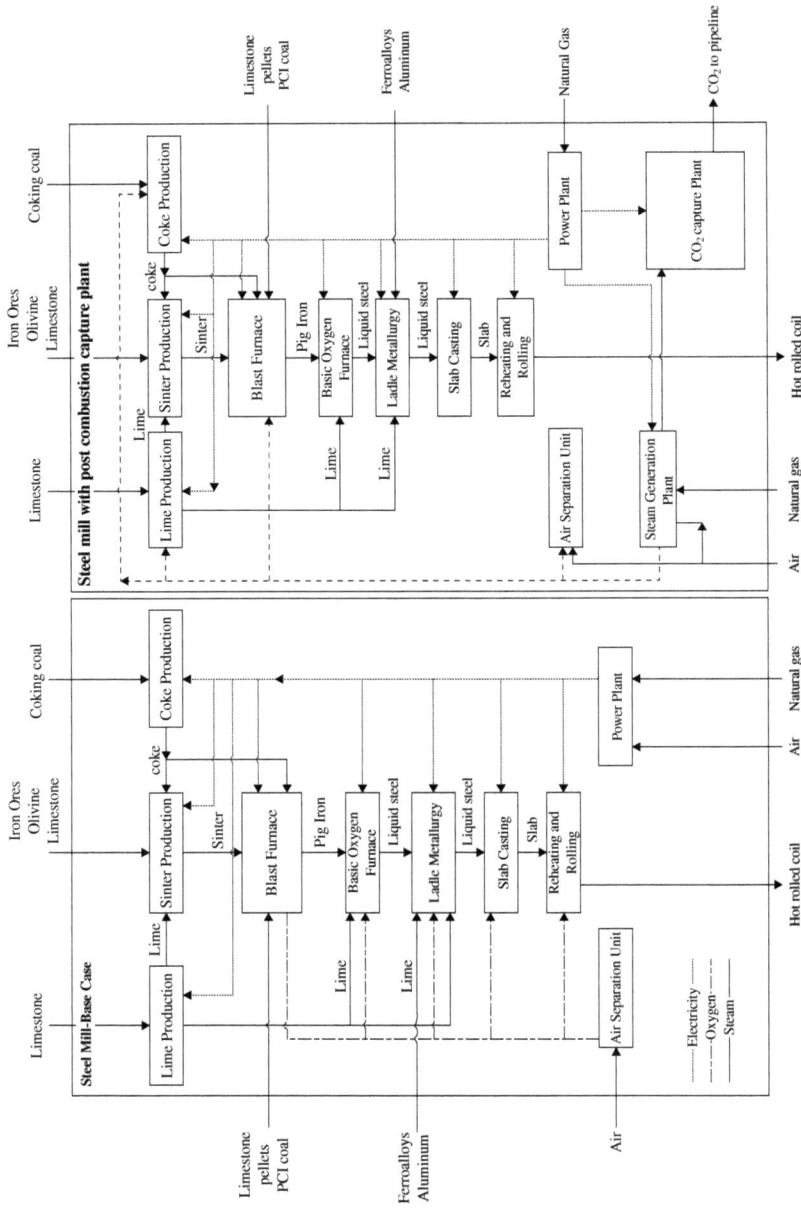

Fig. A.2.   Mass and energy flows in the steel mill with and without the capture plant.

Table A.1. SimaPro inputs/LCI for the steel mill — base case.

Sinter Plant

Total sinter production: 1,111.4 kg/t HRC

Inputs from nature
Air   12,645,530 t/y

Inputs from the Technosphere
Iron Ore (66.2% Fe) — Brazil   730.8 kg/t HRC
Iron Ore (69.77%) — Sweden   61.6 kg/t HRC
Iron Ore (57.95% Fe) — Australia   88.1 kg/t HRC
Olivine — Norway   91,261 t/y
Limestone — UK   523,103 t/y
Lime   44,456 t/y
Coke   231,540 t/y
Electricity   35.6 kWh/t HRC
Water for mixing   1,324,616 m³/y

Emissions to the air
CO_2 250,829.5 gr/t HRC
CO   24,559.92 gr/t HRC
O_2 564,784.4 gr/t HRC
N_2 2,411,480 gr/t HRC
H_2O 147,287.4 gr/t HRC

Coke making

Total coke production: 407.6 kg/t HRC

Inputs from nature
Air   2,402,738 t/y

Inputs from the Technosphere
Hard coal — RNA   2,277,702 t/y
Electricity   14.3 kWh/t HRC
Water   2,931,317 m³/y

Emissions to the air
CO_2 195,020.537 gr/t HRC

(*Continued*)

Table A.1. (*Continued*)

O_2	47,131.19 gr/t HRC
H_2O	57,117.7 gr/t HRC
N_2	573,439.1 gr/t HRC

Lime Production

Total lime production: 86.6 kg/t HRC

Inputs from nature

Air 736,814 t/y

Inputs from the Technosphere

Limestone — UK 591,361 t/y

Electricity 2.6 kWh/t HRC

Emissions to the air

CO_2	71,656.97595 gr/t HRC
O_2	20,850.51715 gr/t HRC
H_2O	19,010.62871 gr/t HRC
N_2	141,557.3 gr/t HRC

Blast Furnace

Total pig iron production: 992.2 kg/t HRC

Inputs from nature

Air 5,060,744 t/y

Inputs from the Technosphere

Limestone — UK	53,027 t/y
Hard coal — RNA	655,708 t/y
Iron Ore (57.95% Fe) — Australia	5,233,041 t/y
Iron Ore (66.2% Fe) — Brazil	1,424,671 t/y
Sinter	4,445,559 t/y
Coke	1,466,917 t/y
Electricity	102.9 kWh/t HRC
Water	3,483,639 m^3/y

Emissions to the air

CO_2	435,068.91 gr/t HRC
O_2	8,846.64 gr/t HRC

Table A.1.	(*Continued*)
H_2O	39,730.95 gr/t HRC
N_2	634,473.9 gr/t HRC

Basic Oxygen Furnace

Total liquid steel production:	1,080.8 kg/t HRC

Inputs from the Technosphere

Pig Iron	3,894,263 t/y
Lime	280,230 t/y
Iron Ore (66.2% Fe) — Brazil	21,983 t/y
Dolomite	47,640 t/y
Electricity	21.6 kWh/t HRC
Water	1,673,679 m³/y
O_2	318,514 t/y

Emissions to air

CO_2	200,882.8 t/y

Ladle Metallurgy

Total refined steel production:	1,086.3 kg/t HRC

Inputs from the Technosphere

Liquid steel	4,323,327 t/y
Lime	21,726 t/y
Aluminum	6,518 t/y
Ferromanganese	47,798 t/y
O_2	13,030 t/y
Electricity	27.2 kWh/t HRC

Emissions to the air

CO_2	3,204.8 t/y

Slab casting

Total slab production:	1,052.6 kg/t HRC

Inputs from the Technosphere

Refined steel	4,345,228 t/y
O_2	12,409 t/y
Electricity	10.9 kWh/t HRC

(*Continued*)

Table A.1.   (*Continued*)

Water	3,655,576 m³/y
Emissions to the air	
CO_2	3,188.2 t/y
Hot rolling	
Total hot rolled coil production:	1,000 kg/t HRC
Inputs from nature	
Air	2,985,364 t/y
Inputs from the Technosphere	
Slab	4,210,526 t/y
O_2	15,031 t/y
Electricity	105.3 kWh/t HRC
Water	8,000,000 m³/y
Emissions to the air	
CO_2	57,781.51 gr/t HRC
O_2	65,831.65 gr/t HRC
H_2O	84,134.54 gr/t HRC
N_2	575,361.8 gr/t HRC
Power Plant	
Total electricity production:	400.1 kWh/t HRC
Inputs from nature	
Air	5,130,503 t/y
Inputs from the Technosphere	
Natural gas	0.052 Nm³/kWh
Emissions to the air	
CO_2	982,741.7 gr/t HRC
O_2	19,189.5 gr/t HRC
H_2O	1,559,230 gr/t HRC
N_2	106,238.2 gr/t HRC
Air Separation Unit	
Total oxygen Production:	693,464 t/y

Table A.1.   (*Continued*)

Inputs from nature	
Air	4,142,730 t/y
Inputs from the Technosphere	
Steam	29,139 t/y
Electricity	66.8 kWh/t HRC

When the CO_2 capture plant is added to the steel mill, the power plant needs to be larger in size and capacity because the CO_2 capture and compression needs more electricity. A steam generation plant is also needed. The CO_2 capture plant has an assumed capture rate of 90%. Other plants will remain as before.

Table A.2.   SimaPro inputs/LCI for the steel mill with the capture and compression plant.

Power Plant	
Total electricity production:	621.7 kWh/t HRC
Inputs from nature	
Air	5,130,503 t/y
Inputs from the Technosphere	
Natural gas	0.1565 Nm³/kWh
Emissions to the air	
CO_2	227,302.9 gr/t HRC
O_2	592,271.2 gr/t HRC
H_2O	194,985.4 gr/t HRC
N_2	2,986,571 gr/t HRC
Steam Generation Plant	
Total steam production:	4,891.6 MJ/t HRC
Inputs from the nature	
Air	6,185,164 t/y

(*Continued*)

Table A.2.   (*Continued*)

Inputs from the Technosphere

Natural gas   0.0084 Nm^3/MJ

Electricity   25.3 kWh/t HRC

Emissions to the air

CO_2 1,059,046.995 gr/t HRC

O_2 21,869.33 gr/t HRC

H_2O 144,268.4 gr/t HRC

N_2 1,817,959 gr/t HRC

CO_2 Capture and Compression Plant

Total captured CO_2:   6,131,310 t/y

Inputs from nature

Air   6,185,164 t/y

Inputs from the Technosphere

Steam   4,769.9 MJ/t HRC

Electricity   196.3 kWh/t HRC

Water   91.7 t/t CO_2 captured

Nomenclature

Climate change	CC
Land use	LU
Water consumption	WC
Climate change potential	CCP
Stratospheric ozone depletion	ODP
Mineral resources scarcity	MRS
Fossil resources scarcity	FRS
Ionizing radiation potential	IRP
Mineral resource scarcity potential	MRSP
Particulate matter formation potential	PMFP
Photochemical oxidant formation potential	POFP
Human health ozone formation potential	HOFP
Ecosystem ozone formation potential	EOFP
Acidification potential	AP
Human toxicity	HT
Freshwater ecotoxicity	FET
Freshwater eutrophication potential	FEP
Marine ecotoxicity	MET
Marine eutrophication potential	MEP
Terrestrial acidification potential	TAP
Terrestrial ecotoxicity potential	TEP

References

1. P.R. Shukla, J. Skea, R. Slade, A. Al Khourdajie, R. van Diemen, *et al.*, Climate change 2022: Mitigation of climate change, the Working Group III contribution, Intergovernmental Panel on Climate Change (2022).
2. V. Masson-Delmotte, P. Zhai, H.-O. Pörtner, D. Roberts, J. Skea, *et al.*, Global warming of 1.5°C. An IPCC Special Report on the impacts of global warming of 1.5°C above pre-industrial levels and related global greenhouse gas emission pathways, in the context of strengthening the global response to the threat of climate change, sustainable development, and efforts to eradicate poverty, Intergovernmental Panel on Climate Change (2018).
3. Intergovernmental Panel on Climate Change, Climate change 2014: Synthesis report, Intergovernmental Panel on Climate Change (2014).
4. M. Bui, C.S. Adjiman, A. Bardow, E.J. Anthony, A. Boston, *et al.*, Carbon capture and storage (CCS): The way forward, *Energy Environ. Sci.* **11**, 1062–1176 (2018).
5. V. Sick, K. Armstrong, G. Cooney, L. Cremonese, A. Eggleston, *et al.*, The need for and path to harmonized life cycle assessment and techno-economic assessment for carbon dioxide capture and utilization, *Energy Technol.* **8** (2020).
6. P. Gabrielli, M. Gazzani and M. Mazzotti, The role of carbon capture and utilization, carbon capture and storage, and biomass to enable a net-zero-CO_2 emissions chemical industry, *Ind. Eng. Chem. Res.* **59**, 7033–7045 (2020).
7. P. Stegmann, V. Daioglou, M. Londo, D.P. van Vuuren and M. Junginger, Plastic futures and their CO_2 emissions, *Nature* **612**, 272–276 (2022).
8. D. Sutter, M. Van Der Spek and M. Mazzotti, 110th anniversary: Evaluation of CO_2-based and CO_2-free synthetic fuel systems using a net-zero-CO_2-emission framework, *Ind. Eng. Chem. Res.* **58**, 19958–19972 (2019).
9. C. Fernández-Dacosta, M. van der Spek, C. Roxanne Hung, G.D. Oregionni, R. Skagestad, *et al.*, Prospective techno-economic and environmental assessment of carbon capture at a refinery and CO_2 utilisation in polyol synthesis, *J. CO$_2$ Util.* **21**, 405–422 (2017).
10. W. Schakel, G. Oreggioni, B. Singh, A. Strømman and A. Ramírez, Assessing the techno-environmental performance of CO_2 utilization via dry reforming of methane for the production of dimethyl ether, *J. CO$_2$ Util.* **16**, 138–149 (2016).
11. J.B. Wevers, L. Shen and M. van der Spek, What does it take to go net-zero-CO_2? A life cycle assessment on long-term storage of intermittent renewables with chemical energy carriers, *Front Energy Res.* **8** (2020).
12. C.A. Hendriks, E. Worrell, D. De Jager, K. Blok and P. Riemer, Emission reduction of greenhouse gases from the cement industry. http://www.iea-green.org.uk/prghgt42.htm

13. H. Mefteh, O. Kebaïli, H. Oucief, L. Berredjem and N. Arabi, Influence of moisture conditioning of recycled aggregates on the properties of fresh and hardened concrete, *J. Clean. Prod.* **54**, 282–288 (2013).
14. J. An, R.S. Middleton and Y. Li, Environmental performance analysis of cement production with CO_2 capture and storage technology in a life-cycle perspective, *Sustainability* **11** (2019).
15. M. Bacatelo, F. Capucha, P. Ferrão and F. Margarido, Selection of a CO_2 capture technology for the cement industry: An integrated TEA and LCA methodological framework, *J. CO_2 Util.* **68** (2023).
16. O. Cavalett, M.D.B. Watanabe, K. Fleiger, V. Hoenig and F. Cherubini, LCA and negative emission potential of retrofitted cement plants under oxyfuel conditions at high biogenic fuel shares, *Sci. Rep.* **12** (2022).
17. S.C. Galusnyak, L. Petrescu and C.C. Cormos, Environmental impact assessment of post-combustion CO_2 capture technologies applied to cement production plants, *J. Environ. Manage.* **320** (2022).
18. D. García-Gusano, D. Garraín, I. Herrera, H. Cabal and Y. Lechón, Life cycle assessment of applying CO_2 post-combustion capture to the Spanish cement production, *J. Clean. Prod.* **104**, 328–338 (2015).
19. A. Rolfe, Y. Huang, M. Haaf, A. Pita, S. Rezvani, *et al.*, Technical and environmental study of calcium carbonate looping versus oxy-fuel options for low CO_2 emission cement plants, *Int. J. Greenh. Gas Control* **75**, 85–97 (2018).
20. W. Schakel, C.R. Hung, L.-A. Tokheim, A.H. Strømman, E. Worrell, *et al.*, Impact of fuel selection on the environmental performance of post-combustion calcium looping applied to a cement plant, *Appl. Energy* **210**, 75–87 (2018).
21. R. Chauvy, L. Dubois, D. Thomas and G. De Weireld, Techno-economic and environmental assessment of carbon capture at a cement plant and CO_2 utilization in production of synthetic natural gas (2021). https://ssrn.com/abstract=3811432
22. M.T. Ho, A. Bustamante and D.E. Wiley, Comparison of CO_2 capture economics for iron and steel mills, *Int. J. Greenh. Gas Control* **19**, 145–159 (2013).
23. Á.A. Ramírez-Santos, C. Castel and E. Favre, A review of gas separation technologies within emission reduction programs in the iron and steel sector: Current application and development perspectives, *Sep. Purif. Technol.* **194**, 425–442 (2018).
24. Z. Fan and S.J. Friedmann, Low-carbon production of iron and steel: Technology options, economic assessment, and policy, *Joule* **5**, 829–862 (2021).
25. S.E. Tanzer, K. Blok and A. Ramírez, Can bioenergy with carbon capture and storage result in carbon negative steel? *Int. J. Greenh. Gas Control* **100** (2020).

26. L. Petrescu, D.-A. Chisalita, C.-C. Cormos, G. Manzolini, P. Cobden and H.A.J. van Dijk, Life cycle assessment of SEWGS technology applied to integrated steel plants, *Sustainability* **11** (2019).
27. D.-A. Chisalita, L. Petrescu, P. Cobden, E. van Dijk, A.-M. Cormos and C.-C. Cormos, Assessing the environmental impact of an integrated steel mill with post-combustion CO_2 capture and storage using the LCA methodology, *J. Clean. Prod.* **211**, 1015–1025 (2019).
28. L. Rigamonti and E. Brivio, Life cycle assessment of methanol production by a carbon capture and utilization technology applied to steel mill gases, *Int. J. Greenh. Gas Control* **115** (2022).
29. N. Thonemann, D. Maga and C. Petermann, Handling of multi-functionality in life cycle assessments for steel mill gas based chemical production, *Chem. Ing. Tech.* **90**, 1576–1586 (2018).
30. C. Fernández-Dacosta, V. Stojcheva and A. Ramirez, Closing carbon cycles: Evaluating the performance of multi-product CO_2 utilisation and storage configurations in a refinery, *J. CO₂ Util.* **23**, 128–142 (2018).
31. B. Young, M. Krynock, D. Carlson, T.R. Hawkins, J. Marriott, *et al.*, Comparative environmental life cycle assessment of carbon capture for petroleum refining, ammonia production, and thermoelectric power generation in the United States, *Int. J. Greenh. Gas Control* **91** (2019).
32. K. Volkart, C. Bauer and C. Boulet, Life cycle assessment of carbon capture and storage in power generation and industry in Europe, *Int. J. Greenh. Gas Control* **16**, 91–106 (2013).
33. L. Petrescu, D. Bonalumi, G. Valenti, A.M. Cormos and C.C. Cormos, Life cycle assessment for supercritical pulverized coal power plants with post-combustion carbon capture and storage, *J. Clean. Prod.* **157**, 10–21 (2017).
34. P. Viebahn, J. Nitsch, M. Fischedick, A. Esken, D. Schüwer, *et al.*, Comparison of carbon capture and storage with renewable energy technologies regarding structural, economic, and ecological aspects in Germany, *Int. J. Greenh. Gas Control* **1**, 121–133 (2007).
35. J. Koornneef, T. van Keulen, A. Faaij and W. Turkenburg, Life cycle assessment of a pulverized coal power plant with post-combustion capture, transport and storage of CO_2, *Int. J. Greenh. Gas Control* **2**, 448–467 (2008).
36. N.A. Odeh and T.T. Cockerill, Life cycle analysis of UK coal fired power plants, *Energy Convers. Manag.* **49**, 212–220 (2008).
37. M. Pehnt and J. Henkel, Life cycle assessment of carbon dioxide capture and storage from lignite power plants, *Int. J. Greenh. Gas Control* **3**, 49–66 (2009).
38. A. Schreiber, P. Zapp and W. Kuckshinrichs, Environmental assessment of German electricity generation from coal-fired power plants with amine-based carbon capture, *Int. J. Life Cycle Assess.* **14**, 547–559 (2009).

39. S. Fadeyi, H.A. Arafat and M.R.M. Abu-Zahra, Life cycle assessment of natural gas combined cycle integrated with CO_2 post combustion capture using chemical solvent, *Int. J. Greenh. Gas Control* **19**, 441–452 (2013).

40. B. Singh, A.H. Strømman and E.G. Hertwich, Comparative life cycle environmental assessment of CCS technologies, *Int. J. Greenh. Gas Control* **5**, 911–921 (2011).

41. H.H. Khoo, P.N. Sharratt, J. Bu, T.Y. Yeo, A. Borgna, *et al.*, Carbon capture and mineralization in Singapore: Preliminary environmental impacts and costs via LCA, *Ind. Eng. Chem. Res.* **50**, 11350–11357 (2011).

42. S. Asante-Okyere, T. Daqing, E. Enemuoh and S. Kwofie, Life cycle assessment of supercritical coal power plant with carbon capture and sequestration in China, *Asian J. Environ. Sci.* **1**, 1–8 (2016).

43. L. Tang, T. Yokoyama, H. Kubota and A. Shimota, Life cycle assessment of a pulverized coal-fired power plant with CCS technology in Japan, *Energy Procedia* **63**, 7437–7443 (2014).

44. N.S. Matin and W.P. Flanagan, Life cycle assessment of amine-based versus ammonia-based post combustion CO_2 capture in coal-fired power plants, *Int. J. Greenh. Gas Control* **113** (2022).

45. Z. Nie, A. Korre and S. Durucan, Life cycle modelling and comparative assessment of the environmental impacts of oxy-fuel and post-combustion CO_2 capture, transport and injection processes, *Energy Procedia* **4**, 2510–2517 (2011).

46. A. Korre, Z. Nie and S. Durucan, Life cycle modelling of fossil fuel power generation with post-combustion CO_2 capture, *Int. J. Greenh. Gas Control* **4**, 289–300 (2010).

47. N. Von Der Assen and A. Bardow, Life cycle assessment of polyols for polyurethane production using CO_2 as feedstock: Insights from an industrial case study, *Green Chem.* **16**, 3272–3280 (2014).

48. L.J. Müller, A. Kätelhön, M. Bachmann, Ar. Zimmermann, A. Sternberg and A. Bardow, A guideline for life cycle assessment of carbon capture and utilization, *Front. Energy Res.* **8** (2020).

49. K. de Kleijne, S.V. Hanssen, L. van Dinteren, M.A.J. Huijbregts, R. van Zelm and H. de Coninck, Limits to Paris compatibility of CO_2 capture and utilization, *One Earth* **5**, 168–185 (2022).

50. W. Schakel, C. Fernández-Dacosta, M. Van Der Spek and A. Ramírez, New indicator for comparing the energy performance of CO_2 utilization concepts, *J. CO_2 Util.* **22**, 278–288 (2017).

51. R. W. Howarth and M. Z. Jacobson, How green is blue hydrogen? *Energy Sci. Eng.* **9**, 1676–1687 (2021).

52. C. Bauer, K. Treyer, C. Antonini, J. Bergerson, M. Gazzani, *et al.*, On the climate impacts of blue hydrogen production, *Sustain. Energy Fuels* **6**, 66–75 (2022).

53. M. Pérez-Fortes, J.C. Schöneberger, A. Boulamanti, G. Harrison and E. Tzimas, Formic acid synthesis using CO_2 as raw material: Techno-economic and environmental evaluation and market potential, *Int. J. Hydrog. Energy* **41**, 16444–16462 (2016).
54. M. Pérez-Fortes, J.C. Schöneberger, A. Boulamanti and E. Tzimas, Methanol synthesis using captured CO_2 as raw material: Techno-economic and environmental assessment, *Appl. Energy* **161**, 718–732 (2016).
55. L.J. Müller, A. Kätelhön, S. Bringezu, S. McCoy, S. Suhet, *et al.*, The carbon footprint of the carbon feedstock CO_2, *Energ. Environ. Sci.* **13**, 2979–2992 (2020).
56. V. Becattini, P. Gabrielli and M. Mazzotti, Role of carbon capture, storage, and utilization to enable a net-zero-CO_2-emissions aviation sector, *Ind. Eng. Chem. Res.* **60**, 6848–6862 (2021).
57. J. Romero, J. Geden, O. Bronwyn Hayward and F. E. L. Otto, *The Gambia,* Aïda Diongue-Niang (Senegal), David Dodman.
58. S.J. Davis, N.S. Lewis, M. Shaner, S. Aggarwal, D. Arent, *et al.*, Net-zero emissions energy systems, *Science* **360**(6396), eaas9793 (2018).
59. E.G. Hertwich, T. Gibon, E.A Bouman and A. Arvesen, Integrated life-cycle assessment of electricity-supply scenarios confirms global environmental benefit of low-carbon technologies, *Proc. Natl. Acad. Sci. U.S.A.* **112**, 6277–6282 (2015).
60. International Energy Agency Environmental Projects (IEAGHG), Iron and Steel CCS study (Techno-economics integrated steel mill), **53** (2013).
61. Y. Qiu, P. Lamers, V. Daioglou, N. McQueen, H.-S. de Boer, *et al.*, Environmental trade-offs of direct air capture technologies in climate change mitigation toward 2100, *Nat. Commun.* **13**, 1–13 (2022).
62. N. Thonemann, A. Schulte D. Maga, How to conduct prospective life cycle assessment for emerging technologies? A systematic review and methodological guidance, *Sustainability* **12**(3), 1192 (2020).
63. T. Beaufils, H. Ward, M. Jakob and L. Wenz, Assessing different European Carbon Border Adjustment Mechanism implementations and their impact on trade partners, *Commun. Earth Environ.* **4** (2023).

Chapter 16

Power Plant Integrated with Carbon Capture and Utilization: Potential Carbon-based Fuels and Chemicals in Singapore

**S.C. Lenny KOH[a,b], Hsien Hui KHOO[c,*]
and Moein SHAMOUSHAKI[a,b]**

[a]*Sheffield University Management School,
The University of Sheffield, Sheffield,
S10 1FL, United Kingdom*
[b]*Energy Institute, The University of Sheffield,
Sheffield, S10 2TN, United Kingdom*
[c]*Institute of Sustainability for Chemicals,
Energy and Environment (ISCE2),
Agency for Science, Technology and Research (A*STAR),
Singapore 627833*

*khoo_hsien_hui@isce2.a-star.edu.sg

In this chapter, we explore the potential of a power system incorporating Carbon Capture and Utilization (CCU) as a promising strategy to mitigate carbon dioxide (CO_2) emissions into the atmosphere. Furthermore, we delve into the generation of eight different CO_2-based products. These products encompass methanol, ethanol, carbon monoxide, methane, formic acid, ethylene, polycarbonate, and gasoline/jet fuel. We assess the carbon footprint of these products by comparing their CO_2 emissions

343

throughout their lifecycles and production phases, drawing on previously conducted Life Cycle Assessments. The literature review highlights that the environmental impacts of producing these fuels can fluctuate based on various factors, including the chosen production methods, materials, and energy sources. A comparison was made between the carbon emissions of CO_2-based and fossil-based fuel production. Environmental assessments have revealed that the production of CO_2-based fuels has the potential to save carbon emissions. The use of CO_2-based fuels shows promise in reducing overall greenhouse gas emissions compared to traditional fossil fuels, thereby making a valuable contribution to the ongoing fight against climate change. To conclude this chapter, we provide recommendations for further research and identify current gaps that warrant attention in future investigations.

1. Introduction

Rapid growth in global chemical and petrochemical industries significantly depend on the availability and use of raw materials, energy, and fuels. At present, conventional energy resources are supplied using coal, oil, and natural gas. The overreliance on fossil energy results in many environmental problems, including climate change. Excessive utilization of fossil fuels has led to severe energy crisis and a sharp rise in carbon dioxide (CO_2) concentration in the atmosphere [1, 2]. Climate change has become a main driver for institutional change, prompting the need for comprehensive adaptation strategies. As an important move towards the transition to a low-carbon nation, Singapore aims to explore various ways to reduce and utilize carbon emissions.

Due to the lack of fossil fuel resources, Singapore relies entirely on fuel imports to meet its growing energy demands, and it is used mainly for electrical power generation and petrochemical production [3]. Other efforts have also recently been made to mitigate climate change and, at the same time, resolve resource scarcity problems. The preparation of valuable fuels or chemicals from CO_2 has attracted great attention, with the potential towards simultaneously achieving long-term targets for greenhouse gas (GHG) reductions and the transition towards low-carbon technologies [4, 5].

Various discussions for the transformation of CO_2 to play a key role in the decarbonization of chemicals and fuels were detailed in Chapters 1 and 2. This chapter will present an overview of ongoing and emerging carbon capture and utilization (CCU) research trends to produce CO_2-based products for use in chemical and petrochemical industries and the overall corresponding GHG reduction strategies for potential application in Singapore. The next subsections will take into account the primary source of fossil fuel, namely natural gas (NG), delivered to power plants in Singapore, and next, a suggested installation of carbon capture. The last few sections discuss CCU for the following eight CO_2-based products: (i) methanol (MeOH), (ii) ethanol (EtOh), (iii) carbon monoxide (CO), (iv) methane (CH4), (v) formic acid (FA), (vi) ethylene, (vii) polycarbonate, and (viii) gasoline/jet fuel. Finally, carbon footprint results for all these CO_2-based products are presented, and they present circular-economy-led energy technologies in practice.

1.1. *From fuel to a natural gas power plant*

Singapore imports most of its NG from South Sumatra, Indonesia, for the supply of power production and other energy demands. After extraction, natural gas has a wide range of acid gas concentrations. The unprocessed natural gas, or "sour gas", contains other small hydrocarbon molecules such as ethane, propane, butane, pentane, and hexane. These impurities have to be removed before the gas is accepted for export to Singapore [3]. An overall schematic of natural gas extraction and processing is illustrated in Fig. 1(a). The stages show the processes of gas extraction, production, treatment (to remove impurities compression), and finally, export via pipelines. The gas supply from Indonesia is piped through a 477 km

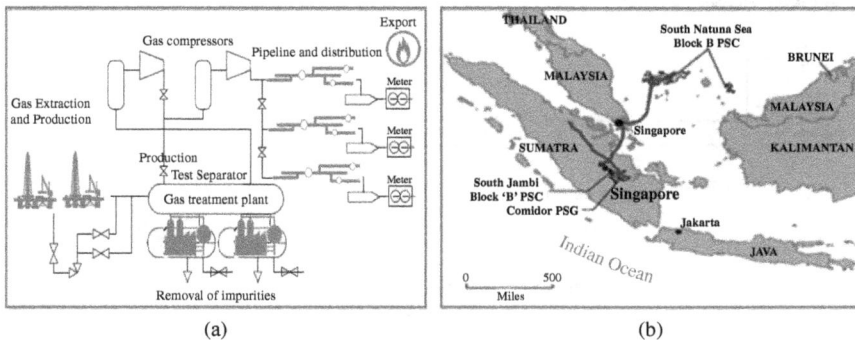

(a)   (b)

Fig. 1.   (a) Overall processes of gas extraction, production, treatment, compression, pipeline, and distribution. (b) Network of gas pipelines from Indonesia to Singapore.

undersea pipeline operated by Transgasindo, a unit of Indonesia's state-owned gas transmission company, Perusahaan Gas Negara, from South Sumatra to Singapore, passing through Batam (Fig. 1(b)).

Figure 2(a) shows the carbon footprint of 1 kg of NG delivered from Malaysia. Figure 2(b) summarizes the carbon footprint of 1 kg of NG

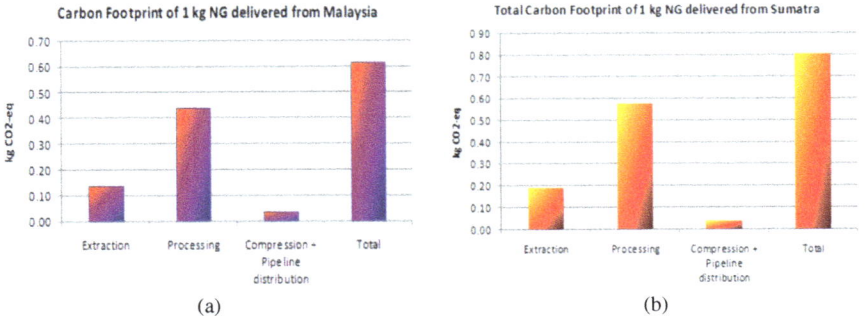

(a)

(b)

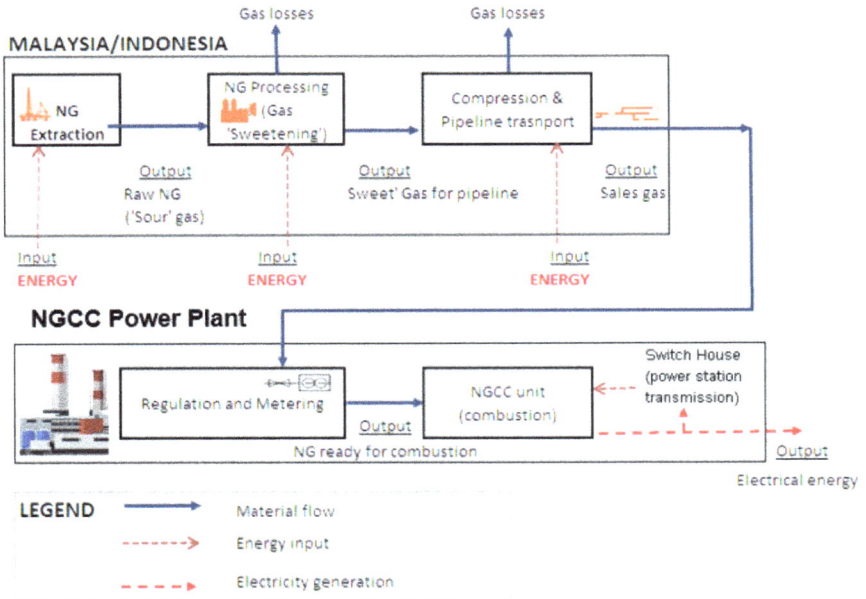

(c)

Fig. 2.   Life cycle stages for the the natural gas combined cycle with (a) natural gas (NG) sourced from Malaysia, (b) NG sourced from Indonesia, and (c) NG production to power.

delivered from Sumatra. They both include upstream operations — fuel extraction, processing, compression, and pipeline delivery — and show that NG from Malaysia is more environmentally sustainable than from Sumatra. Despite this, at the NG power plant, the source of NG comes from Sumatra (80%), and the rest from Malaysia (20%). In the final stages, NG is combusted in the natural gas combined cycle (NGCC) to generate electricity (Fig. 2(c)).

1.2. *Power plant with carbon dioxide capture technology*

Carbon dioxide (CO_2) capture with monoethanolamine (MEA) is recommended for a high CO_2 capture rate. This chemical absorption method of CO_2 capture relies on a reaction between CO_2 and the MEA chemical solvent. It is assessed that the amine method can capture approximately 85–95% carbon from power plant flue gas with a purity above 99.95%, with an energy penalty of 3.8–4.2 MJ/kg CO_2 captured for use [6–8].

The preliminary results of total CO_2 from NG extraction, delivery to fuel combustion, and carbon capture are described in Fig. 3.

The CO_2 results are aligned with Khoo in Ref. [3], where "cradle-to-gate" reported for NG power was to be 390 kg CO_2 per 1 MWh for 2020.

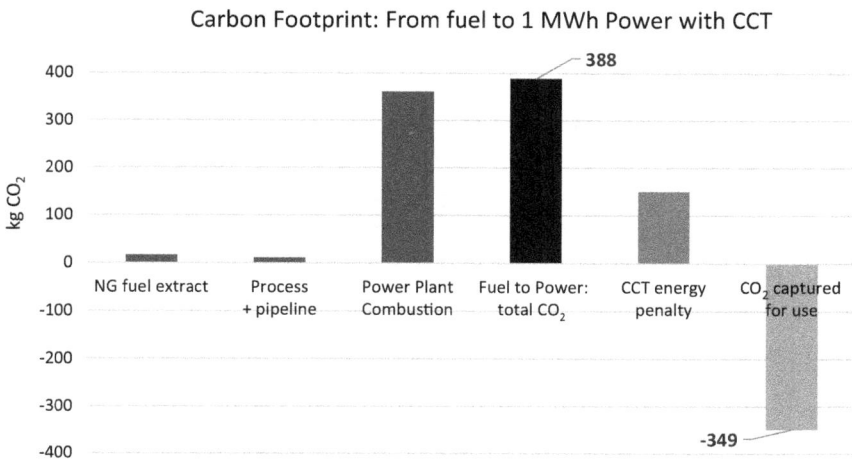

Fig. 3.   Preliminary fuel to power carbon dioxide (CO_2) results (natural gas extraction + delivery + combustion = 388 kg CO_2 per 1 MWh).

2. Carbon Capture and Utilization Methods — Emerging Areas

Carbon dioxide (CO_2), a non-toxic, abundant, cheap, and easily obtained carbon source, arouses great interest among scientists all over the world, to decrease massive CO_2 in the atmosphere. With the aim of climate change mitigation, the utilization of CO_2 has attracted attention with the aim of reducing carbon emissions by converting CO_2 into high-value chemicals or energy. Among all the proposed methods for converting CO_2, catalytic hydrogenation of CO_2 is one of the most promising applications due to its possibility of massively converting CO_2 into fuels and building blocks for the chemical industry [9–12].

In the research arena involving CCU, electrochemical CO_2 reduction reaction (CO2RR) is one of the main scientific pathways to achieve carbon-based products. CO2RR involves multiple electron/proton transfer processes, where CO_2 can be reduced into various gaseous and liquid products, including formic acid (HCOOH), CO, hydrocarbons (CH4 and C2H4), and alcohols (CH3OH and C2H5OH) [8].

CO_2 is a thermodynamically very stable molecule and, thus, a substantial input of energy combined with effective reaction conditions and active catalysts are required for its conversion (Fig. 4) [8].

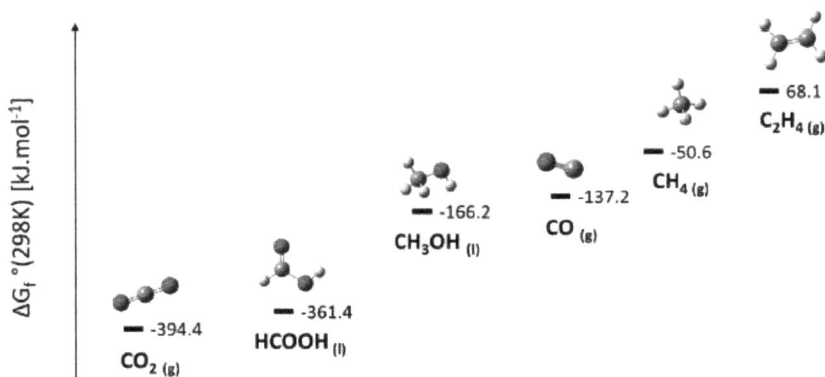

Standard Gibbs free energy of formation (at 298 K), expressed in kJ mol^{-1}, of CO2 and possible reduction products

Fig. 4.   The Gibbs free energy of formation at 298 K in kJ per mole for carbon dioxide and its potential reduction byproducts [8].

It was recommended in various chapters in this book that the potential net carbon reduction benefits of emerging CO_2 reduction technologies be investigated by a Life Cycle Assessment (LCA). The LCA is internationally well recognized for its ability to provide important evaluation in a broader context (e.g., double role of CO_2 as emission and feedstock) [13]. Recently, various research cases covering the CCU arena applied LCA to analyze the environmental impacts of the production of a number of new CO_2-based products [11, 14, 15]. In most CCU cases, high energetic co-reactants such as hydrogen (H2) [16] are often needed to activate the chemically inert CO_2. The production of these high energetic co-reactants, however, is associated with high CO_2 impacts. The following CO_2-based products will be reviewed and evaluated via an LCA (Fig. 5).

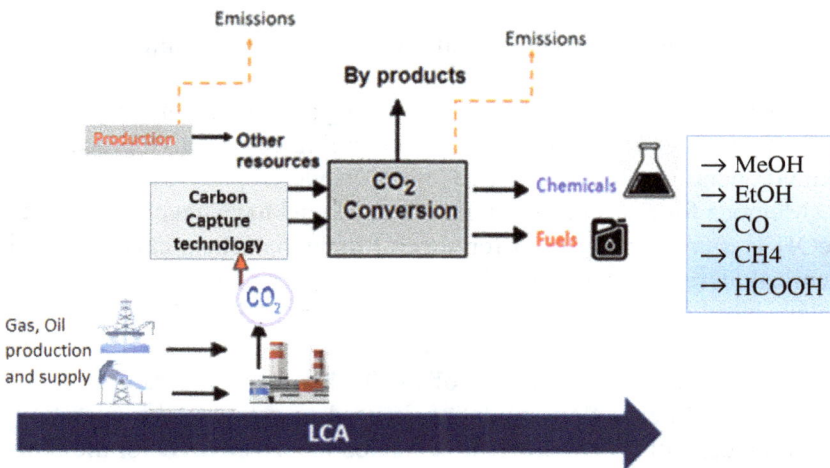

Fig. 5.   Life Cycle Assessment model for carbon dioxide-based products.

2.1. *Carbon dioxide to methanol*

Methanol (MeOH) occupies a critical position in the chemical industry as a highly versatile building block in many innovative applications of various products (Fig. 6) [17].

MeOH is also a promising liquid energy carrier that can be used for the synthesis of heavy alcohols and gasoline. As a hydrogen and a carbon

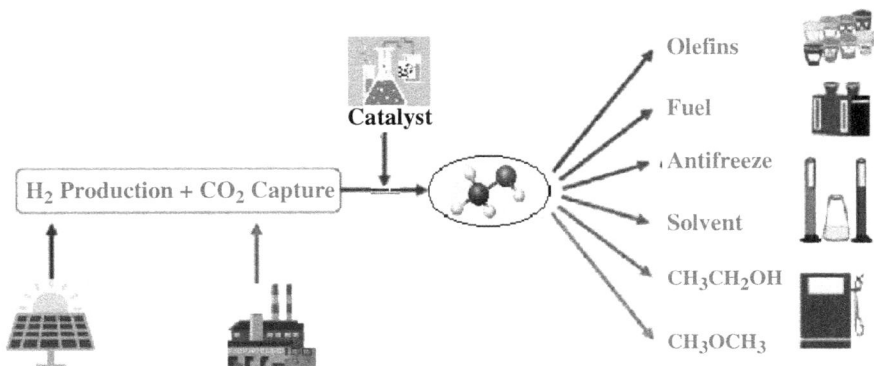

Fig. 6.   Carbon dioxide hydrogenation to methanol, supported with catalyst use [17].

carrier, MeOH is considered one of the most promising among the CO_2-conversion products.

MeOH synthesis is a mature technology; therefore, making MeOH production via CO_2 hydrogenation an emerging competitive process, but optimal operating conditions have to be considered [18, 19].

Methods for hydrogenation of CO_2 to MeOH have developed rapidly. MeOH is a commodity used for several industrial chemicals. The main chemical derivatives produced are formaldehyde, acetic acid, methyl tertiary-butyl ether (MTBE), and dimethyl ether (DME) [9].

In the presence of catalysts like Cu/ZnO/Al2O3, CO_2 reacts with H2 to form MeOH at a pressure of 5–10 MPa and temperature of 210–270°C [20]. Process flow modeling is used to estimate the operational performance, with a flowsheet implemented in CHEMCAD for the evaluation of a MeOH plant producing 440 ktMeOH/yr. In another case, Li and colleagues [21] stated that to produce a ton of MeOH, 0.2 ton of H2, 1.5 ton of CO_2 and 0.28 MWh of power are needed.

More efforts on the electrochemical reduction of CO_2 to MeOH can be found in Refs. [22, 23].

2.2. *Carbon dioxide to ethanol*

Besides MeOH, EtOh is also an important industrial chemical that can be used as a solvent and in the synthesis of other organic chemicals [24]. One of the interests in EtOh (C2H5OH) lies in the research area of generating

H2 via EtOh–water mixtures through catalytic steam reforming (e.g., Ref. [25]).

Bio-based EtOh can be produced by anaerobic fermentation of biomass materials. However, the production chain of bio-EtOh's dependence on agricultural systems, land management, and geographical limitations has sparked global environmental concern [26, 27].

Electrochemical reduction of CO_2 to EtOh, a clean and renewable liquid fuel with high heating value, is an attractive strategy for global warming mitigation and resource utilization. Moreover, $C2H5OH$ produced from CO_2 can be used as a fuel or a fuel additive with higher energy density than MeOH from the same C1 source. In one example, Mo and colleagues [28] carried out an LCA study to evaluate the potential environmental impacts and benefits of an innovative electrochemical process versus conventional biochemical and thermochemical processes towards the sustainable synthesis of 1 kg of EtOH.

Alternatively, Lee [29] evaluated the use of high-purity CO_2 — generated from a bio-based fermentation process — as the feedstock for EtOH production through a gas fermentation process plus an electrochemical reduction process (Fig. 7).

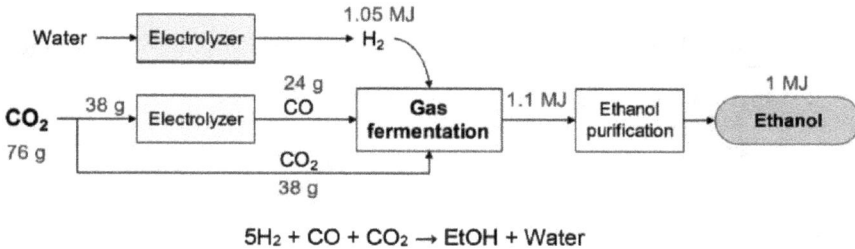

$$5H_2 + CO + CO_2 \rightarrow EtOH + Water$$

Fig. 7. Case of carbon dioxide (CO_2) reduction into carbon monoxide (CO); next, CO_2 and CO is fed into the gas fermentation process, along with hydrogen produced from an on-site water electrolyzer [29].

As the conversion of CO_2 to EtOH gains momentum due to its potential for industrial uses as a chemical solvent and in the synthesis of organic compounds, a few scientific challenges still need to be solved. CO_2-to-$C2H5OH$ conversion yield has been rather low, and the underlying catalytic mechanism remains vague or unexplored in most cases [30]. Part of the problem lies in the low activity, poor product selectivity, and stability of electrocatalysts [31].

2.3. *Carbon dioxide to carbon monoxide*

Carbon dioxide (CO_2) conversion to CO is the first step in the synthesis of more complex carbon-based fuels and feedstocks and holds great significance for the chemical industry. Basically, CO_2-to-CO conversion involves two-electron/proton transfer and thus is regarded as kinetically fast. Among various developed CO_2-to-CO reduction electrocatalysts, transition metal/N-doped carbon (M-N-C) catalysts are attractive due to their low cost and high activity [12].

Alternatively, noble metal-based electrocatalysts such as Au, Ag, and Pd were regarded as one of the major classes for efficient CO2RR to CO with high selectivity (>80%) and activity because they could stabilize COOH intermediate (Fig. 8) [32].

In another work, Nabil and colleagues [33] carried out comparative cradle-to-gate LCA study of one and two-step electrochemical conversion of CO_2 to eight major value-added products, including a one-step production of CO via:

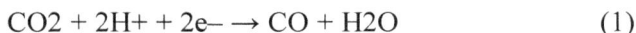

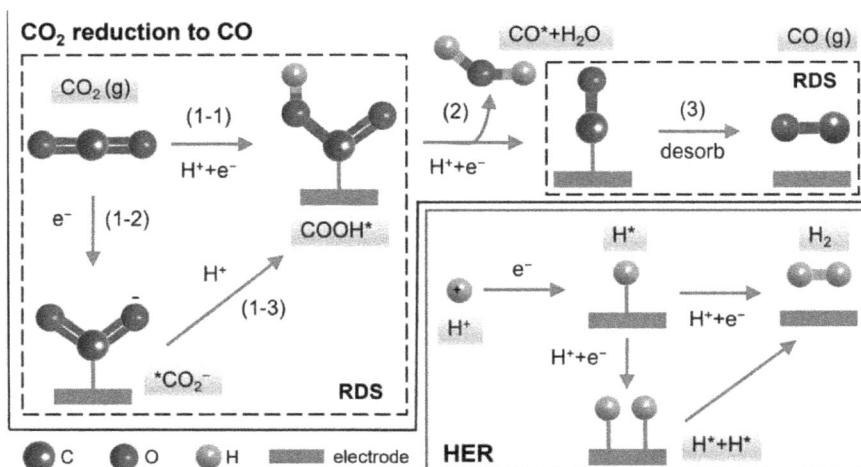

$$CO2 + 2H+ + 2e- \rightarrow CO + H2O \qquad (1)$$

Fig. 8. Proposed reaction pathway for carbon dioxide (CO_2) reduction to carbon monixide (CO) and HER (C atoms: gray; O atoms: red; H atoms: orange; and electrode: blue) [32].

CO$_2$ reduction to CO (as depicted in Eq. (1)) gained interest since CO can play many roles in the petrochemical industry, including the interest as a component of syngas, a H2/CO mixture for useful liquid fuel synthesis (e.g., Ref. [34]). Further research areas in the development of affordable electrocatalysts that can drive the reduction of CO$_2$ to CO with high selectivity, efficiency, and large current densities remain a critical step forward for the sustainable production of liquid carbon-based fuels [35].

2.4. *Carbon dioxide to methane*

As a high-valuable fuel that has many applications in the chemical and petrochemical industry, CH4, several production processes from different feedstock (typically CO$_2$ and H2) are ongoing. CH4 can be produced through renewables-powered electrolytic routes such as alkaline electrolyzers, proton exchange membrane electrolyzers, and solid oxide electrolyzers, commonly known as solid oxide electrolysis cells (SOECs) (Fig. 9) [36].

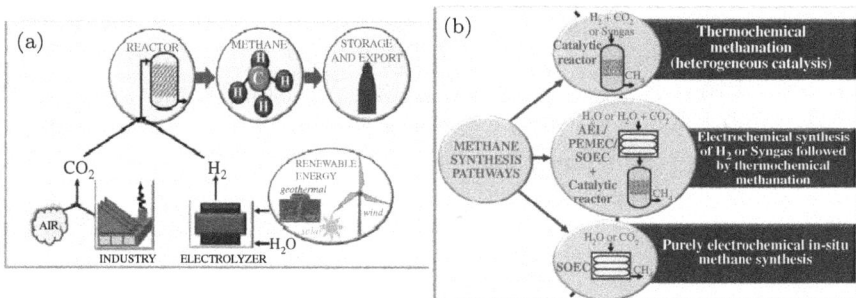

Fig. 9.   (a) Methane synthesis through renewables-powered electrochemical reactors. (b) Three major pathways of methane synthesis [36].

Among these, thermochemical methanation is perhaps the most conventional application in industry and extensively studied (e.g., Ref. [37]). Other research studies intensively investigated focus on catalyst development to support CO$_2$ hydrogenation to CH4 [38]. Many works have focused on catalysts used to promote CO$_2$ methanation. Discussions and reviews on catalytic materials for enhanced structure–activity relationships for CO_2 methanation have been extensively done [39]. Apart from the view of catalyst design and development, the environmental aspects of CO_2 conversion have also been evaluated [8, 11, 33]. A noteworthy study was performed by Federici and colleagues [40] on the life cycle environmental impact of CH4 production using a combined alkaline electrolyzer and methanation reaction. Their work demonstrated that energy and reactants consumed, plus construction materials, are all relevant to the environmental performance of the processes in different impact categories.

2.5. *Carbon dioxide to formic acid*

Formic acid (FA) is a fundamental chemical feedstock in the organic chemical industry. It has been used in the perfume industry as a mordant in the dyeing industry, a neutralizer in tanning, and a disinfectant and preservative agent in sanitary stations. As a raw material in the chemical industry, it has promoted the production of formate esters, which have been used to produce a variety of organic derivatives, such as aldehydes, ketones, carboxylic acids, and amides.

Currently, commercial processes for FA production are mainly the result of hydrolysis of methyl formate and direct synthesis from CO and water (Fig. 10). As these traditional methods consume a large amount of energy and produce hazardous waste, the development of a clean method of FA synthesis is a high priority. The hydrogenation of CO_2, as shown in

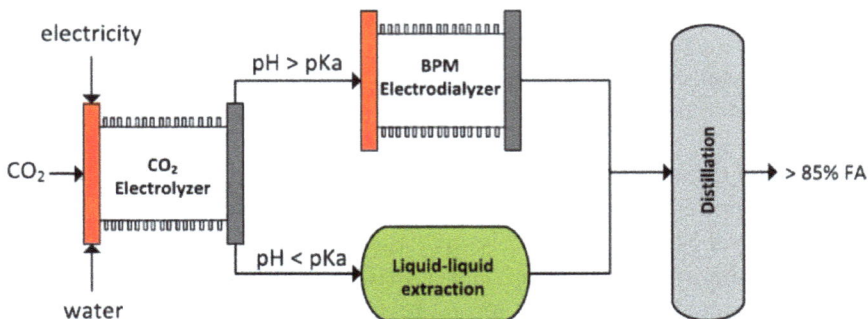

Fig. 10.   High-pressure semicontinuous batch electrolyzer converting carbon dioxide to formic acid [42].

the equation below, provides a possible technique to synthesize FA using CO_2 as a raw material [23]. We have following reaction [41]:

$$CO2 + H2 \rightarrow HCOOH \qquad (2)$$

2.6. Carbon dioxide to ethylene

One route for CO_2 electro-conversion is via the reduction of CO_2 and water (H_2O) in an aqueous electrolyte to produce ethylene (C2H4):

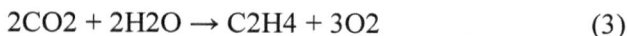

$$2CO2 + 2H2O \rightarrow C2H4 + 3O2 \qquad (3)$$

Among the various CO_2 hydrogenation reaction products, light olefins, such as ethylene and propylene, are very important intermediates in the chemical industry. For example, direct and selective hydrogenation of CO_2 to ethylene and propene by bifunctional catalysts [43] enables electrochemical CO_2 conversion to C2H4 is in line with research and industry [44] because ethylene is an important base chemical for the chemical and plastics industry (Fig. 11).

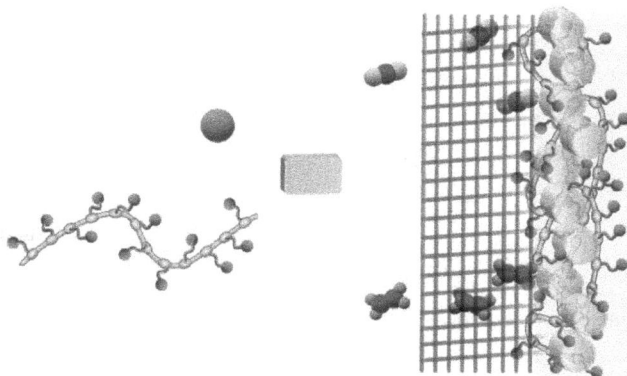

Fig. 11. Preparing the Cu-polymer catalyst involves the co-electroplating of P1 and copper onto the Gas Diffusion Layer (GDL), as depicted in the schematic illustration [45].

2.7. Carbon dioxide to poly (propylene carbonate)

One of the economically efficient methods involves using CO_2 to create a range of biodegradable polymers with customized characteristics [46].

Polycarbonate is a thermoplastic polymer that comprises carbonate segments [47].

The challenges faced in substituting polyolefin plastics with biodegradable polymers arise from the expensive source materials. Hence, the application of CO_2, a valuable resource, becomes evident in the production of affordable, eco-friendly, and degradable polycarbonates. The conversion of waste CO_2 into poly (propylene carbonate) (PPC) represents a notable accomplishment, given its minimal environmental impact on polymer production. CO_2 offers a direct route to CO_2-derived plastics that are completely degradable, showcasing considerable potential and diverse uses. PPC is applied in various areas, including barrier substances, foam materials, electrolytes, and more [1].

To facilitate the production of these CO_2-based polyols, the typical approach involves co-polymerizing CO_2 with epoxides, usually conducted alongside chain transfer agents (CTAs). The inclusion of CTAs substantially reduces the amount of catalyst required and enables the creation of polyols with precisely controlled molecular weight and functional properties [48]. Currently, the most widely used and effective method for producing aliphatic PPC involves the copolymerization of CO_2 and epoxide [49] (Fig. 12).

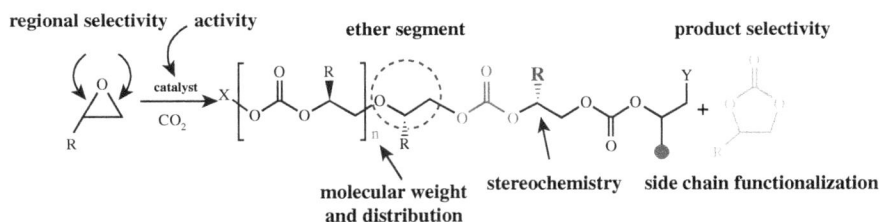

Fig. 12.   Combining carbon dioxide and epoxide through copolymerization [50].

2.8. *Carbon dioxide to gasoline/jet fuel*

2.8.1. *Carbon dioxide to gasoline*

The process of directly creating liquid fuels through the hydrogenation of CO_2 has garnered substantial attention due to its crucial contributions to

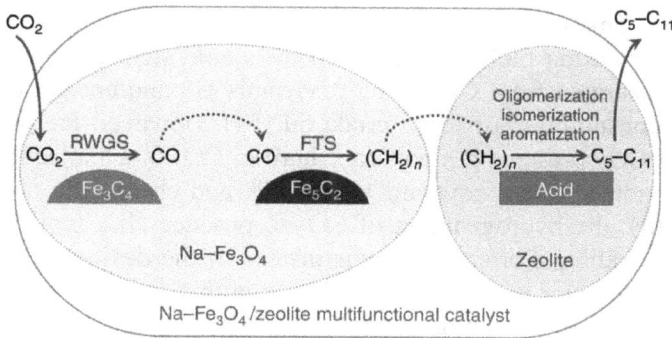

Fig. 13. Diagram outlining the process of transforming carbon dioxide through hydrogenation into hydrocarbons suitable for gasoline [51].

lessening CO_2 emissions and decreasing reliance on petrochemicals [51] (Fig. 13). Selecting the appropriate catalyst and creating a reactor for producing gasoline fuel is merely the initial phase of an industrial process design project. In order to maintain competitiveness, industrial facilities must also be effectively optimized concerning the movement of materials and energy. As Wei and colleagues [51] have indicated, CO_2 molecules are transformed into CO through a reversed water–gas shift (RWGS) reaction as described by Eq. (1). In simpler terms, the Fischer–Tropsch synthesis (FTS) process can be described as a polymerization method that creates hydrocarbon (HC) chains by combining a carbon source and hydrogenating CO [52]. Following this, certain amounts of CO are subject to hydrogenation through FTS, resulting in the creation of HCs, as represented by Eq. (2) [53].

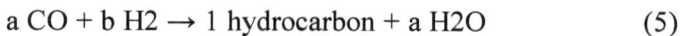

$$CO2 + H2 \rightarrow CO + H2O \ (RWGS) \tag{4}$$

$$a \ CO + b \ H2 \rightarrow 1 \ hydrocarbon + a \ H2O \tag{5}$$

2.8.2. *Carbon dioxide to jet fuel*

Amid increasing apprehensions about climate change, the conversion or utilization of CO_2 into environmentally friendly synthetic HC fuels, primarily intended for transportation needs, remains a topic of global fascination. This is especially pertinent in the quest for sustainable alternatives for aviation fuels (SAFs). Such fuels hold significant promise as they are generated from CO_2 using renewable H2 and energy, diverging from the consumption of fossil crude oil [54]. However, the process of activating CO_2 presents a significant hurdle. CO_2 is a molecule that is entirely oxidized, thermodynamically stable, and chemically unreactive. Additionally, the hydrogenation of CO_2 to produce HCs typically leans toward generating shorter HC chains instead of the desired longer ones. As a result, a major portion of the research in this field has concentrated on selectively hydrogenating CO_2 to yield CH4, oxygenates like CH3OH and HCOOH, as well as light olefins (C2–C4 olefins) [54–56].

Converting CO_2 directly with the use of environmentally friendly H2 represents a sustainable strategy for producing jet fuel. Nonetheless, attaining a remarkable level of effectiveness remains a significant hurdle, primarily due to CO_2's unreactive nature and its limited capacity for forming subsequent C–C bonds [57] (Fig. 14).

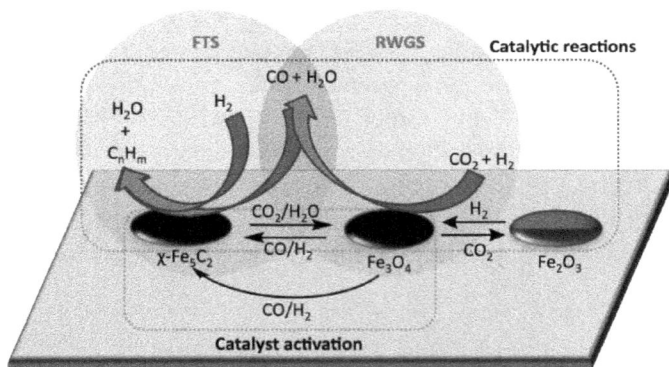

Fig. 14. Illustration of the process for converting carbon dioxide into hydrocarbons within the jet fuel range [54].

Table 1 presents the comparison summary of eight CO_2-based products generation considered in this chapter.

Table 1.   Summary of the eight CO_2-based products considered in this chapter, based on a review of the literature.

CO_2 Converted Product	Method	Amount of CO_2 Converted	Other Solvent/ Chemical Use	Product Yield	Energy Use	Ref.
MeOH	Process flow modeling implemented in CHEMCAD to obtain mass and energy flows.	1.46 t	0.199 H2	1 t MeOH	1.5 MWh	[9]
	The MeOH CCU-plant studied can utilize about 21.5% of CO_2 from 550 MW coal power plant. Reaction of CO_2 with H2 to form MeOH at 5–10 MPa and 210–270°C. (Catalysts: Cu/ZnO/Al2O3)	1,400 kg CO_2	~200 kg H2, ~1,700 kg H2O	1,000 kg MeOH	10–11 MWh	[20]
	Electrolysis of water to H$_2$ and hydrogenation of CO_2 to MeOH	1.5 t CO_2	0.2 t H$_2$	1 ton MeOH	0.28 MWh	[21]
EtOh	LCA comparison of innovative electrochemical process vs. biochem. and thermochem. to synthesis EtOH	4.58 kg CO_2	2.07 kg H2O 9.99×10^{-16} kg KHCO3	1 kg EtOh	8.89 kWh	[28]
	LCA of CO_2 from corn biorefineries for EtOH production	76.5 g CO_2	0.08 L H2O	1 MJ EtOh	1.98 (total) MJ	[29]
CO	Comparative cradle-to-gate LCA of electrochemical conversion of CO_2 to value-added products (one-step): $CO_2 + 2H+ + 2e- \longrightarrow CO + H2O$	1.57 kg CO_2	Not reported	1 kg CO	Not stated	[33]
	CO_2 conversion to CO using chemical looping over Co-In oxide. CO_2 splitting rate done at T = 723–823 K. CO_2 to CO conversion (ca. 80%)	1 mol CO_2	1 mol H2	80% conversion from one mol CO_2	134.4 kJ per mol activation energy	[35]

(Continued)

3. Life Cycle Assessment Results

Based on an analysis of existing literature and environmental assessments of these technologies, the carbon footprint of each approach varies depending on the chosen CO_2 to product method, the energy resources used to meet process demands, and the materials employed. Figure 15 summarizes the Global Warming Impact (GWI) values for the mentioned CO_2-based product generation process in comparison to three fossil-based fuels. The CO_2-based products are shown in blue bars and fossil-based fuels are presented in black bars. The LCA results of conducted environmental assessments indicate that the highest GWI is linked to the production of MeOH, while the lowest GWI is associated with CO production. Ethanol production ranks as the second-highest contributor to carbon footprint consequences. CH4 and gasoline production impacts are fairly similar, and jet fuel production results in approximately the same level of CO_2 emissions as the FA generation process. The extent of the GWI through the electrochemical route is greatly influenced by the emissions associated with electricity production. This approach becomes

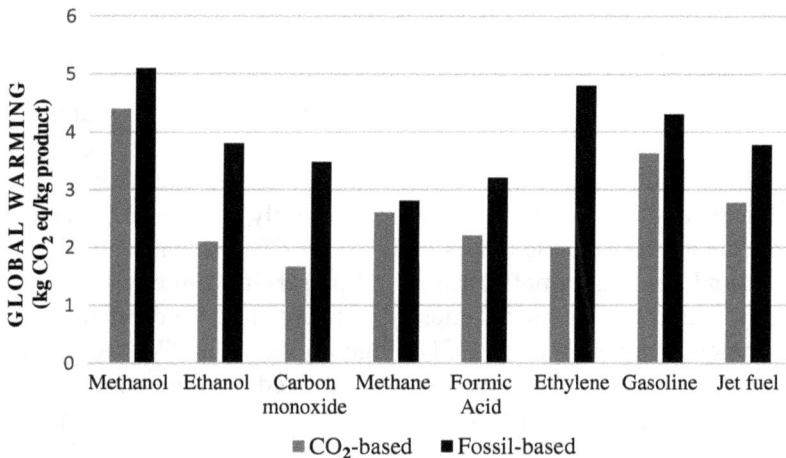

Fig. 15. Global Warming Impact of different CO_2-based and fossil-based products (mCO_2 and fossil-based ethanol [60], CO_2-based ethanol [61], fossil-based ethanol [62], CO_2-based carbon monoxide [33], fossil-based carbon monoxide [60], CO_2-based methane [33], fossil-based methane [63], CO_2-based and fossil-based formic acid [64], CO_2-based ethylene [33], fossil-based ethylene [65], CO_2-based gasoline [66], fossil-based gasoline [67], CO_2-based jet fuel [59], fossil-based jet fuel [68]).

particularly attractive in comparison to existing methods only when used in conjunction with low-emission intensity electricity [33].

When we examined products derived from CO_2 and compared them to their fossil-based counterparts, a significant decrease in the carbon footprint of fuels was observed. The results indicated that the most substantial reduction in carbon emissions was associated with the production of ethylene from CO_2, resulting in a 58.3% reduction in carbon emissions when compared to the use of fossil fuels for ethylene production. On the other hand, the lowest reduction in carbon emissions was observed in the case of CH4 production, where there was only a 7.1% reduction in CO_2 emissions when CH4 was produced from CO_2. Additionally, the generation of CO and EtOh from CO_2 had a notable impact on global warming, with reductions of 52.1% and 44.7%, respectively.

Discussions

With the goal of addressing global warming and meeting the increasing need for sustainable products and energy, academia, the chemical industry, and the energy sector are investing significant research endeavors into the development of technologies capable of utilizing CO_2 as a raw material [69].

Various approaches are being implemented to reduce CO_2 emissions. One strategy involves Carbon Capture and Storage (CCS), which primarily focuses on retaining CO_2 [70]. Another promising alternative is CCU, which is particularly appealing because it not only eliminates CO_2 but also transforms it into valuable compounds [71]. CCU is seen as a technology aimed at addressing climate change and preserving finite resources, particularly within the chemical industry. It is crucial to understand that the environmental advantages of CCU hinge on variables like the particular technology utilized, the energy source employed, and the comprehensive evaluation of the products created throughout their life cycle. Furthermore, achieving their environmental potential will heavily rely on the broad adoption and scalability of CCU technologies.

The production of alternative products derived from CO_2 has the potential to significantly reduce the economy's dependence on fossil resources. Earlier research indicated that incorporating CO_2 into chemical syntheses could offer a chance to attain GHG reductions and promote a low-carbon economy [72]. For instance, based on a previous study [73], when generated through CO_2 hydrogenation, the production of formic acid can result in a 95.01% reduction in its GWI. The environmental consequences of CO_2-based production techniques fluctuate based on several factors. Although these methods have the potential to cut down on carbon emissions and advance sustainability, their effectiveness in addressing environmental concerns hinges on how they are precisely executed and their reliance on clean energy sources.

LCA was done to quantify the environmental effects of the CO_2-based products from cradle-to-gate and encompasses all types of environmental impacts. LCA is a powerful tool that promotes environmental sustainability, informs decision-making, reduces resource consumption, and contributes to a more eco-friendly and responsible approach to product development and resource management. Assessing different CO_2-based product generation is important to understand the environmental impact of the whole life cycle of each product to find the most sustainable products. Among the considered products in this chapter, according to the previous studies, the environmental assessments conducted have revealed that MeOH production has the most significant GHG impact, whereas CO production is associated with the least impact on the GWI. Nevertheless, it is important to acknowledge that numerous variables can influence the GWI stemming from each approach and CO_2-based products. It is crucial to take into account the complete lifecycle of CO_2-based fuel manufacturing, encompassing the energy sources employed, the sourcing of feedstock, and the distribution process. The environmental advantages of CO_2-based fuels hinge on these aspects and their seamless integration into a comprehensive strategy for lessening environmental consequences. The utilization of CO_2-based fuels can result in decreased total GHG emissions in contrast to conventional fossil fuels, thereby aiding in the effort to combat climate change.

Applying renewable energies to provide the required demand for CO_2-based fuel products could significantly decrease the life cycle environmental impact. In addition, it can reduce the required energy for running the processes and consequently decline the production costs. While the utilization of renewable resources like solar energy systems has the potential to create CO_2-based fuels, such as gasoline, with a reduced environmental footprint compared to the traditional petroleum refining process, certain critical factors that enhance their environmental advantages and sustainability need enhancement. Altering the procedure to minimize losses could lead to improved environmental performance by enhancing process energy efficiency. This, in turn, could reduce the overall impact on the LCA through more effective heat integration within the primary facility and the adoption of advanced CO_2 capture technologies. These enhancements have the potential to yield better economic outcomes in addition to their environmental advantages [66].

To sum it up, harnessing CO_2 as a resource for making a range of products offers the potential to cut down on carbon emissions, foster sustainability, and advance the concept of a circular carbon economy. Nevertheless, the successful application of these technologies on a significant scale demands additional research, technological progress, and careful assessment of economic and environmental factors.

4. Conclusions

Carbon Capture and Utilization (CCU) allows for the creation of CO_2-based products, including synthetic fuels, chemicals, and materials, offering viable substitutes for conventional products derived from fossil fuels. Numerous investigations have explored CCU technology and methods for generating products from CO_2. However, there is still a need for conducting environmental impact assessments to gain a more accurate understanding of these technologies. There are some challenges in the LCA of some products, such as propylene carbonate. The conducted LCA has revealed that the production of CO_2-based products can take various forms and that the resulting carbon footprint can vary significantly based on factors such as the chosen CO_2 to product synthesis, use of chemicals, energy consumption, and its source, among others. Nonetheless, innovative interventions are expected to be implemented to mitigate the environmental impact of CO_2-derived fuels, given their substantial potential to reduce the detrimental environmental impacts associated with these products.

Enhancing the carbon-saving potential and reducing the carbon footprint is crucial for the commercialization of new systems, methods, and technologies. Comparing the potential of producing eight CO_2 and fossil-based fuels reveals a significant advantage in reducing the carbon footprint and carbon emissions. Assessing the CO_2-based products in comparison to their fossil-based equivalents, it is clear that there is substantial potential for carbon emissions reduction by producing CO_2-based fuels. The most noteworthy reduction in GHG emissions is observed in the production of CO_2-based ethylene, CO, and EtOh, with reductions of 58.3%, 52.1%, and 44.7%, respectively. On the other hand, CO_2-based CH4 production offers the smallest carbon savings, around 7.1%, in comparison to the production of CH4 from fossil sources.

While CO_2-derived fuels present these benefits, it is crucial to recognize that their effective deployment hinges on considerations such as the energy source employed, the efficiency of the technology, and the capacity for scalability. Additionally, the environmental advantages they offer are tied to the comprehensive evaluation of their entire life cycle, which encompasses the energy source utilized during their production.

5. Recommendations/Future Work

Additional investigations are required to evaluate various production methods and the application of different processes. Moreover, the potential adoption of diverse renewable energy sources and resources based on regional availability could be explored to assess the reduction in environmental impact associated with CO_2-based product generation. Furthermore, there is an evident need for a comprehensive comparative LCA study encompassing all eight production approaches, allowing for a comprehensive evaluation across different impact categories and indicators. Such a study would provide a broader perspective on the potential air, land, and water pollution implications of these products. Additionally, novel production techniques or approaches could be considered for each technology to enhance the sustainability of the resulting products.

References

1. F. Gasbarro, F. Iraldo and T. Daddi, The drivers of multinational enterprises' climate change strategies: A quantitative study on climate-related risks and opportunities, *J. Clean. Prod.* **160**, 8–26 (2017).
2. K.E. Diehl, J. Bachinger and S.K. Hamadeh, Requisite variety in adaptation strategies: Case studies from two regions prone to climate change, Brandenburg, Germany and semi-arid Bekaa, Lebanon, *Procedia Environ. Sci.* **29**,132–133 (2015).
3. H.H. Khoo, LCA of mixed generation systems in Singapore: Implications for national policy making, *Energies* **15**(24), 9272 (2022).

4. X. Tan, H. Li, J. Guo, B. Gu and Y. Zeng, Energy-saving and emission-reduction technology selection and CO_2 emission reduction potential of China's iron and steel industry under energy substitution policy, *J. Clean. Prod.* **222**, 823–834 (2019).

5. Q. Wang, B. Su, J. Sun, P. Zhou and D. Zhou, Measurement and decomposition of energy-saving and emissions reduction performance in Chinese cities, *Appl. Energy* **151**, 85–92 (2015).

6. M. Vaccarelli, R. Carapellucci and L. Giordano, Energy and economic analysis of the CO_2 capture from flue gas of combined cycle power plants, *Energy Procedia* **45**, 1165–1174 (2014).

7. B. Xue, Y. Yu, J. Chen, X. Luo and M. Wang, A comparative study of MEA and DEA for post-combustion CO_2 capture with different process configurations, *Int. J. Coal Sci. Technol.* **4**, 15–24 (2017).

8. C.A. Pappijn, M. Ruitenbeek, M.-F. Reyniers and K.M. Van Geem, Challenges and opportunities of carbon capture and utilization: Electrochemical conversion of CO_2 to ethylene, *Front. Energy Res.* **8**, 557466 (2020).

9. M. Pérez-Fortes, J.C. Schöneberger, A. Boulamanti and E. Tzimas, Methanol synthesis using captured CO_2 as raw material: Techno-economic and environmental assessment, *Appl. Energy* **161**, 718–732 (2016).

10. S. Nitopi, E. Bertheussen, S.B. Scott, X. Liu, A.K. Engstfeld, S. Horch, *et al.*, Progress and perspectives of electrochemical CO_2 reduction on copper in aqueous electrolyte, *Chem. Rev.* **119**(12), 7610–7672 (2019).

11. H.H. Khoo, I. Halim and A.D. Handoko, LCA of electrochemical reduction of CO_2 to ethylene, *J. CO2 Util.* **41**, 101229 (2020).

12. F.-Y. Gao, R.-C. Bao, M.-R. Gao and S.-H. Yu, Electrochemical CO_2-to-CO conversion: Electrocatalysts, electrolytes, and electrolyzers, *J. Mater. Chem. A* **8**(31), 15458–15478 (2020).

13. J. Artz, T.E. Müller, K. Thenert, J. Kleinekorte, R. Meys, A. Sternberg, *et al.* Sustainable conversion of carbon dioxide: An integrated review of catalysis and life cycle assessment, *Chem. Rev.* **118**(2), 434–504 (2018).

14. G. Garcia, M.C, Fernandez, K. Armstrong, S. Woolass, and P. Styring, Analytical review of life-cycle environmental impacts of carbon capture and utilization technologies, *ChemSusChem.* **14**(4), 995–1015 (2021).

15. N. Thonemann and A. Schulte, From laboratory to industrial scale: A prospective LCA for electrochemical reduction of CO_2 to formic acid, *Environ. Sci. Technol.* **53**(21), 12320–12329 (2019).

16. L.J. Müller, A. Kätelhön, M. Bachmann, A. Zimmermann, A. Sternberg and A. Bardow, A guideline for life cycle assessment of carbon capture and utilization, *Front. Energy Res.* **8**, 15 (2020).

17. X. Zhang, G. Zhang, C. Song and X. Guo, Catalytic conversion of carbon dioxide to methanol: Current status and future perspective, *Front. Energy Res.* **8**, 413 (2021).

18. F. Studt, I. Sharafutdinov, F. Abild-Pedersen, C.F. Elkjær, J.S. Hummelshøj, S. Dahl, *et al.*, Discovery of a Ni-Ga catalyst for carbon dioxide reduction to methanol, *Nat. Chem.* **6**(4), 320–324 (2014).
19. Q. Smejkal, U. Rodemerck, E. Wagner and M. Baerns, Economic assessment of the hydrogenation of CO_2 to liquid fuels and petrochemical feedstock, *Chemie Ingenieur Technik* **86**(5), 679–686 (2014).
20. P.R.N. Becht, Methanol from CO_2: A technology and outlook overview. Optimal capture of CO_2 towards methanol production compels development of sustainable renewable solutions like green methanol, https://www.digit-alrefining.com/article/1002891/methanol-from-co2-a-technology-and-outlook-overview
21. P. Li, S. Gong, C. Li and Z. Liu, Analysis of routes for electrochemical conversion of CO_2 to methanol, *Clean Energy* **6**(1), 202–210 (2022).
22. L. Lu, X. Sun, J. Ma, D. Yang, H. Wu, B. Zhang, *et al.*, Highly efficient electroreduction of CO_2 to methanol on palladium–copper bimetallic aerogels, *Angewandte Chemie* **130**(43), 14345–14349 (2018).
23. M.G. Mazzotta, M. Xiong and M.M. Abu-Omar, Carbon dioxide reduction to silyl-protected methanol catalyzed by an oxorhenium pincer PNN complex, *Organometallics* **36**(9), 1688–1691 (2017).
24. D. Xu, H. Zhang, H. Ma, W. Qian and W. Ying, Effect of Ce promoter on Rh-Fe/TiO_2 catalysts for ethanol synthesis from syngas, *Catal. Commun.* **98**, 90–93 (2017).
25. A. Casanovas, C. de Leitenburg, A. Trovarelli and J. Llorca, Ethanol steam reforming and water gas shift reaction over Co–Mn/ZnO catalysts, *Chem. Eng. J.* **154**(1–3), 267–273 (2009).
26. J.B. Dunn, S. Mueller, H.-Y. Kwon and M.Q. Wang, Land-use change and greenhouse gas emissions from corn and cellulosic ethanol, *Biotechnol. Biofuels* **6**, 1–13 (2013).
27. L. Bhatia, S. Johri and R. Ahmad, An economic and ecological perspective of ethanol production from renewable agro waste: A review, *Amb Express* **2**(1), 1–19 (2012).
28. W. Mo, X.-Q. Tan and W.-J. Ong, Prospective life cycle assessment bridging biochemical, thermochemical, and electrochemical CO_2 reduction toward sustainable ethanol synthesis, *ACS Sustain. Chem. Eng.* **11**(14), 5782–5799 (2023).
29. U. Lee, T.R. Hawkins, E. Yoo, M. Wang, Z. Huang and L. Tao, Using waste CO_2 from corn ethanol biorefineries for additional ethanol production: Life-cycle analysis. *Biofuel Bioprod. Biorefin.* **15**(2), 468–480 (2021).
30. Y. Li, Y. Chen, T. Chen, G. Shi, L. Zhu, Y. Sun, *et al.*, Insight into the electrochemical CO2-to-ethanol conversion catalyzed by Cu2S nanocrystal-decorated Cu nanosheets, *ACS Appl. Mater. Interfaces.* **15**(15), 18857–18866 (2023).

31. Y. Liu, Y, Zhang, K. Cheng, X. Quan, X. Fan, Y. Su, *et al.*, Selective electrochemical reduction of carbon dioxide to ethanol on a boron- and nitrogen-Co-doped nanodiamond, *Angewandte Chemie* **129**(49), 15813–15817 (2017).

32. S. Liang, L. Huang, Y. Gao, Q. Wang and B. Liu, Electrochemical reduction of CO_2 to CO over transition metal/N-doped carbon catalysts: The active sites and reaction mechanism, *Adv. Mat. Sci.* **8**(24), 2102886 (2021).

33. S.K. Nabil, S. McCoy and M.G. Kibria, Comparative life cycle assessment of electrochemical upgrading of CO_2 to fuels and feedstocks, *Green Chem.* **23**(2), 867–880 (2021).

34. J. Medina-Ramos, R.C. Pupillo, T.P. Keane, J.L. DiMeglio and J. Rosenthal, Efficient conversion of CO_2 to CO using tin and other inexpensive and easily prepared post-transition metal catalysts, *J. Am. Chem. Soc.* **137**(15), 5021–5027 (2015).

35. J.-I. Makiura, S. Kakihara, T. Higo, N. Ito, Y. Hiranob and Y. Sekine, Efficient CO_2 conversion to CO using chemical looping over Co–In oxide, *Chemical Comm.* **58**(31), 4837–4840 (2022).

36. S. Biswas, A.P. Kulkarni, S. Giddey and S. Bhattacharya1, A review on synthesis of methane as a pathway for renewable energy storage with a focus on solid oxide electrolytic cell-based processes, *Front. Energy Res.* **8**, 570112 (2020).

37. C. Acar and I. Dincer, Review and evaluation of hydrogen production options for better environment, *J. Clean. Prod.* **218**, 835–849 (2019).

38. F. Hu, X. Chen, Z. Tu, Z.-H. Lu, G. Feng and R. Zhang, Graphene aerogel supported Ni for CO_2 hydrogenation to methane, *Ind. Eng. Chem. Res.* **60**(33), 12235–12243 (2021).

39. F. Hu, R. Ye, Z.-H. Lu, R. Zhang and G. Feng, Structure–activity relationship of Ni-based catalysts toward CO_2 methanation: Recent advances and future perspectives, *Energy Fuels* **36**(1), 156–169 (2021).

40. F. Federici, J. Puna, T.M. Mata and A.A. Martins, Life cycle analysis of a combined electrolysis and methanation reactor for methane production, *Energy Rep.* **8**, 554–560 (2022).

41. C. Hao, S. Wang M. Li, L. Kang and X. Ma, Hydrogenation of CO_2 to formic acid on supported ruthenium catalysts, *Catal. Today* **160**(1), 184–190 (2011).

42. M. Ramdin, A.R.T. Morrison, M. de Groen, R. van Haperen, R. de Kler, E. Irtem, *et al.*, High-pressure electrochemical reduction of CO_2 to formic acid/formate: Effect of pH on the downstream separation process and economics, *Ind. Eng. Chem. Res.* **58**(51), 22718–22740 (2019).

43. J. Gao, C. Jia and B. Liu, Direct and selective hydrogenation of CO_2 to ethylene and propene by bifunctional catalysts, *Catal. Sci. Technol.* **7**(23), 5602–5607 (2017).

44. R.I. Masel, Z. Liu, H. Yang, J.J. Kaczur, D. Carrillo, S. Ren, *et al.*, An industrial perspective on catalysts for low-temperature CO_2 electrolysis, *Nat. Nanotechnol.* **16**(2), 118–128 (2021).

45. X. Chen, J. Chen, N.M. Alghoraibi, D.A. Henckel, R. Zhang, U.O. Nwabara, *et al.*, Electrochemical CO_2-to-ethylene conversion on polyamine-incorporated Cu electrodes, *Nat. Catal.* **4**(1), 20–27 (2021).
46. S. Ye, S. Wang, L. Lin, M. Xiao and Y. Meng, CO_2 derived biodegradable polycarbonates: Synthesis, modification and applications, *Adv. Ind. Eng. Polym. Res.* **2**(4), 143–160 (2019).
47. L. Fu, Z. Ren, W. Si, Q. Ma, W. Huang, K. Liao, *et al.*, Research progress on CO_2 capture and utilization technology, *J. CO2 Util.* **66**,102260 (2022).
48. C. Chen, Y. Gnanou and X. Feng, Ultra-productive upcycling CO_2 into polycarbonate polyols via borinane-based bifunctional organocatalysts, *Macromolecules* **56**(3), 892–898 (2023).
49. O. Coulembier, J. Huang, J.C. Worch and A.P. Dove, Update and challenges in CO_2-based polycarbonate synthesis, *ChemSusChem* **13**(3), 469–487 (2020).
50. Y. Qin, X. Wang and F. Wang, Synthesis and properties of carbon dioxide based copolymers, *Sci. Sin. Chim.* **48**, 883–893 (2018).
51. J. Wei, Q. Ge, R. Yao, Z. Wen, C. Fang, Li. Guo, *et al.*, Directly converting CO_2 into a gasoline fuel, *Nat. Commun.* **8**(1), 15174 (2017).
52. A.D.N. Kamkeng and M. Wang, Technical analysis of the modified Fischer-Tropsch synthesis process for direct CO_2 conversion into gasoline fuel: Performance improvement via ex-situ water removal, *Chem. Eng. J.* **462**, 142048 (2023).
53. M.J. Fernández-Torres, W. Dednam and J.A. Caballero, Economic and environmental assessment of directly converting CO_2 into a gasoline fuel, *Energ. Convers. Manag.* **252**, 115115 (2022).
54. B. Yao, T. Xiao, O.A. Makgae, X. Jie, S. Gonzalez-Cortes, S. Guan, *et al.*, Transforming carbon dioxide into jet fuel using an organic combustion-synthesized Fe-Mn-K catalyst, *Nat. Commun.* **11**(1), 6395 (2020).
55. H. Tian, C. Lu, P. Ciais, A.M. Michalak, J.G. Canadell, E. Saikawa, *et al.*, The terrestrial biosphere as a net source of greenhouse gases to the atmosphere, *Nature* **531**(7593), 225–228 (2016).
56. R.-P. Ye, J. Ding, W. Gong, M.D. Argyle, Q. Zhong, Y. Wang, *et al.*, CO_2 hydrogenation to high-value products via heterogeneous catalysis, *Nat. Commun.* **10**(1), 5698 (2019).
57. L. Zhang, Y. Dang, X. Zhou, P. Gao, A.P. van Bavel, H. Wang, *et al.*, Direct conversion of CO_2 to a jet fuel over CoFe alloy catalysts, *The Innovation* **2**(4), 100170 (2021).
58. E. Hernández, R. Santiago, A. Belinchón, G.M. Vaquerizo, C. Moya, P. Navarro, *et al.*, Universal and low energy-demanding platform to produce propylene carbonate from CO_2 using hydrophilic ionic liquids, *Sep. Purif. Technol.* **295**, 121273 (2022).
59. C. Falter, V. Batteiger and A. Sizmann, Climate impact and economic feasibility of solar thermochemical jet fuel production. *Environ. Sci. Technol.* **50**(1), 470–477 (2016).

60. J. Kim, J.E. Miller, C.T. Maravelias and E.B. Stechel, Comparative analysis of environmental impact of S2P (Sunshine to Petrol) system for transportation fuel production, *Appl. Energy* **111**, 1089–1098 (2013).
61. N. Mac Dowell, P.S. Fennell, N. Shah and G.C. Maitland, The role of CO_2 capture and utilization in mitigating climate change, *Nat. Clim. Change* **7**(4), 243–249 (2017).
62. J. Gibbins and H. Chalmers, Carbon capture and storage, *Energ. Policy* **36**(12), 4317–4322 (2008).
63. F.M. Stuardi, F. MacPherson, and J. Leclaire, Integrated CO_2 capture and utilization: A priority research direction, *Curr. Opin. Green Sustain. Chem.* **16**, 71–76 (2019).
64. W. Hoppe, N. Thonemann, and S. Bringezu, Life cycle assessment of carbon dioxide–based production of methane and methanol and derived polymers, *J. Ind. Ecol.* **22**(2), 327–340 (2018).
65. N. Thonemann, Environmental impacts of CO2-based chemical production: A systematic literature review and meta-analysis, *Appl. Energy* **263**, 114599 (2020).

Index

378   *Towards Net-Zero Carbon Initiatives: A Life Cycle Assessment Perspective*

waste treatment, 133, 140

waste-to-chemicals, 133, 143

waste-to-energy, 115, 118, 142, 285,

well-to-wheels (WTW), 224

WOWnature, 252

zero carbon emissions, 2, 22

zero CO_2, 252

zero waste, 155

www.ingramcontent.com/pod-product-compliance
Lightning Source LLC
Chambersburg PA
CBHW050536190326
41458CB00007B/1805